AC Electrical Circuit Analysis

Mehdi Rahmani-Andebili

AC Electrical Circuit Analysis

Practice Problems, Methods, and Solutions

Springer

Mehdi Rahmani-Andebili
Engineering Technology
State University of New York
Buffalo, NY, USA

ISBN 978-3-030-60988-7 ISBN 978-3-030-60986-3 (eBook)
https://doi.org/10.1007/978-3-030-60986-3

This Springer imprint is published by the registered company Springer Nature Switzerland AGThe registered company address is: Gewerbestrasse 11, 6330 Cham, Switzerland

Preface

Electrical Circuit Analysis is one of the most fundamental subjects of Electrical Engineering which is taught in two courses in successive semesters under the names of "Electrical Circuit Analysis I" and "Electrical Circuit Analysis II" or under the names of "DC Electrical Circuit Analysis" and "AC Electrical Circuit Analysis" in universities and colleges all over the world.

This textbook, like the previously published *DC Electrical Circuit Analysis*, includes basic and advanced exercises of AC Electrical Circuit Analysis with very detailed and multiple methods of solutions. The textbook can be used as a practicing textbook by students and as a supplementary teaching source by instructors.

To help students study the textbook in the most efficient way, the exercises have been categorized in nine different levels. In this regard, for each problem of the textbook, a difficulty level (easy, normal, or hard) and a calculation amount (small, normal, or large) have been assigned. Moreover, in each chapter, problems have been ordered from the easiest problem with the smallest calculations to the most difficult problem with the largest calculations. Therefore, students are suggested to start studying the textbook from the easiest problems and continue practicing until they reach the normal and then the hardest ones. On the other hand, this classification can help instructors choose their desired problems to conduct a quiz or a test. Moreover, the classification of computation amount can help students manage their time during future exams and instructors give the appropriate problems based on the exam duration.

Since the problems have very detailed solutions and some of them include multiple methods of solutions, the textbook can be useful for the underprepared students. In addition, the textbook is beneficial for knowledgeable students because it includes advanced exercises.

In the preparation of problem solutions, use of typical methods of Electrical Circuit Analysis has been tried to present the textbook as an instructor-recommended one. In other words, the heuristic methods have never been used as the first method of problem solution. By considering this key point, the textbook is in the direction of instructors' lectures, and the instructors will not see any untaught problem solutions in their students' answer sheets.

The Iranian University Entrance Exams for the Master's and PhD degrees of Electrical Engineering major is the main reference of the textbook; however, all the problem solutions have been provided by me. The Iranian University Entrance Exam is one of the most competitive university entrance exams in the world that allows only 10% of the applicants to get into prestigious and tuition-free Iranian universities.

Buffalo, NY, USA

Mehdi Rahmani-Andebili

Contents

About the Author

Mehdi Rahmani-Andebili is an Assistant Professor in the Engineering Technology Department at State University of New York, Buffalo State. He received his first M.Sc. and Ph.D. degrees in Electrical Engineering (Power System) from Tarbiat Modares University and Clemson University in 2011 and 2016, respectively, and his second M.Sc. degree in Physics and Astronomy from the University of Alabama in Huntsville in 2019. Moreover, he was a Postdoctoral Fellow at Sharif University of Technology during 2016–2017. As a professor, he has taught many courses such as Essentials of Electrical Engineering Technology, Electrical Circuits Analysis I, Electrical Circuits Analysis II, Electrical Circuits and Devices, Industrial Electronics, and Renewable Distributed Generation and Storage. Dr. Rahmani-Andebili has more than 100 single-author publications including textbooks, books, book chapters, journal papers, and conference papers. His research areas include Smart Grid, Power System Operation and Planning, Integration of Renewables and Energy Storages into Power System, Energy Scheduling and Demand-Side Management, Plug-in Electric Vehicles, Distributed Generation, and Advanced Optimization Techniques in Power System Studies.

Problems: Sinusoidal Steady-State Analysis

1

Abstract

This chapter helps both groups of underprepared and knowledgeable students taking courses in AC electrical circuit analysis. In this chapter, the basic and advanced problems of important subjects of AC circuit analysis, that is, sinusoids and phasors, sinusoidal steady-state analysis, and AC power analysis, are presented. The problems of sinusoids and phasors include complex numbers; rectangular, polar, and exponential forms of phasors; phasor relationships for circuit elements; impedance and admittance and their combinations; resonance frequency; and bandwidth of frequency response of series and parallel RLC circuits. The problems of sinusoidal steady-state analysis include Kirchhoff's laws in frequency domain; nodal and mesh analyses in frequency domain; sinusoidal steady-state response; superposition theorem; source transformation theorem; Thevenin and Norton theorems; and maximum average power transfer theorem. The problems of AC power analysis are concerned with root mean square (rms) and peak quantities; average power, active and reactive powers, apparent power, complex power; and lagging, unity, and leading power factors. In this chapter, the problems are categorized in different levels based on their difficulty levels (easy, normal, and hard) and calculation amounts (small, normal, and large). Additionally, the problems are ordered from the easiest problem with the smallest computations to the most difficult problems with the largest calculations.

1.1. In the circuit of Figure 1.1, what must be the resistance of the purely resistive load to absorb the maximum average power [1]?

Difficulty level ● Easy ○ Normal ○ Hard

Calculation amount ● Small ○ Normal ○ Large

1) 5 Ω
2) 7 Ω
3) 7.5 Ω
4) 4 Ω

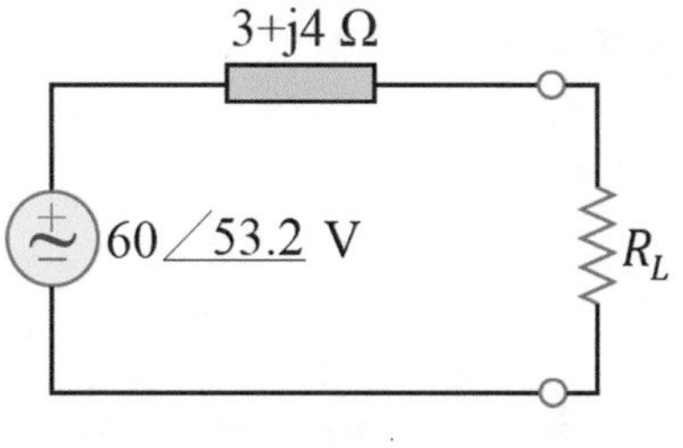

Figure 1.1 The circuit of problem 1.1

1.2. In the circuit of Figure 1.2, calculate the resistance of R so that it can absorb the maximum average power.

Difficulty level ● Easy ○ Normal ○ Hard

Calculation amount ● Small ○ Normal ○ Large

M. Rahmani-Andebili, *AC Electrical Circuit Analysis*, https://doi.org/10.1007/978-3-030-60986-3_1

1) 1 Ω
2) 3 Ω
3) 9 Ω
4) $1 + 2\sqrt{2}\,\Omega$

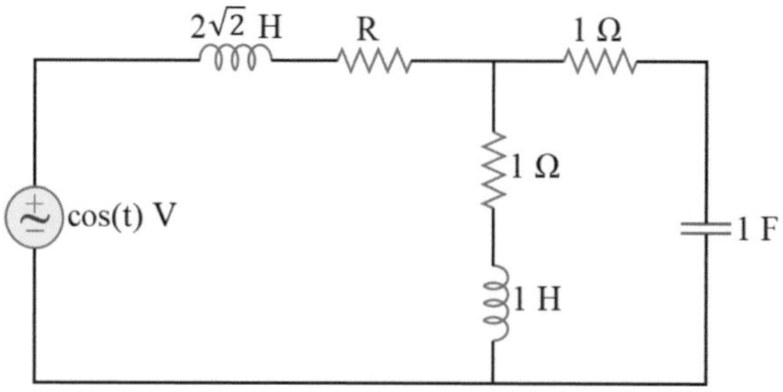

Figure 1.2 The circuit of problem 1.2

1.3. In the circuit of Figure 1.3, calculate the value of $v_i(t)$ if the sinusoidal steady-state response of $v_o(t)$ is equal to $2\cos(0.5t - 30^\circ)\ V$.

Difficulty level ● Easy ○ Normal ○ Hard

Calculation amount ● Small ○ Normal ○ Large

1) $2\sqrt{2}\cos\left(0.5t - 15^\circ\right)\ V$
2) $2\sqrt{2}\cos\left(0.5t + 15^\circ\right)\ V$
3) $2\cos(0.5t + 30^\circ)\ V$
4) $2\sqrt{2}\cos\left(0.5t - 30^\circ\right)\ V$

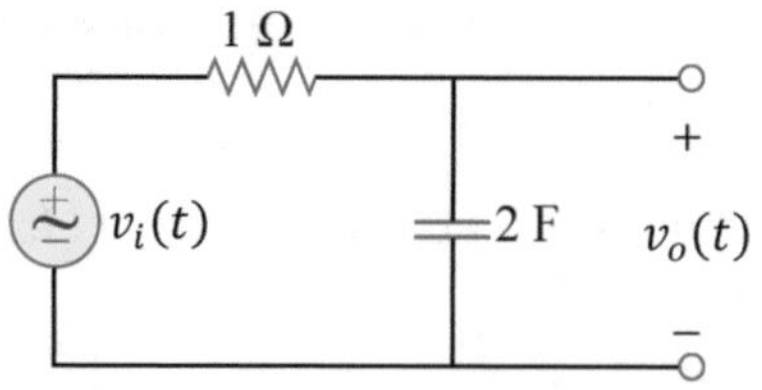

Figure 1.3 The circuit of problem 1.3

1.4. In the circuit of Figure 1.4, calculate the sinusoidal steady-state response of $v(t)$.

Difficulty level ● Easy ○ Normal ○ Hard

Calculation amount ● Small ○ Normal ○ Large

1) $\frac{5}{2}\cos\left(2t + 45^\circ\right)\ V$
2) $5\cos(2t - 45^\circ)\ V$
3) $-5\cos(2t - 45^\circ)\ V$
4) $\frac{5\sqrt{2}}{2}\cos\left(2t - 45^\circ\right)\ V$

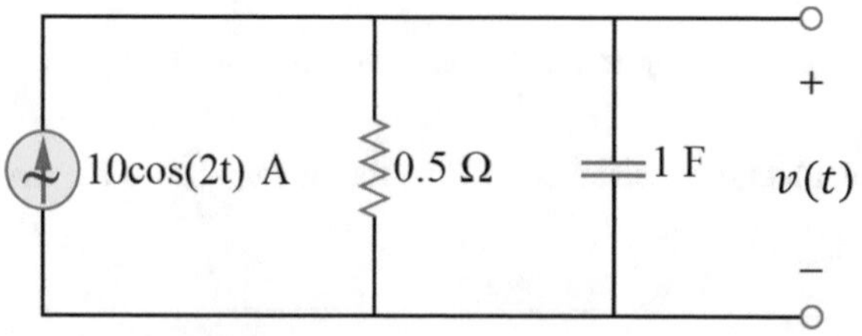

Figure 1.4 The circuit of problem 1.4

1.5. The circuit of Figure 1.5 is in its sinusoidal steady state. Calculate the phasor of the voltage of the inductor.

Difficulty level ● Easy ○ Normal ○ Hard
Calculation amount ● Small ○ Normal ○ Large

1) $\sqrt{2}e^{-j135}\ V$
2) $\frac{\sqrt{2}}{2}e^{-j135}\ V$
3) $-\frac{\sqrt{2}}{2}e^{-j45}\ V$
4) $\sqrt{2}e^{-j45}\ V$

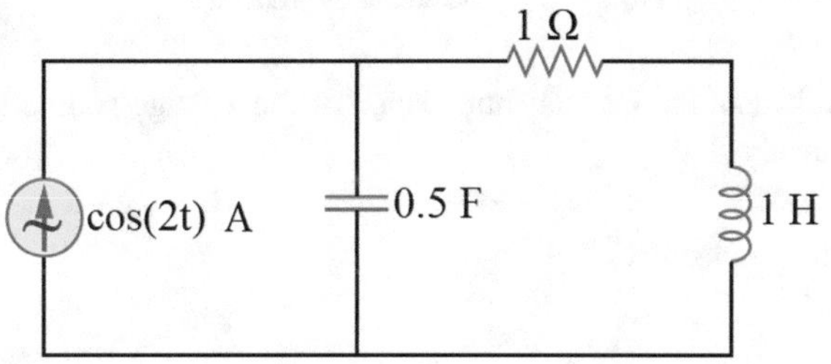

Figure 1.5 The circuit of problem 1.5

1.6. In the circuit of Figure 1.6, what impedance must be connected to terminal a–b to absorb the maximum average power in sinusoidal steady state?

Difficulty level ● Easy ○ Normal ○ Hard
Calculation amount ● Small ○ Normal ○ Large

1) $\left(\frac{4}{5}+j\frac{2}{5}\right)\ \Omega$
2) $\left(\frac{4}{5}-j\frac{2}{5}\right)\ \Omega$
3) $\left(\frac{2}{5}+j\frac{4}{5}\right)\ \Omega$
4) $\left(\frac{2}{5}-j\frac{4}{5}\right)\ \Omega$

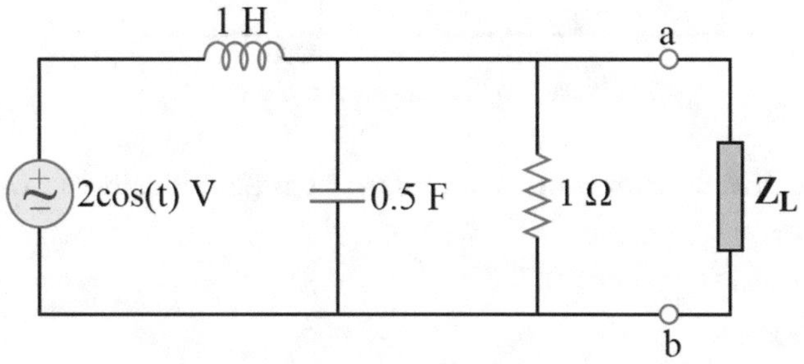

Figure 1.6 The circuit of problem 1.6

1.7. In the circuit of Figure 1.7, calculate the sinusoidal steady-state voltage of the inductor.

Difficulty level ● Easy ○ Normal ○ Hard
Calculation amount ● Small ○ Normal ○ Large

1) $2\sin(2t)\ A$
2) $4\sin(2t)\ A$
3) $-2\sin(2t)\ A$
4) $-4\sin(2t)\ A$

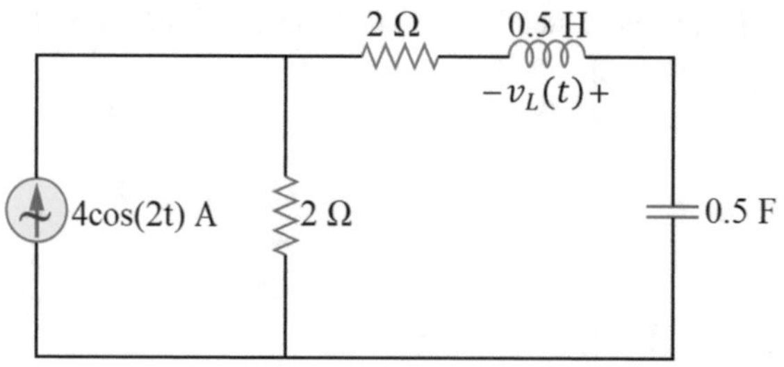

Figure 1.7 The circuit of problem 1.7

1.8. In the circuit of Figure 1.8, calculate the internal impedance of the voltage source ($\mathbf{Z_S}$) that can absorb the maximum average power in sinusoidal steady state.

Difficulty level ● Easy ○ Normal ○ Hard

Calculation amount ● Small ○ Normal ○ Large

1) 4 Ω
2) 5 Ω
3) $(4 - j10)\ \Omega$
4) $(5 + j10)\ \Omega$

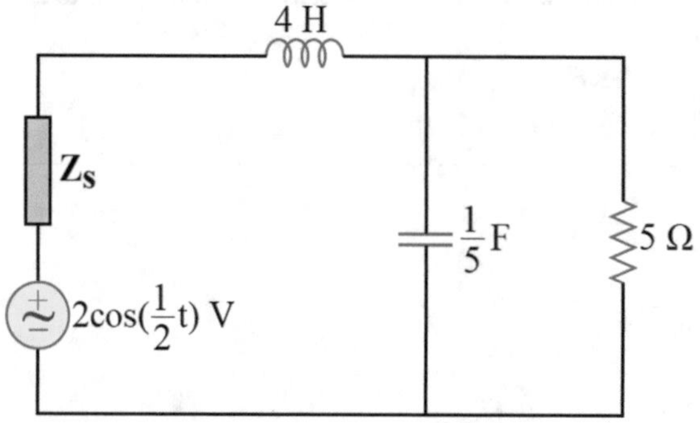

Figure 1.8 The circuit of problem 1.8

1.9. Calculate the average power that the current of $2\cos(10t) - 3\cos(20t)\ A$ delivers to a 4 Ω resistor.

Difficulty level ● Easy ○ Normal ○ Hard

Calculation amount ● Small ○ Normal ○ Large

1) 26 W
2) 13 W
3) 6.5 W
4) 18 W

1.10. Determine the resonance frequency of the circuit shown in Figure 1.9.

Difficulty level ● Easy ○ Normal ○ Hard

Calculation amount ● Small ○ Normal ○ Large

1) 5000 *rad/sec*
2) 1000 *rad/sec*
3) 500 *rad/sec*
4) 100 *rad/sec*

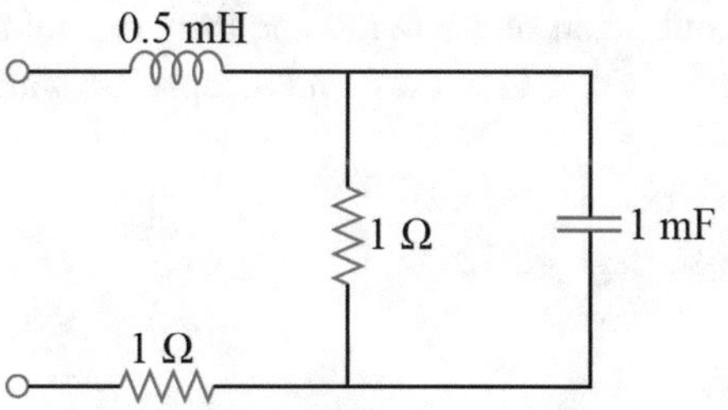

Figure 1.9 The circuit of problem 1.10

1.11. In the circuit of Figure 1.10, calculate the voltage of the capacitor in sinusoidal steady state.

Difficulty level ● Easy ○ Normal ○ Hard

Calculation amount ● Small ○ Normal ○ Large

1) $45.8\cos(100t - 24.3^\circ)\ V$
2) $36.3\cos(1000t - 18.4^\circ)\ V$
3) $4.58\cos(100t - 24.3^\circ)\ V$
4) $3.63\cos(1000t - 18.4^\circ)\ V$

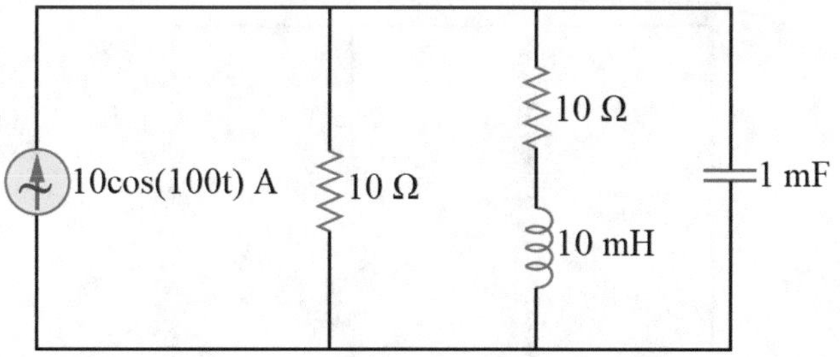

Figure 1.10 The circuit of problem 1.11

1.12. In the circuit of Figure 1.11, $v(t) = A\cos(t)\ V$. Calculate the sinusoidal steady-state response of $i(t)$.

Difficulty level ● Easy ○ Normal ○ Hard

Calculation amount ● Small ○ Normal ○ Large

1) $A\cos(t)\ A$
2) $-A\cos(t)\ A$
3) $\frac{1}{2}A\cos(t)\ A$
4) $\frac{1}{4}A\cos(t)\ A$

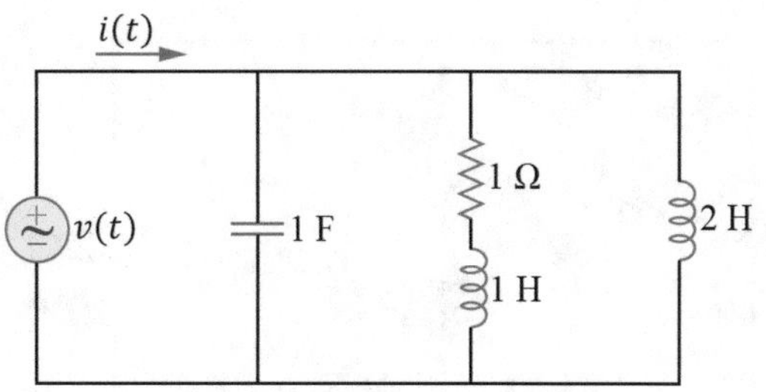

Figure 1.11 The circuit of problem 1.12

1.13. In a certain frequency, the series connection of a 2 Ω resistor and a capacitor with the capacitance of C has the same impedance as the parallel connection of a 4 Ω resistor and a capacitor with the capacitance of $\frac{1}{40}$ F. Determine the capacitance of the capacitor (C).

Difficulty level ● Easy ○ Normal ○ Hard

Calculation amount ○ Small ● Normal ○ Large

1) $\frac{1}{50}$ F
2) $\frac{1}{20}$ F
3) $\frac{1}{10}$ F
4) $\frac{1}{5}$ F

1.14. Which of the following statement is correct for the circuit of Figure 1.12 in sinusoidal steady state?

Difficulty level ○ Easy ● Normal ○ Hard

Calculation amount ● Small ○ Normal ○ Large

1) Increasing the frequency of the current source will increase the magnitude of the voltage of terminal a–b.
2) Increasing the frequency of the current source will decrease the magnitude of the voltage of terminal a–b.
3) Increasing the frequency of the current source will not change the magnitude of the voltage of terminal a–b.
4) Increasing the frequency of the current source will increase the phase angle of the voltage of terminal a–b.

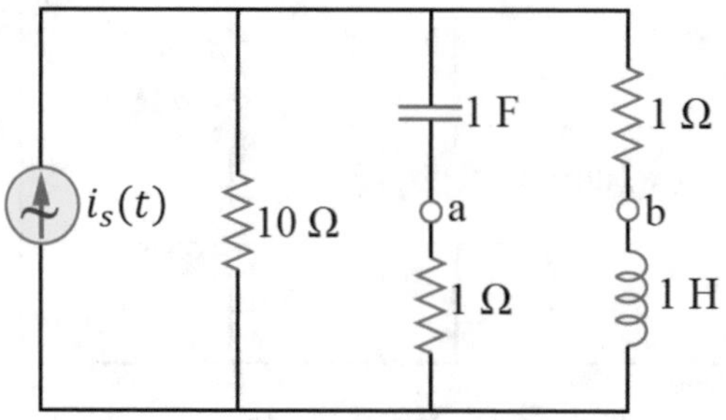

Figure 1.12 The circuit of problem 1.14

1.15. The circuit of Figure 1.13 is in sinusoidal steady state. Calculate the current of the inductor ($\mathbf{I_L}$).

Difficulty level ○ Easy ● Normal ○ Hard

Calculation amount ● Small ○ Normal ○ Large

1) $(7 - j4)$ A
2) $(7 + j4)$ A
3) $(2 - j4)$ A
4) $(2 + j4)$ A

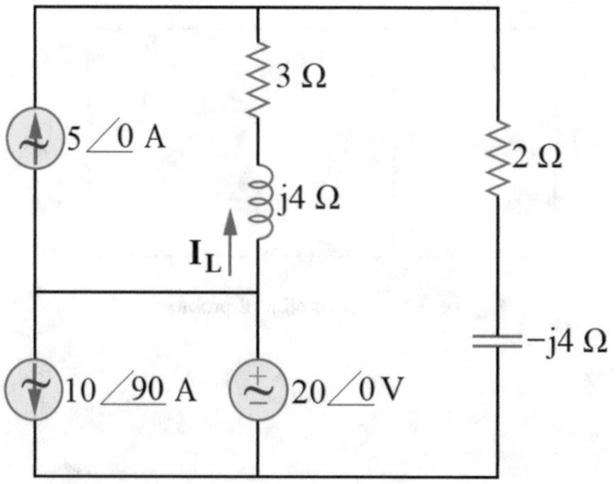

Figure 1.13 The circuit of problem 1.15

1.16. In the circuit of Figure 1.14, calculate the peak value of the amplitude of $\mathbf{I_2}$ if the average power of R_1 is 1.5 W.

Difficulty level ○ Easy ● Normal ○ Hard

Calculation amount ● Small ○ Normal ○ Large

1) $\frac{\sqrt{2}}{2}\ A$
2) $1\ A$
3) $\sqrt{\frac{7}{2}}A$
4) Impossible, since it depends on the other parameters of the circuit that are unknown.

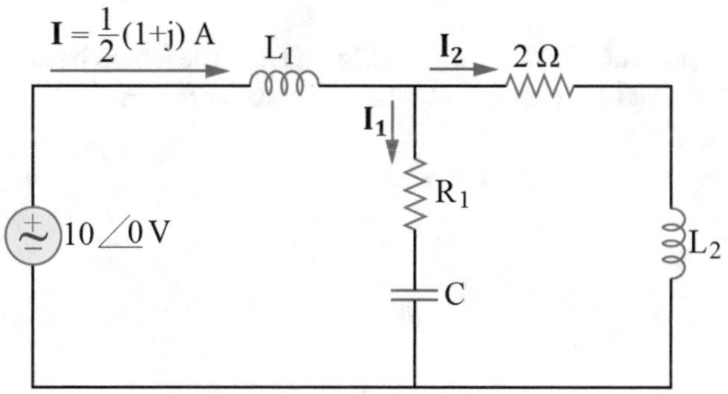

Figure 1.14 The circuit of problem 1.16

1.17. Determine the size of the inductor and capacitor in the circuit of Figure 1.15 to adjust the resonance frequency and $-3\ dB$ bandwidth at 50 kHz and 500 Hz, respectively.

Difficulty level ○ Easy ● Normal ○ Hard

Calculation amount ● Small ○ Normal ○ Large

1) $L = 18.4\ mH,\ C = 6.4\ nF$
2) $L = 13.1\ mH,\ C = 64\ nF$
3) $L = 15.9\ mH,\ C = 64\ nF$
4) $L = 15.9\ mH,\ C = 0.64\ nF$

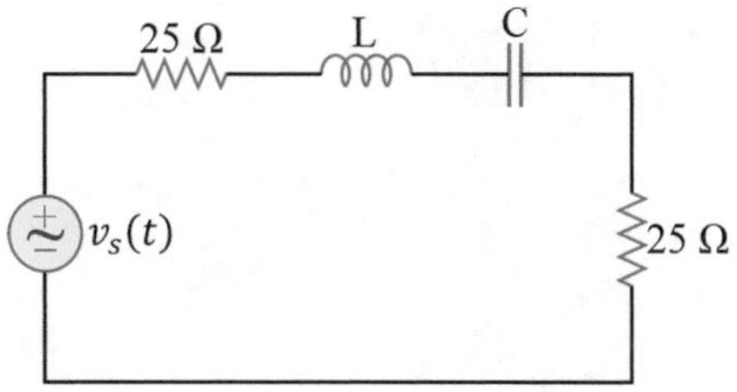

Figure 1.15 The circuit of problem 1.17

1.18. Parametrically calculate the phasor of the voltage of the terminal indicated in the circuit of Figure 1.16. Herein, "$\angle$" is the symbol of phase angle.

Difficulty level ○ Easy ● Normal ○ Hard

Calculation amount ● Small ○ Normal ○ Large

1) $\frac{1+LC\omega^2}{1-LC\omega^2}\ V$
2) $\frac{1+LC\omega^2}{1-LC\omega^2}\angle 90^\circ\ V$
3) $\frac{1-LC\omega^2}{1+LC\omega^2}\ V$
4) $\frac{1-LC\omega^2}{1+LC\omega^2}\angle 90^\circ\ V$

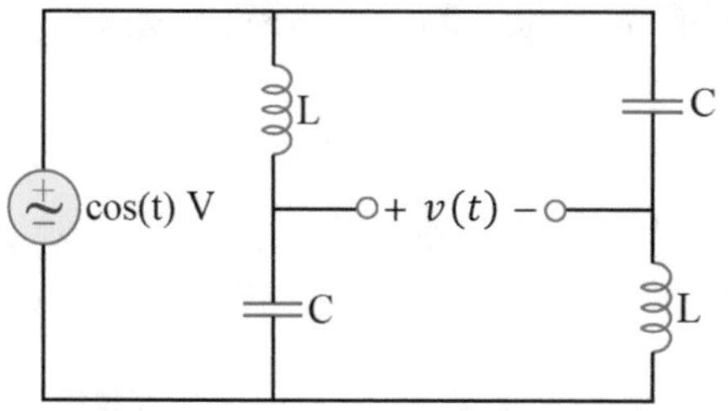

Figure 1.16 The circuit of problem 1.18

1.19. Consider the parallel and series RLC circuits shown in Figure 1.17. The −3 dB bandwidth of the parallel RLC circuit is about 1 Hz. Determine C' so that both circuits have the same quality factor.

Difficulty level ○ Easy ● Normal ○ Hard

Calculation amount ● Small ○ Normal ○ Large

1) 2 F
2) 0.5 F
3) 4 F
4) 0.25 F

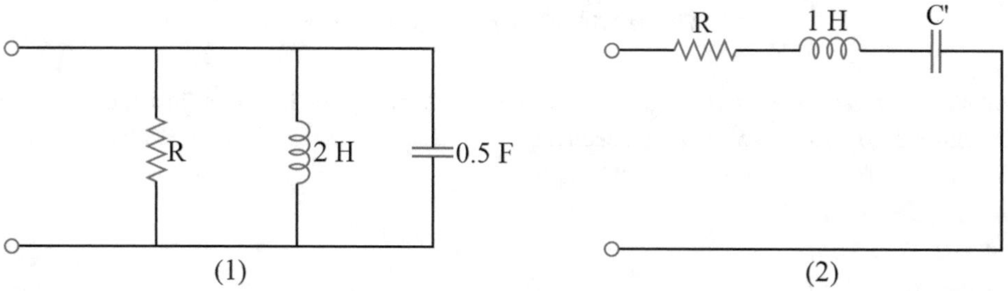

Figure 1.17 The circuit of problem 1.19

1.20. In the circuit of Figure 1.18, calculate $v_3(t)$ in sinusoidal steady state.

Difficulty level ○ Easy ● Normal ○ Hard

Calculation amount ● Small ○ Normal ○ Large

1) $-4\sin(4t)V$
2) $\sin(4t)V$
3) $-2\sin(4t)V$
4) $2\sin(4t)V$

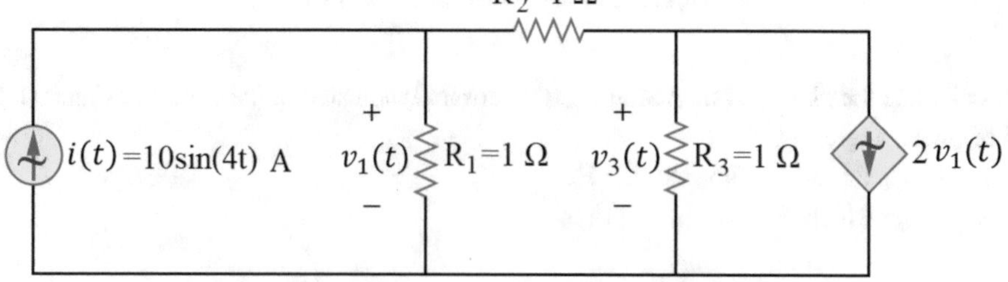

Figure 1.18 The circuit of problem 1.20

1.21. The circuit of Figure 1.19 is in its sinusoidal steady state. The root mean square (rms) value of the voltage is about 100 V. The complex powers of the load "A" and the load "B" are $100e^{j30}$ kVA and $100e^{-j30}$ kVA, respectively. Calculate the rms value of the current.

Difficulty level ○ Easy ● Normal ○ Hard

Calculation amount ● Small ○ Normal ○ Large

1) $0\,A$
2) $200\sqrt{3}\,A$
3) $500\sqrt{3}\,A$
4) $1000\sqrt{3}\,A$

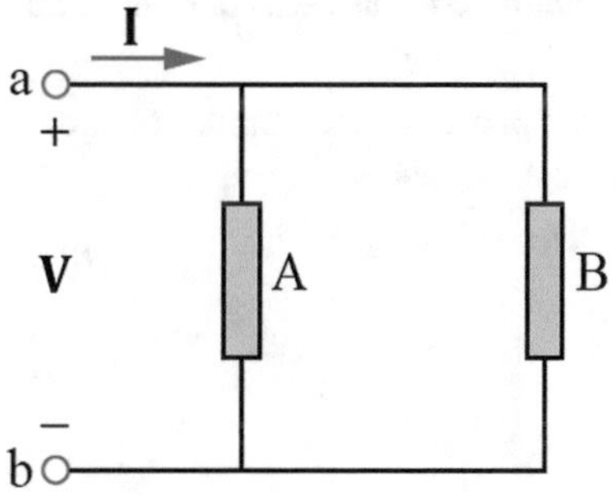

Figure 1.19 The circuit of problem 1.21

1.22. Determine the resonance frequency of the circuit illustrated in Figure 1.20.

Difficulty level ○ Easy ● Normal ○ Hard

Calculation amount ● Small ○ Normal ○ Large

1) $\sqrt{\frac{1}{LC} - \left(\frac{R_2}{L}\right)^2}$
2) $\sqrt{\frac{1}{LC}}$
3) $\sqrt{\frac{1}{LC}\left(\frac{R_2}{L}\right)^2}$
4) $\sqrt{\frac{1}{LC-R_2}}$

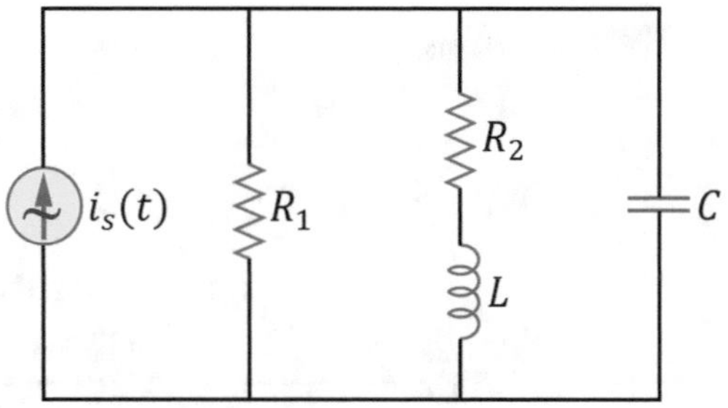

Figure 1.20 The circuit of problem 1.22

1.23. In the circuit of Figure 1.21, calculate the impedance of $\mathbf{Z_x}$ so that it can absorb the maximum average power.

Difficulty level ○ Easy ● Normal ○ Hard

Calculation amount ● Small ○ Normal ○ Large

1) $j2\ \Omega$
2) $1\ \Omega$
3) $(1 + j0.5)\ \Omega$
4) $(1 - j0.5)\ \Omega$

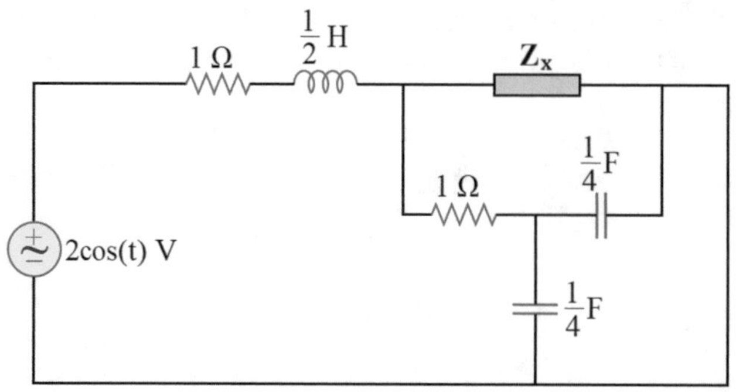

Figure 1.21 The circuit of problem 1.23

1.24. In the circuit of Figure 1.22, calculate the phasor of the output voltage.

Difficulty level ○ Easy ● Normal ○ Hard

Calculation amount ● Small ○ Normal ○ Large

1) $\frac{\sqrt{2}}{4}e^{-j45}\ V$
2) $\frac{\sqrt{2}}{2}e^{-j45}\ V$
3) $\frac{\sqrt{2}}{4}e^{\,j45}\ V$
4) $\frac{\sqrt{2}}{2}e^{\,j45}\ V$

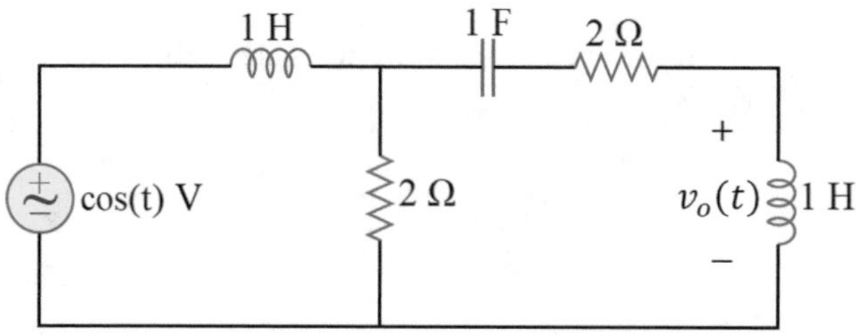

Figure 1.22 The circuit of problem 1.24

1.25. In the circuit of Figure 1.23, calculate the sinusoidal steady-state response of the voltage of the capacitor.

Difficulty level ○ Easy ● Normal ○ Hard

Calculation amount ● Small ○ Normal ○ Large

1) $63.2\cos(100t - 18.4^\circ)\ V$
2) $63.2\cos(100t + 71.6^\circ)\ V$
3) $70.7\cos(100t - 71.6^\circ)\ V$
4) $70.7\cos(100t + 18.4^\circ)\ V$

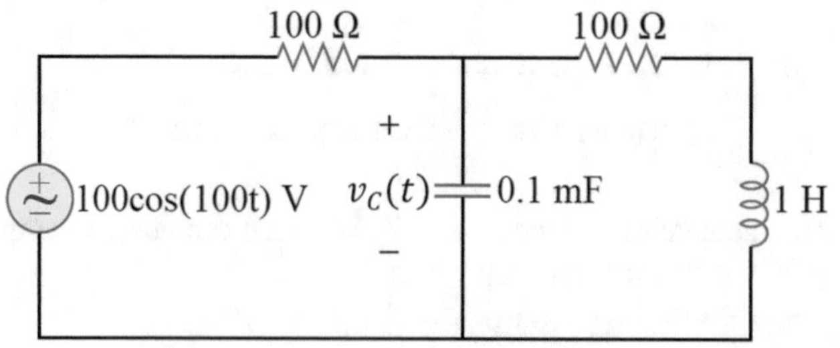

Figure 1.23 The circuit of problem 1.25

1.26. In the circuit of Figure 1.24, the resistor (R) is consuming the average power of 1000 W and the impedance ($\mathbf{Z_L}$) has the apparent power of 1000 VA with the lagging power factor of 0.6. Calculate the phasor (rms value) of the current of the circuit.

Difficulty level ○ Easy ● Normal ○ Hard
Calculation amount ● Small ○ Normal ○ Large

1) $(1.79\angle -26.5^\circ)\ A$
2) $(3.58\angle 53^\circ)\ A$
3) $(3.58\angle -26.5^\circ)\ A$
4) $(3.58\angle -53^\circ)\ A$

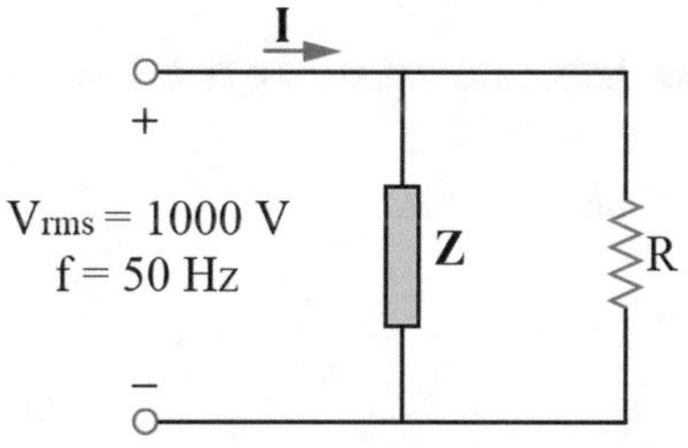

Figure 1.24 The circuit of problem 1.26

1.27. In the circuit of Figure 1.25, calculate the impedance of the load ($\mathbf{Z_L}$) so that it can absorb the maximum average power in sinusoidal steady state.

Difficulty level ○ Easy ● Normal ○ Hard
Calculation amount ● Small ○ Normal ○ Large

1) $(0.3 - j0.9)\ \Omega$
2) $(0.3 + j0.9)\ \Omega$
3) $(0.9 - j0.3)\ \Omega$
4) $(0.9 + j0.3)\ \Omega$

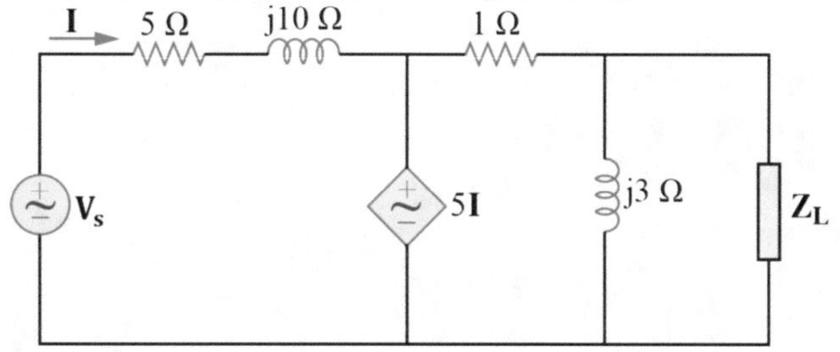

Figure 1.25 The circuit of problem 1.27

1.28. The circuit of Figure 1.26 is its sinusoidal steady state. Calculate the average power of the dependent voltage source supplied only by the AC independent current source.

Difficulty level ○ Easy ● Normal ○ Hard
Calculation amount ● Small ○ Normal ○ Large

1) $\frac{4}{49}\ W$
2) $\frac{2}{49}\ W$
3) $\frac{3}{49}\ W$
4) $\frac{1}{49}\ W$

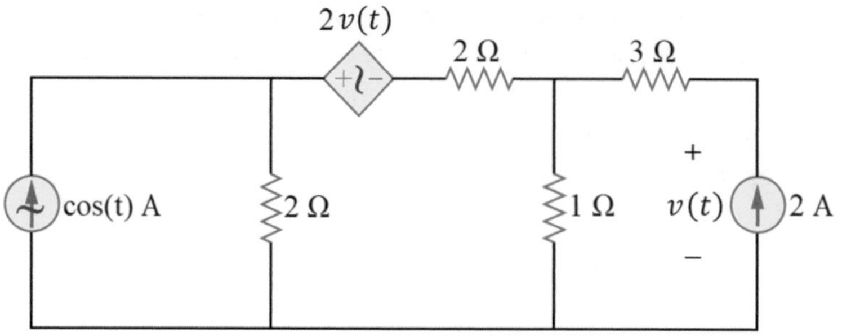

Figure 1.26 The circuit of problem 1.28

1.29. In the circuit of Figure 1.27, determine the sinusoidal steady-state response of $i(t)$.

Difficulty level ○ Easy ● Normal ○ Hard

Calculation amount ○ Small ● Normal ○ Large

1) $\frac{1}{2}\cos(t)\ A$
2) $-\frac{8}{5}\cos(t)\ A$
3) $-\cos(t)\ A$
4) $0\ A$

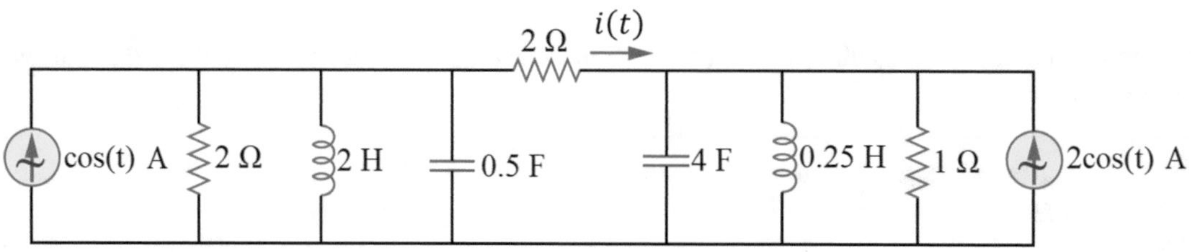

Figure 1.27 The circuit of problem 1.29

1.30. The circuit shown in Figure 1.28 is in sinusoidal steady state. Determine the time-dependent equation of the output voltage ($v_o(t)$).

Difficulty level ○ Easy ● Normal ○ Hard

Calculation amount ○ Small ● Normal ○ Large

1) $1-\sqrt{2}\cos(t)V$
2) $1+\sqrt{2}\cos\left(t+45^\circ\right)V$
3) $2\cos(t+45^\circ)V$
4) $\sqrt{2}\cos\left(t-45^\circ\right)V$

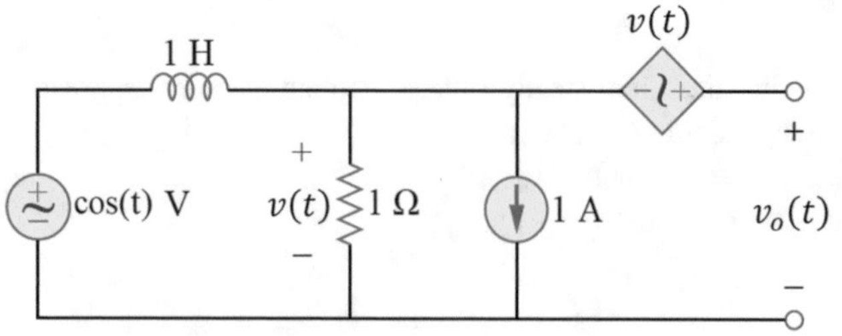

Figure 1.28 The circuit of problem 1.30

1.31. The circuit illustrated in Figure 1.29 is in its sinusoidal steady state. Calculate the root mean square (rms) value of the output voltage if:
$v_s(t) = 4\left(\cos\left(\frac{1}{2}t\right) - \frac{1}{3}\cos\left(\frac{3}{2}t\right) + \frac{1}{5}\cos\left(\frac{5}{2}t\right)\right)\ V$
Difficulty level ○ Easy ● Normal ○ Hard
Calculation amount ○ Small ● Normal ○ Large
1) 1.83 V
2) 2.12 V
3) 2.59 V
4) 3.72 V

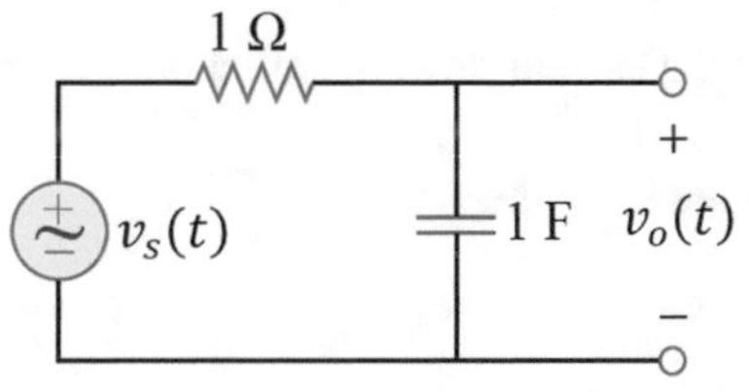

Figure 1.29 The circuit of problem 1.31

1.32. The circuit of Figure 1.30 is in its sinusoidal steady state. What condition needs to be applied on g_m to keep the relation of $|\mathbf{V}_o| \geq |\mathbf{V}_i|$ for all the frequencies?
Difficulty level ○ Easy ● Normal ○ Hard
Calculation amount ○ Small ● Normal ○ Large
1) It is impossible to keep the relation
2) $|g_m| = 1$
3) $|g_m| \leq 0.5$
4) $|g_m| \geq 0.5$

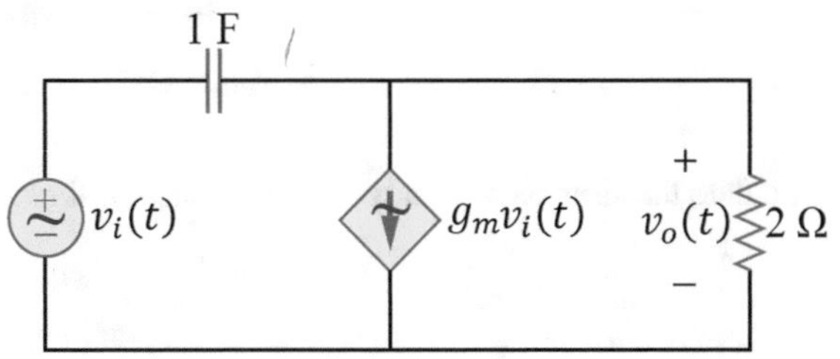

Figure 1.30 The circuit of problem 1.32

1.33. In the circuit of Figure 1.31, determine the value of $\mathbf{Z_L}$ to absorb the maximum average power.
Difficulty level ○ Easy ● Normal ○ Hard
Calculation amount ○ Small ● Normal ○ Large
1) $(1 + j2)\ \Omega$
2) $(1 - j2)\ \Omega$
3) $(1 - j4)\ \Omega$
4) $(1 + j4)\ \Omega$

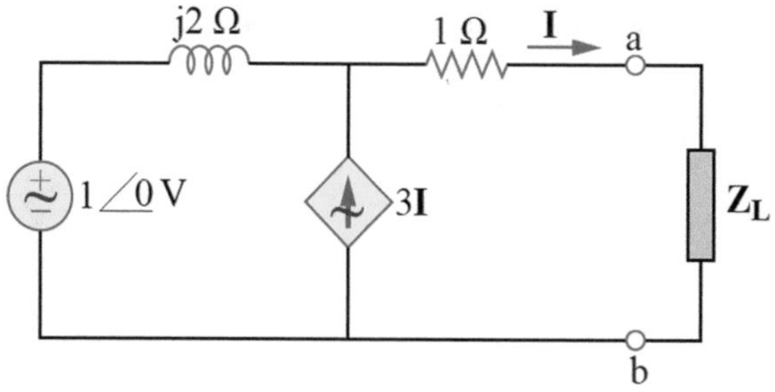

Figure 1.31 The circuit of problem 1.33

1.34. In the circuit of Figure 1.32, calculate the current flowing from node "a" to node "b" if terminal a–b is short-circuited.

Difficulty level ○ Easy ● Normal ○ Hard

Calculation amount ○ Small ● Normal ○ Large

1) $\sqrt{2}\cos\left(3t+45^{\circ}\right)A$
2) $\sqrt{2}\cos\left(3t-45^{\circ}\right)A$
3) $\frac{1}{\sqrt{2}}\cos\left(3t+45^{\circ}\right)A$
4) $\frac{1}{\sqrt{2}}\cos\left(3t-45^{\circ}\right)A$

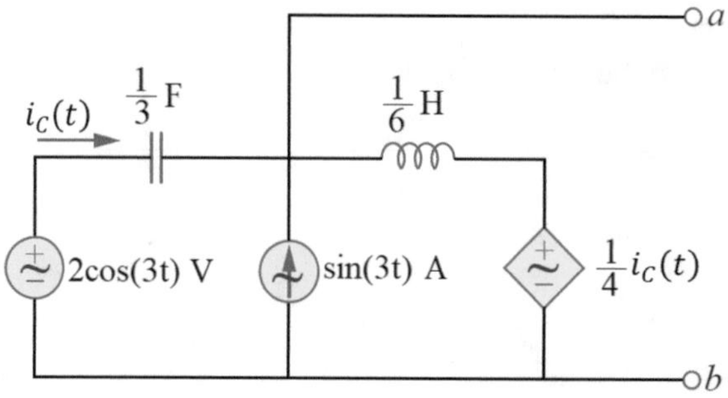

Figure 1.32 The circuit of problem 1.34

1.35. In the circuit of Figure 1.33, calculate the sinusoidal steady-state response of the output voltage ($v_o(t)$).

Difficulty level ○ Easy ● Normal ○ Hard

Calculation amount ○ Small ● Normal ○ Large

1) 3.16 cos (t + 18.43°) + 0.33 cos (2t – 20.54°) V
2) 3.16 cos (t + 161.57°) + 0.66 cos (2t + 20.54°) V
3) 3.16 cos (t + 161.75°) + 0.33 cos (2t + 20.54°) V
4) 3.16 cos (t + 18.43°) + 0.66 cos (2t – 20.54°) V

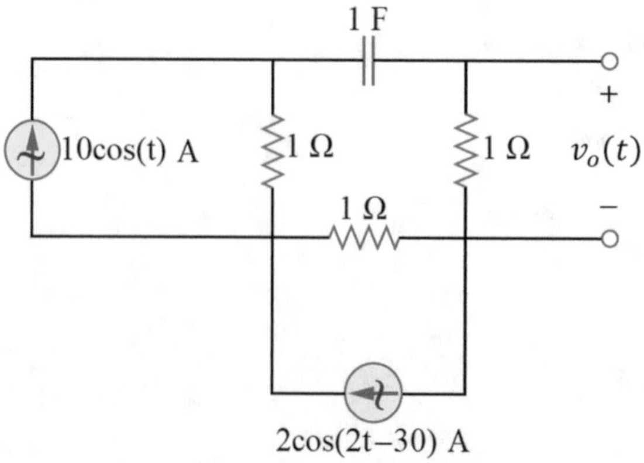

Figure 1.33 The circuit of problem 1.35

1.36. In the circuit of Figure 1.34, what must be the impedance of the load to absorb the maximum average power?

Difficulty level ○ Easy ● Normal ○ Hard

Calculation amount ○ Small ● Normal ○ Large

1) $(1 - j3)\ \Omega$
2) $(0.9 - j0.3)\ \Omega$
3) $(0.9 + j0.3)\ \Omega$
4) $(0.27 - j2.86)\ \Omega$

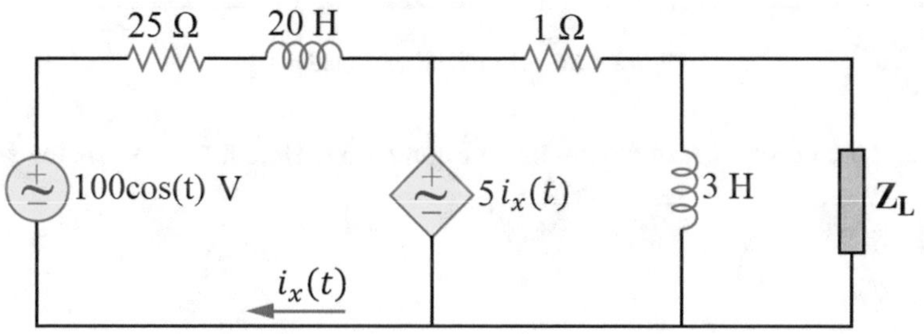

Figure 1.34 The circuit of problem 1.36

1.37. In the circuit of Figure 1.35, determine the impedance of the load ($\mathbf{Z_L}$) so that the load can absorb the maximum average power. In addition, how much is this maximum average power?

Difficulty level ○ Easy ● Normal ○ Hard

Calculation amount ○ Small ● Normal ○ Large

1) $\mathbf{Z_L} = (26.66 + j20)\ \Omega,\ P_{max} = 0.83\ W$
2) $\mathbf{Z_L} = (22.33 + 18.33)\ \Omega,\ P_{max} = 0.73\ W$
3) $\mathbf{Z_L} = (26.66 - j20)\ \Omega,\ P_{max} = 0.83\ W$
4) $\mathbf{Z_L} = (26.66 - j18.33)\ \Omega,\ P_{max} = 0.73\ W$

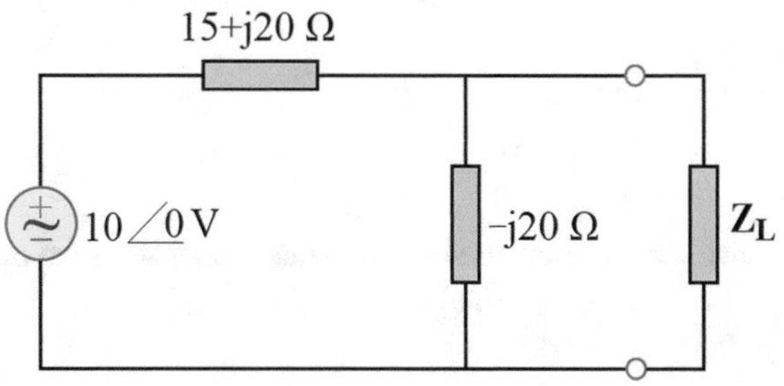

Figure 1.35 The circuit of problem 1.37

1.38. In the circuit of Figure 1.36, calculate the sinusoidal steady-state response of $i(t)$.

Difficulty level ○ Easy ● Normal ○ Hard

Calculation amount ○ Small ● Normal ○ Large

1) $4.5 \sin(1000t + 30^\circ)\ A$
2) $1.5 \cos(1000t + 39^\circ)\ A$
3) $4.5 \sin(1000t + 39^\circ)\ A$
4) $1.5 \sin(1000t + 45^\circ)\ A$

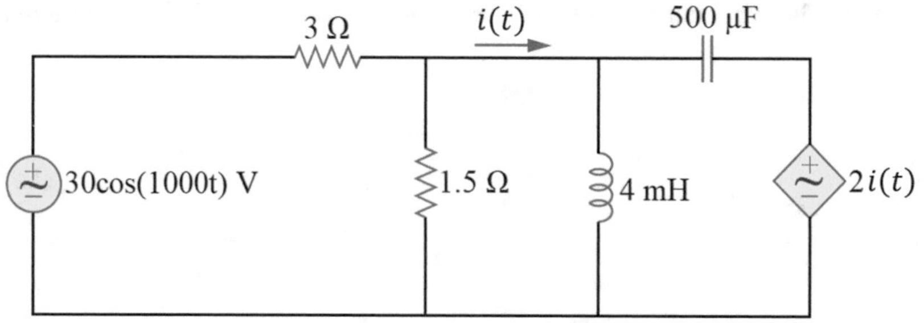

Figure 1.36 The circuit of problem 1.38

1.39. Calculate the phasor of the current $i(t)$ in the circuit of Figure 1.37. Herein "$\angle$" is the symbol of phase angle.

Difficulty level ○ Easy ● Normal ○ Hard
Calculation amount ○ Small ● Normal ○ Large

1) $(4.47\angle 36.9°)\ A$
2) $(4.47\angle 143.1°)\ A$
3) $(0.44\angle 36.9°)\ A$
4) $(0.44\angle -143.1°)\ A$

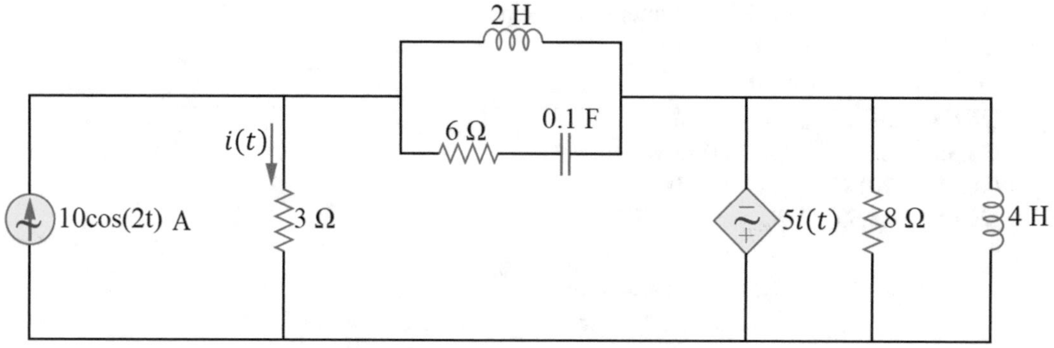

Figure 1.37 The circuit of problem 1.39

1.40. In the circuit of Figure 1.38, calculate the sinusoidal steady-state response of the output voltage ($v_o(t)$).

Difficulty level ○ Easy ● Normal ○ Hard
Calculation amount ○ Small ● Normal ○ Large

1) $\sqrt{10}\sin\left(4t + (\tan^{-1}(3))^{\circ}\right)\ V$
2) $\frac{\sqrt{10}}{2}\sin\left(4t - (\tan^{-1}(\frac{1}{3}))^{\circ}\right)\ V$
3) $\frac{\sqrt{10}}{2}\sin\left(4t + (\tan^{-1}(3))^{\circ}\right)\ V$
4) $\frac{\sqrt{10}}{2}\sin\left(4t - (\tan^{-1}(3))^{\circ}\right)\ V$

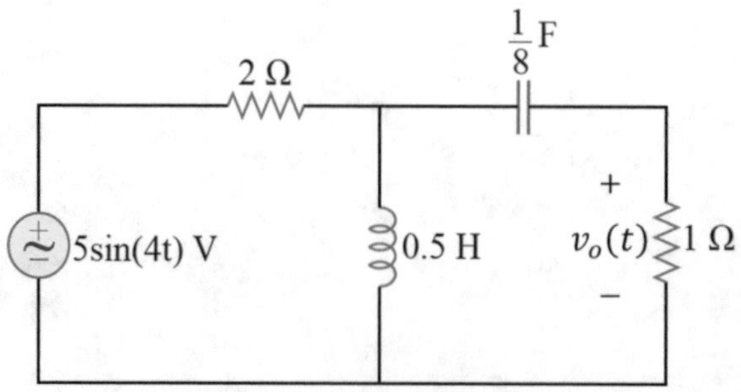

Figure 1.38 The circuit of problem 1.40

1.41. In the circuit of Figure 1.39, $C_1 = \frac{2}{5}\ mF$, $C_2 = \frac{5}{6}\ mF$, and $L = \frac{3}{5}\ H$. Calculate the phasor of $v_x(t)$.

Difficulty level ○ Easy ● Normal ○ Hard

Calculation amount ○ Small ● Normal ○ Large

1) $(150 - j200)\ V$
2) $(150 + j200)\ V$
3) $(-200 + j150)\ V$
4) $(200 - j150)\ V$

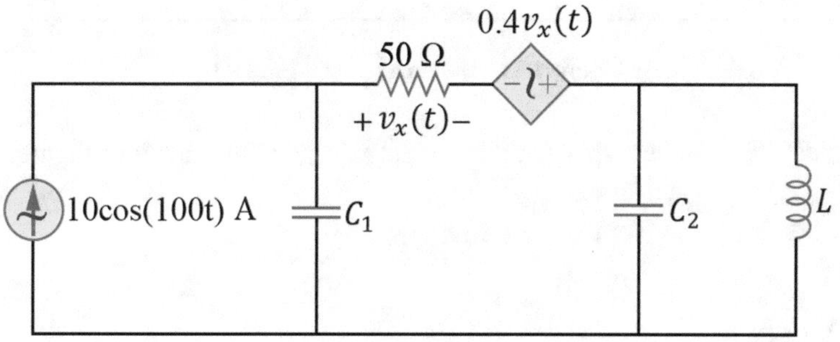

Figure 1.39 The circuit of problem 1.41

1.42. In the circuit of Figure 1.40, calculate the sinusoidal steady-state current indicated by $i(t)$.

Difficulty level ○ Easy ● Normal ○ Hard

Calculation amount ○ Small ● Normal ○ Large

1) $\sqrt{2}\cos\left(10t + 45^\circ\right)\ A$
2) $\sqrt{2}\cos\left(10t + 135^\circ\right)\ A$
3) $\sqrt{2}\cos\left(10t - 45^\circ\right)\ A$
4) $\sqrt{2}\cos\left(10t - 135^\circ\right)\ A$

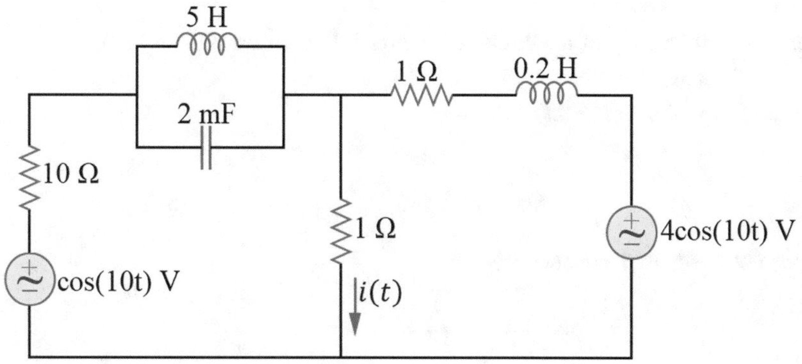

Figure 1.40 The circuit of problem 1.42

1.43. In the circuit of Figure 1.41, determine the sinusoidal steady-state response of $i(t)$.

Difficulty level ○ Easy ● Normal ○ Hard

Calculation amount ○ Small ● Normal ○ Large

1) $5\sqrt{2}\sin\left(1000t + 45^\circ\right)\ A$
2) $5\sqrt{2}\sin\left(1000t - 45^\circ\right)\ A$
3) $10\sqrt{2}\sin\left(1000t - 45^\circ\right)\ A$
4) $10\sqrt{2}\sin\left(1000t + 45^\circ\right)\ A$

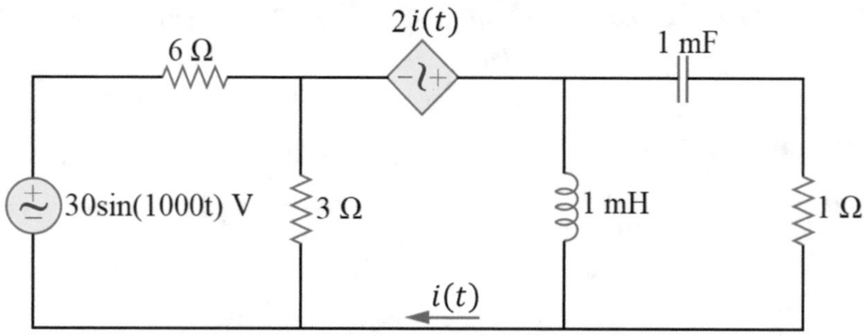

Figure 1.41 The circuit of problem 1.43

1.44. Calculate the sinusoidal steady-state response of the output voltage in the circuit of Figure 1.42.

Difficulty level ○ Easy ● Normal ○ Hard

Calculation amount ○ Small ● Normal ○ Large

1) $2\sqrt{2}\cos\left(t-\frac{\pi}{4}\right)\ V$
2) $2\sqrt{2}\sin\left(t-\frac{\pi}{4}\right)\ V$
3) $2\sqrt{2}\sin\left(t+\frac{3\pi}{4}\right)\ V$
4) $2\sqrt{2}\cos\left(t+\frac{3\pi}{4}\right)\ V$

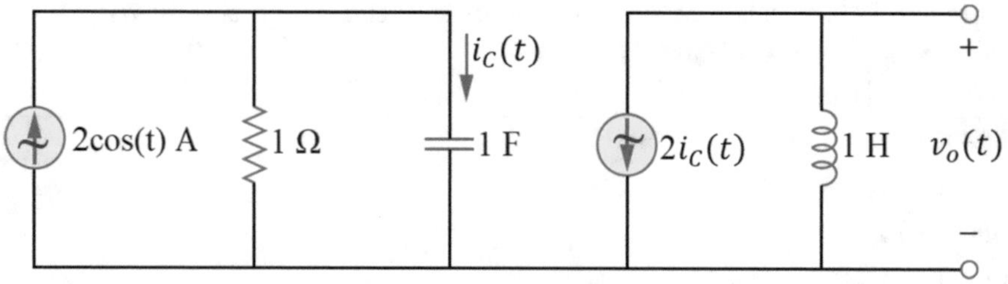

Figure 1.42 The circuit of problem 1.44

1.45. Determine the resonance frequency of the circuit of Figure 1.43 seen from the terminal.

Difficulty level ○ Easy ● Normal ○ Hard

Calculation amount ○ Small ● Normal ○ Large

1) $\frac{\sqrt{3}}{2}$ *rad/sec*
2) $\frac{1}{2}$ *rad/sec*
3) 1 *rad/sec*
4) The circuit does not have any resonance frequency

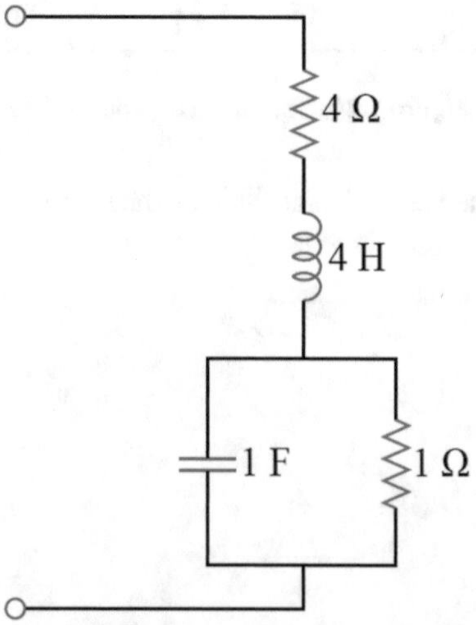

Figure 1.43 The circuit of problem 1.45

1.46. In the circuit of Figure 1.44, determine the sinusoidal steady-state response of $v(t)$.

Difficulty level ○ Easy ● Normal ○ Hard

Calculation amount ○ Small ● Normal ○ Large

1) $\sqrt{2}\cos(2t)\ V$
2) $\sqrt{2}\sin(2t)\ V$
3) $\sqrt{2}\cos\left(2t+\frac{\pi}{4}\right)\ V$
4) $\sqrt{2}\sin\left(2t+\frac{\pi}{4}\right)\ V$

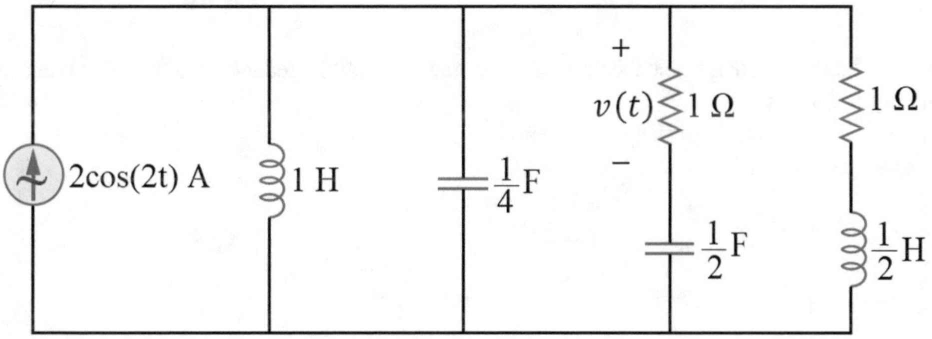

Figure 1.44 The circuit of problem 1.46

1.47. In the circuit of Figure 1.45, $\mathbf{V_1} = (10 + j2)\ V$ and $\mathbf{V_2} = (12 + j12)\ V$. Calculate the voltage of the left-side voltage source, that is, $\mathbf{V_{s1}}$.

Difficulty level ○ Easy ● Normal ○ Hard

Calculation amount ○ Small ● Normal ○ Large

1) $(67 + j59)\ V$
2) $(29 + j59)\ V$
3) $(67 + j57)\ V$
4) $(24 + j54)\ V$

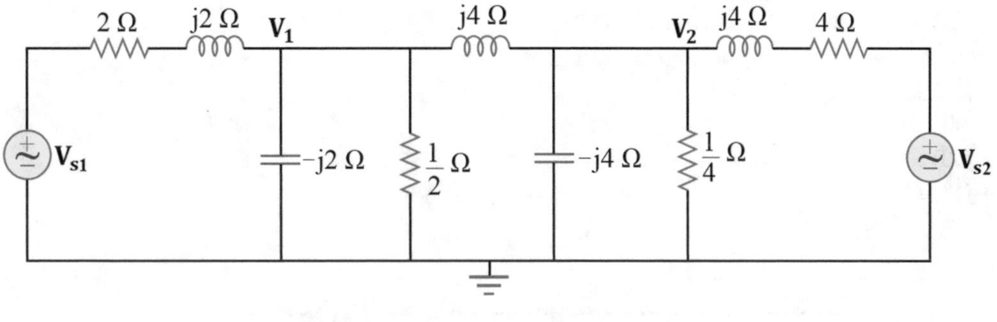

Figure 1.45 The circuit of problem 1.47

1.48. Calculate the average power of the resistor in the circuit of Figure 1.46.

Difficulty level ○ Easy ● Normal ○ Hard

Calculation amount ○ Small ● Normal ○ Large

1) $0.65\ W$
2) $1.1\ W$
3) $2.6\ W$
4) $4.8\ W$

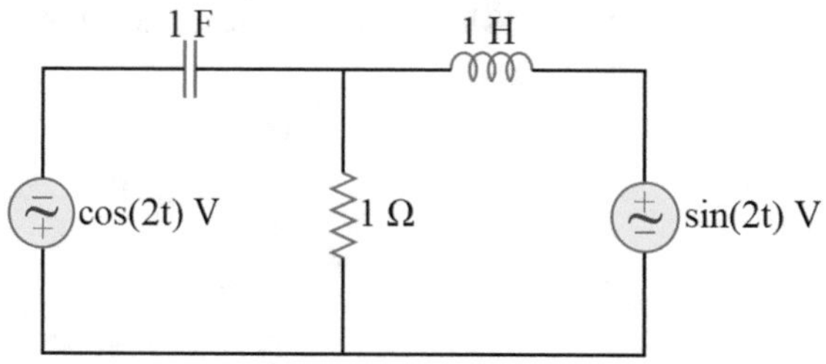

Figure 1.46 The circuit of problem 1.48

1.49. In the circuit of Figure 1.47, determine the value of C so that the input impedance and the input admittance of the circuit are equal in any frequency.

Difficulty level ○ Easy ● Normal ○ Hard

Calculation amount ○ Small ● Normal ○ Large

1) $C = 2\ F$
2) $C = 0.5\ F$
3) $C = 1\ F$
4) No C can be found.

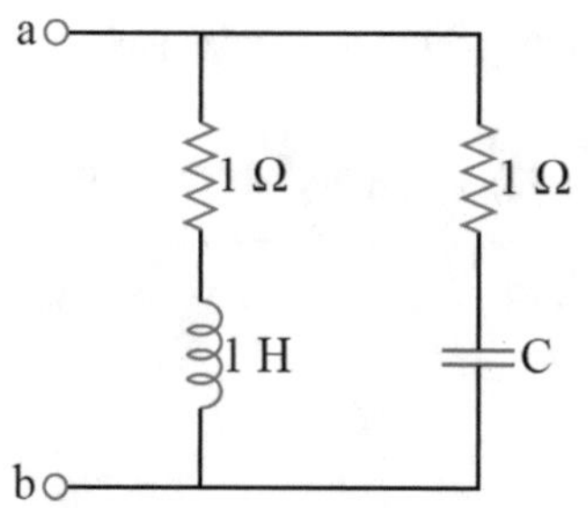

Figure 1.47 The circuit of problem 1.49

1.50. Calculate the phasor of the output voltage ($v_O(t)$) in the circuit of Figure 1.48.

Difficulty level ○ Easy ● Normal ○ Hard

Calculation amount ○ Small ● Normal ○ Large

1) $(50 + j20)\ \Omega$
2) $(54 - j48)\ \Omega$
3) $(20 + j50)\ \Omega$
4) $(54 + j48)\ \Omega$

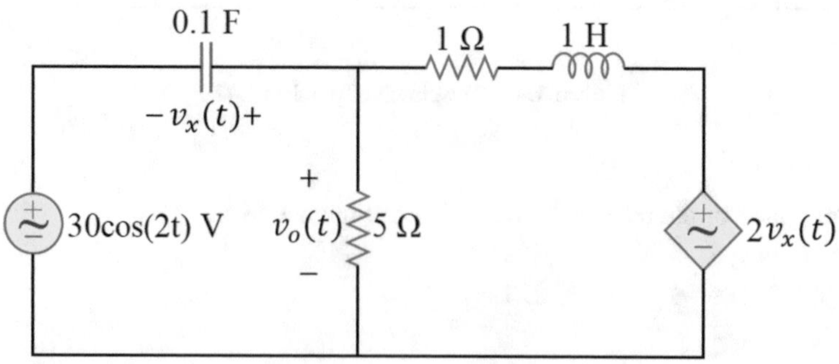

Figure 1.48 The circuit of problem 1.50

1.51. In the circuit of Figure 1.49, calculate the voltage of $v_x(t)$ in sinusoidal steady state.

Difficulty level ○ Easy ● Normal ○ Hard

Calculation amount ○ Small ● Normal ○ Large

1) $125\cos\left(100t - \left(\tan^{-1}\left(\frac{3}{4}\right)\right)^{\circ}\right)\ V$
2) $250\cos\left(100t - \left(\tan^{-1}\left(\frac{4}{3}\right)\right)^{\circ}\right)\ V$
3) $125\cos\left(100t - \left(\tan^{-1}\left(\frac{4}{3}\right)\right)^{\circ}\right)\ V$
4) $250\cos\left(100t - \left(\tan^{-1}\left(\frac{3}{4}\right)\right)^{\circ}\right)\ V$

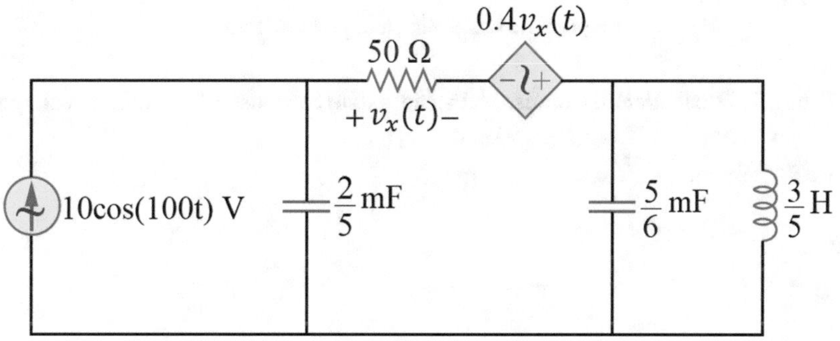

Figure 1.49 The circuit of problem 1.51

1.52. In the circuit of Figure 1.50, calculate the average power of the independent voltage source in sinusoidal steady state.

Difficulty level ○ Easy ● Normal ○ Hard

Calculation amount ○ Small ● Normal ○ Large

1) 1250 W
2) 2500 W
3) −1250 W
4) −2500 W

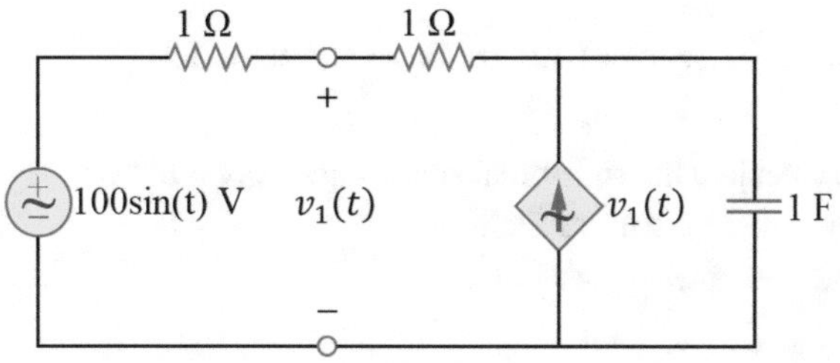

Figure 1.50 The circuit of problem 1.52

1.53. In the circuit of Figure 1.51, determine the value of α in sinusoidal steady state so that $v_0(t) = \sin(t)\ V$.

Difficulty level ○ Easy ● Normal ○ Hard

Calculation amount ○ Small ● Normal ○ Large

1) $\alpha = 0.5$
2) $\alpha = -0.5$
3) $\alpha = -1$
4) $\alpha = 1$

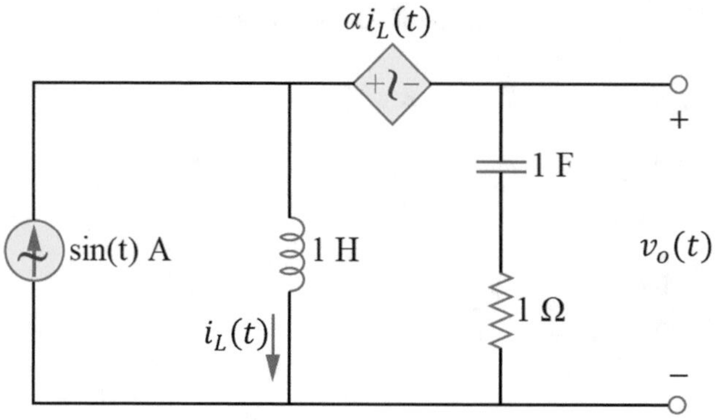

Figure 1.51 The circuit of problem 1.53

1.54. In the circuit of Figure 1.52 calculate the sinusoidal steady-state response of the output voltage, that is, $v_o(t)$.

Difficulty level ○ Easy ● Normal ○ Hard

Calculation amount ○ Small ● Normal ○ Large

1) $3\sin(t)\ V$
2) $3\sqrt{2}\cos(t)\ V$
3) $3\sqrt{2}\sin\left(t-\frac{\pi}{4}\right)\ V$
4) $3\cos\left(t+\frac{\pi}{4}\right)\ V$

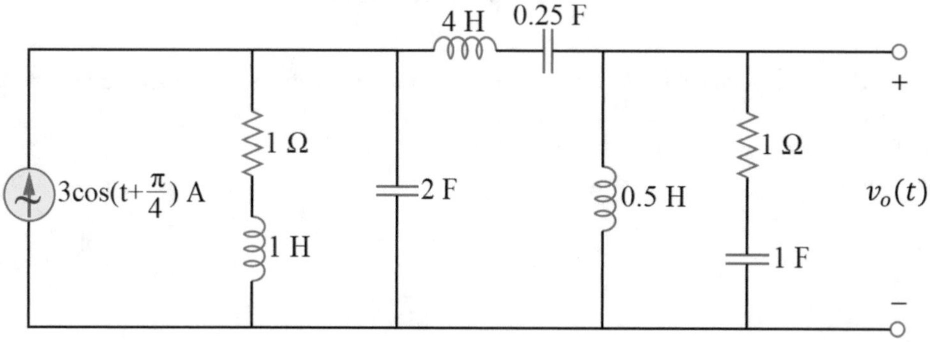

Figure 1.52 The circuit of problem 1.54

1.55. In the circuit of Figure 1.53, calculate the sinusoidal steady-state value of $v(t)$.

Difficulty level ○ Easy ● Normal ○ Hard

Calculation amount ○ Small ● Normal ○ Large

1) $5\cos(10t+45^{\circ})\ V$
2) $5\cos(10t+90^{\circ})\ V$
3) $10\cos(10t+45^{\circ})\ V$
4) $10\cos(10t+90^{\circ})\ V$

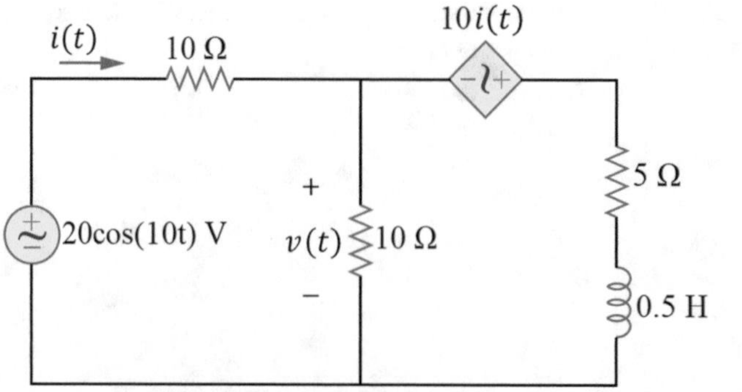

Figure 1.53 The circuit of problem 1.55

1.56. In the circuit of Figure 1.54, what must be the resistance of R to see unity power factor by the voltage source in sinusoidal steady state?

Difficulty level ○ Easy ● Normal ○ Hard

Calculation amount ○ Small ● Normal ○ Large

1) 1 Ω
2) 2 Ω
3) 3 Ω
4) 0 Ω

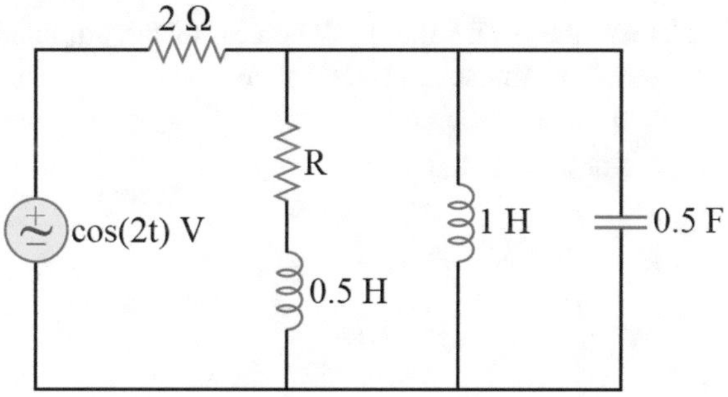

Figure 1.54 The circuit of problem 1.56

1.57. In the circuit of Figure 1.55, calculate the sinusoidal steady-state response of $i(t)$.

Difficulty level ○ Easy ● Normal ○ Hard

Calculation amount ○ Small ● Normal ○ Large

1) $5\sin(2t + 53^{\circ})\ A$
2) $5\sin(2t)\ A$
3) $5\cos(2t)\ A$
4) $\cos(2t)\ A$

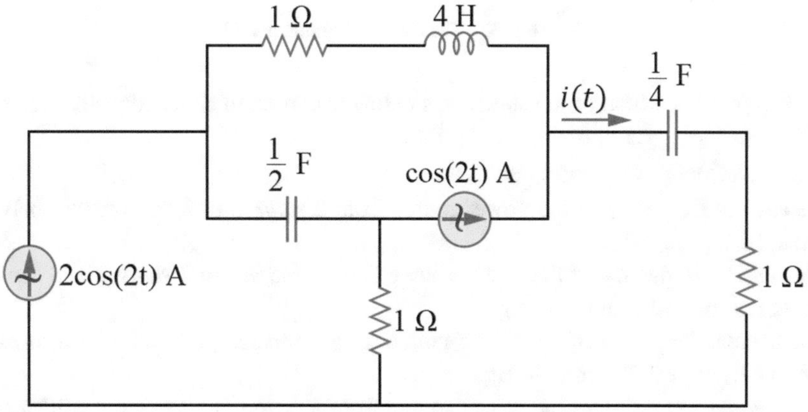

Figure 1.55 The circuit of problem 1.57

1.58. In the circuit of Figure 1.56, determine the complex power supplied by the current source.

Difficulty level ○ Easy ● Normal ○ Hard

Calculation amount ○ Small ● Normal ○ Large

1) $\left(\frac{1}{4} + j\frac{3}{2}\right)\ VA$
2) $\left(\frac{1}{4} - j\frac{3}{2}\right)\ VA$
3) $\left(\frac{1}{2} - j\frac{3}{4}\right)\ VA$
4) $\left(\frac{1}{2} + j\frac{3}{4}\right)\ VA$

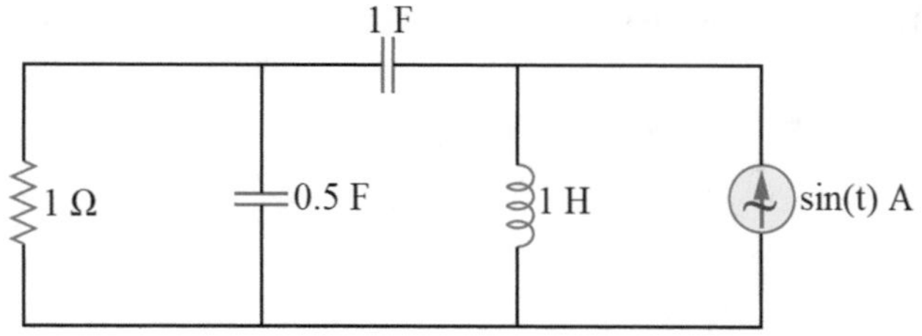

Figure 1.56 The circuit of problem 1.58

1.59. In the circuit of Figure 1.57, calculate the peak value peak values of the amplitude of the output voltage while it is absorbing the maximum average power in sinusoidal steady state.

Difficulty level ○ Easy ● Normal ○ Hard

Calculation amount ○ Small ○ Normal ● Large

1) $\frac{10}{\sqrt{2}}\ V$
2) $10\ V$
3) $\frac{100}{\sqrt{2}}\ V$
4) $100\ V$

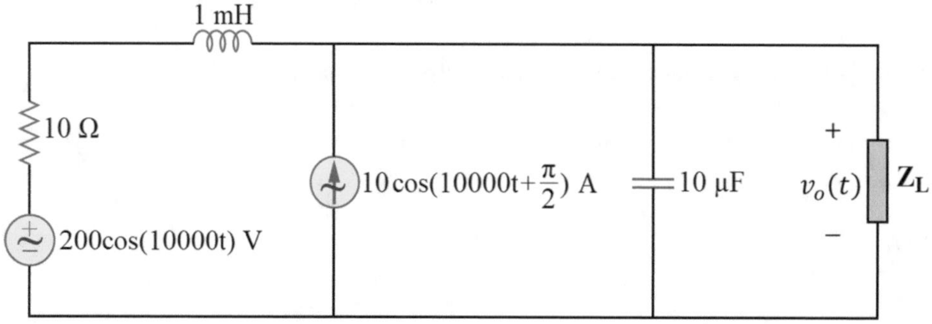

Figure 1.57 The circuit of problem 1.59

1.60. For the circuit of Figure 1.58, which of the statements below is correct in sinusoidal steady state?

Difficulty level ○ Easy ● Normal ○ Hard

Calculation amount ○ Small ○ Normal ● Large

1) Increasing the angular frequency of the power source from 2 *rad/sec* to 3 *rad/sec* will halve the amplitude of the indicated current (|**I**|).
2) Decreasing the angular frequency of the power source from 3 *rad/sec* to 2 *rad/sec* will cause the amplitude of the indicated current (|**I**|) to be 1.5 times as big.
3) Decreasing the angular frequency of the power source from 3 *rad/sec* to 2 *rad/sec* will cause the amplitude of the indicated current (|**I**|) to be 2.25 times as big.
4) Decreasing the angular frequency of the power source from 3 *rad/sec* to 2 *rad/sec* will cause the amplitude of the indicated current (|**I**|) to be 2.12 times as big.

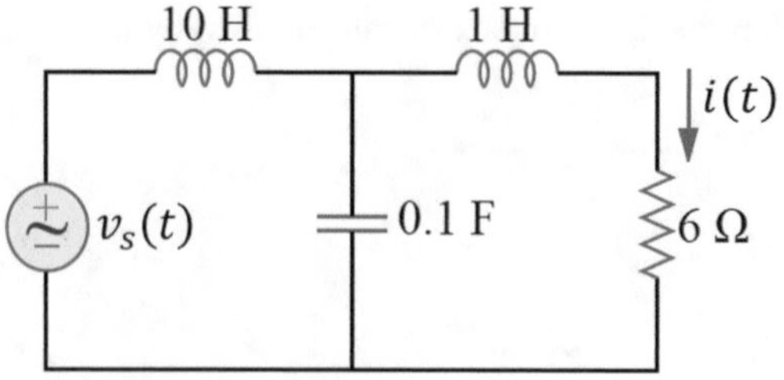

Figure 1.58 The circuit of problem 1.60

1.61. Determine the resonance frequency of the circuit illustrated in Figure 1.59.

Difficulty level ○ Easy ● Normal ○ Hard

Calculation amount ○ Small ○ Normal ● Large

1) $\sqrt{\frac{\alpha-1}{LC}}$
2) $\sqrt{\frac{\alpha}{LC}}$
3) $\sqrt{\frac{1-\alpha}{LC}}$
4) $\sqrt{\frac{1+\alpha}{LC}}$

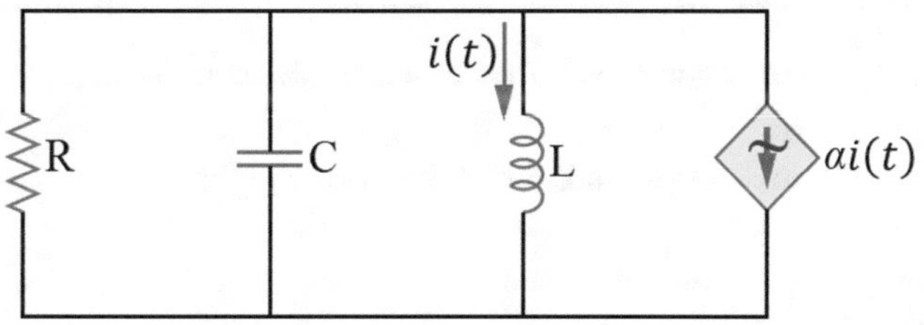

Figure 1.59 The circuit of problem 1.61

1.62. Which of the relations below must be held to calculate an equal resonance frequency in the circuits of Figure 1.60?

Difficulty level ○ Easy ● Normal ○ Hard

Calculation amount ○ Small ○ Normal ● Large

1) $R = \frac{L}{C}$
2) $R = \frac{C}{L}$
3) $R = \sqrt{\frac{L}{C}}$
4) $R = \sqrt{\frac{C}{L}}$

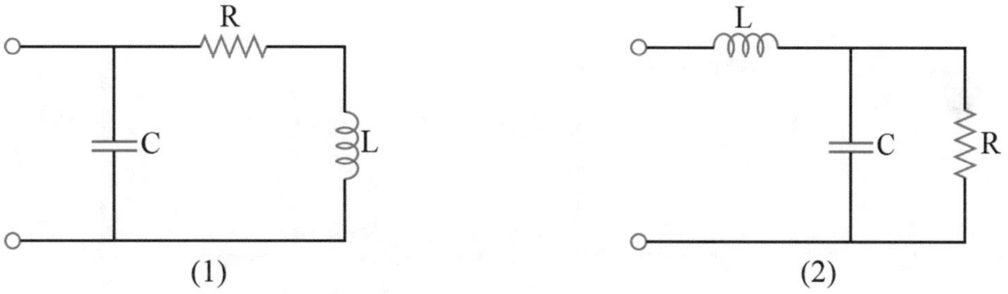

Figure 1.60 The circuit of problem 1.62

1.63. In the circuit of Figure 1.61, calculate the sinusoidal steady-state response of $i_x(t)$.

Difficulty level ○ Easy ● Normal ○ Hard

Calculation amount ○ Small ○ Normal ● Large

1) $-2\cos(t-60^\circ)+0.2\sin(10t)\ A$
2) $2\cos(t-60^\circ)+0.2\sin(10t)\ A$
3) $\sin(10t)-2\cos(t-60^\circ)\ A$
4) $-\sin(10t)+0.2\cos(t-60^\circ)\ A$

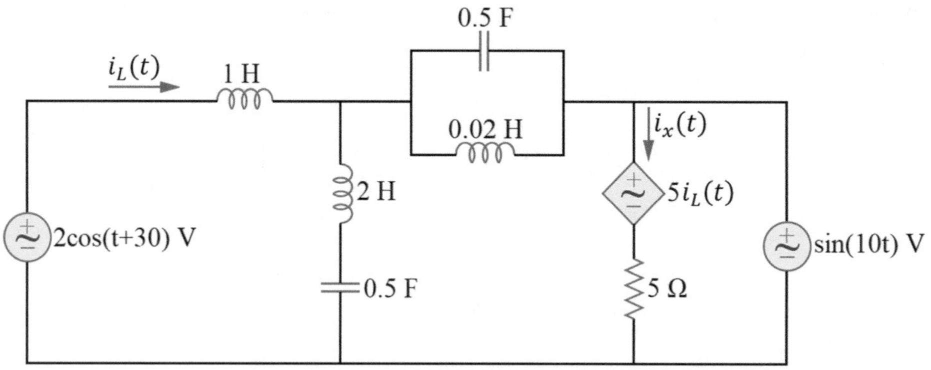

Figure 1.61 The circuit of problem 1.63

1.64. In the circuit of Figure 1.62 determine the impedance of the load ($\mathbf{Z_L}$) that can absorb the maximum average power in sinusoidal steady state.

Difficulty level ○ Easy ● Normal ○ Hard

Calculation amount ○ Small ○ Normal ● Large

1) $(2 + j0.5)\ \Omega$
2) $(-1 + j)\ \Omega$
3) $(2 - j0.5)\ \Omega$
4) $(1 + j)\ \Omega$

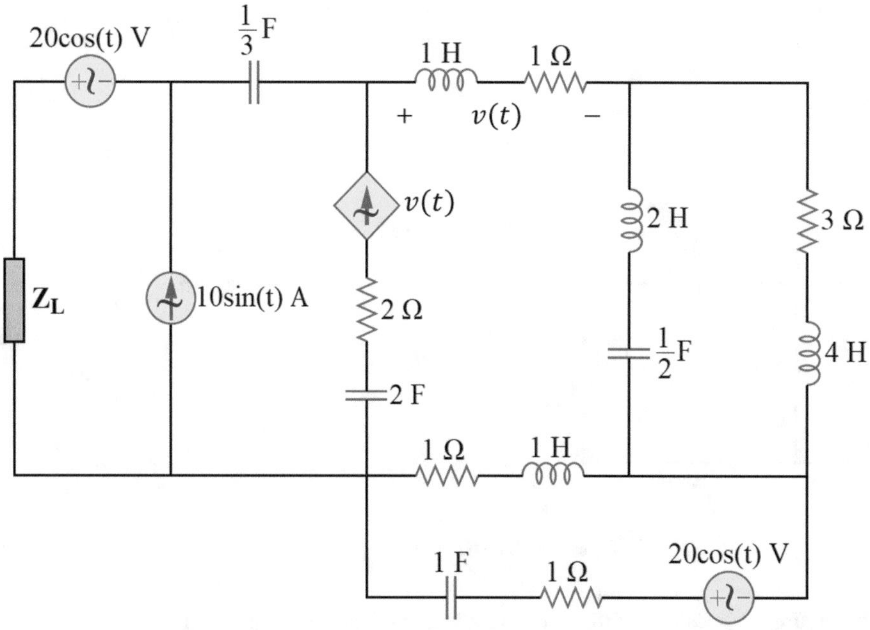

Figure 1.62 The circuit of problem 1.64

1.65. The circuit illustrated in Figure 1.63 is in sinusoidal steady state. Calculate the average power of the resistor located in the horizontal branch.

Difficulty level ○ Easy ● Normal ○ Hard

Calculation amount ○ Small ○ Normal ● Large

1) 16.4 W
2) 8.2 W
3) 14 W
4) 7 W

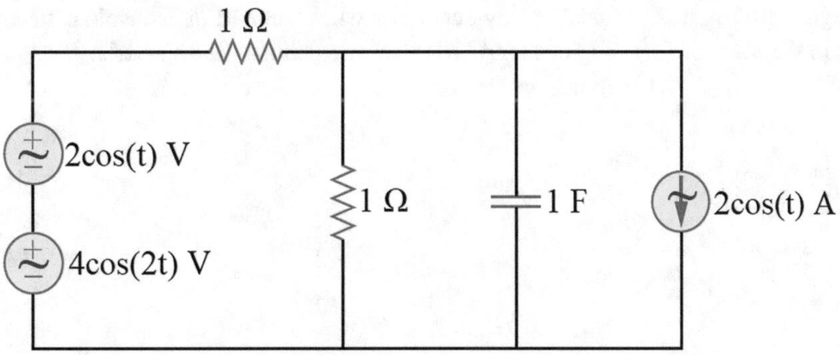

Figure 1.63 The circuit of problem 1.65

1.66. In the circuit of Figure 1.64, calculate sinusoidal steady-state value of $v(t)$.

Difficulty level ○ Easy ● Normal ○ Hard

Calculation amount ○ Small ○ Normal ● Large

1) $\frac{16}{15}\sin(4t) + \frac{2}{3}\cos(2t)\ V$
2) $\frac{16}{15}\cos(4t) + \frac{2}{3}\sin(2t)\ V$
3) $\frac{15}{16}\cos(4t) + \frac{3}{2}\sin(2t)\ V$
4) $\frac{15}{16}\sin(4t) + \frac{3}{2}\cos(2t)\ V$

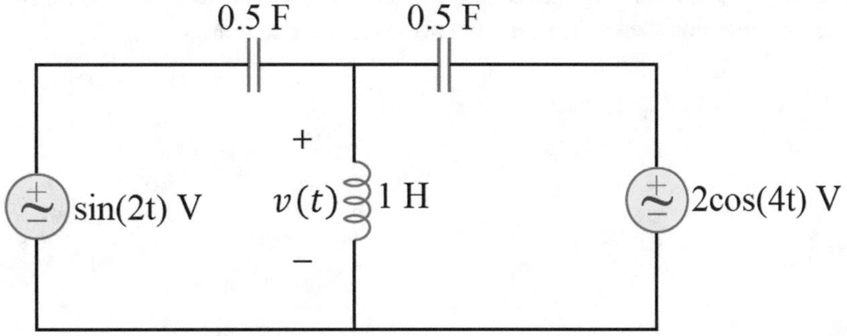

Figure 1.64 The circuit of problem 1.66

1.67. In the circuit of Figure 1.65, the impedance of the load is inductive, has the magnitude of 26 Ω, and absorbs the active power of 13 kW. The voltage source delivers 13.5 kW active power. Determine the reactance of the load.

Difficulty level ○ Easy ○ Normal ● Hard

Calculation amount ● Small ○ Normal ○ Large

1) 15.6 Ω
2) 20.8 Ω
3) 26 Ω
4) The information of the problem is not enough to solve it.

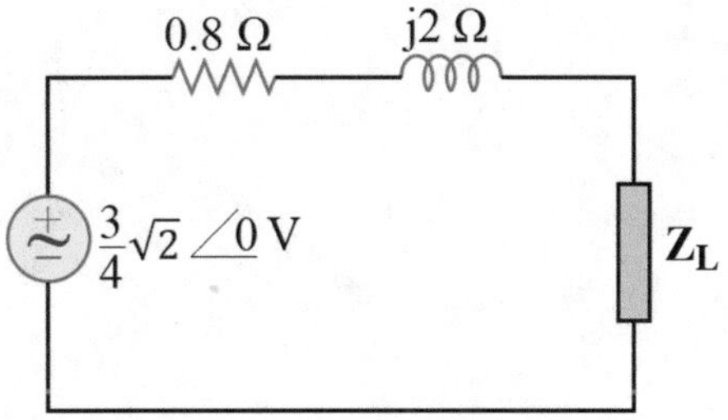

Figure 1.65 The circuit of problem 1.67

1.68. The circuit of Figure 1.66 is in sinusoidal steady state. For what value of φ, sinusoidal steady-state response of the output current is in the form of $i_o(t) = I_m \cos(t)\ A$? Moreover, calculate the value of I_m.

Difficulty level ○ Easy ○ Normal ● Hard

Calculation amount ● Small ○ Normal ○ Large

1) $\varphi = -\frac{\pi}{4}\ rad, \mathrm{I_m} = \sqrt{2}\ A$
2) $\varphi = -\frac{\pi}{4}\ rad, \mathrm{I_m} = \frac{1}{\sqrt{2}}\ A$
3) $\varphi = \frac{\pi}{4}\ rad, \mathrm{I_m} = \frac{1}{\sqrt{2}}\ A$
4) $\varphi = \frac{\pi}{4}\ rad, \mathrm{I_m} = \sqrt{2}\ A$

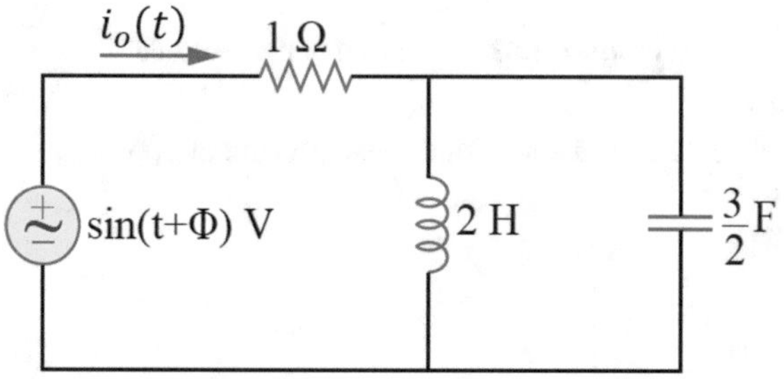

Figure 1.66 The circuit of problem 1.68

1.69. In the circuit of Figure 1.67, determine the phase angle of $\mathbf{Z_a}$ if $|\mathbf{Z_a}| = 100\ \Omega$, $|\mathbf{Z_b}| = 50\ \Omega$, and $\omega = 100\ rad/sec$. The network includes linear time-invariant (LTI) components and power sources.

Difficulty level ○ Easy ○ Normal ● Hard

Calculation amount ○ Small ● Normal ○ Large

1) 36°
2) 47°
3) 61°
4) 85°

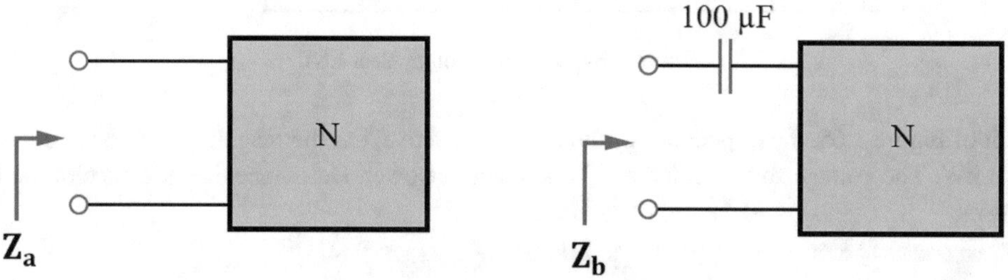

Figure 1.67 The circuit of problem 1.69

1.70. The circuit shown in Figure 1.68 is in sinusoidal steady state. Calculate the root mean square (rms) value of $v(t)$.

Difficulty level ○ Easy ○ Normal ● Hard

Calculation amount ○ Small ● Normal ○ Large

1) $\frac{\sqrt{5}}{2}\ V$
2) $\frac{\sqrt{3}}{2}\ V$
3) $\frac{1+\sqrt{3}}{2}\ V$
4) $\sqrt{\frac{7}{20}}\ V$

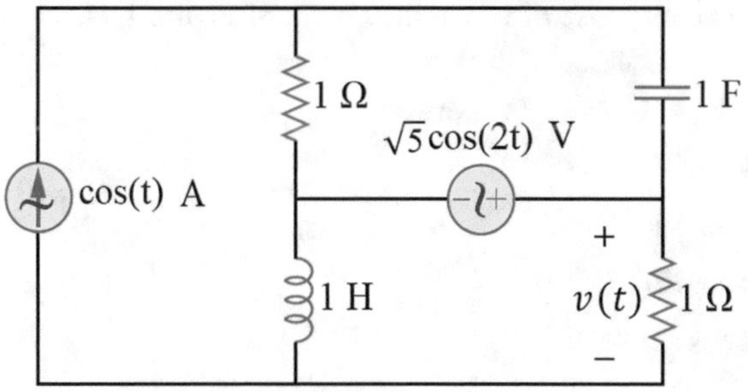

Figure 1.68 The circuit of problem 1.70

1.71. The differential equation for the input of $v_s(t)$ and the output of $i(t)$, related to the circuit of Figure 1.69, is as follows:

$$\frac{d^4}{dt^4}i(t) + 10\frac{d^3}{dt^3}i(t) + 40\frac{d^2}{dt^2}i(t) + 60\frac{d}{dt}i(t) + 784i(t) = 10\frac{d}{dt}v_s(t) + 40v_s(t)$$

Determine an equivalent circuit for sinusoidal steady-state behavior of the linear time-invariant (LTI) circuit in the angular frequency of 4 *rad/sec*.

Difficulty level ○ Easy ○ Normal ● Hard

Calculation amount ○ Small ● Normal ○ Large

1) The circuit is equivalent to a single $\frac{1}{40}$ *H* inductor
2) The circuit is equivalent to a single $\frac{1}{40}$ *F* capacitor
3) The circuit is equivalent to the series connection of a resistor and a capacitor
4) The circuit is equivalent to the series connection of a resistor and an inductor

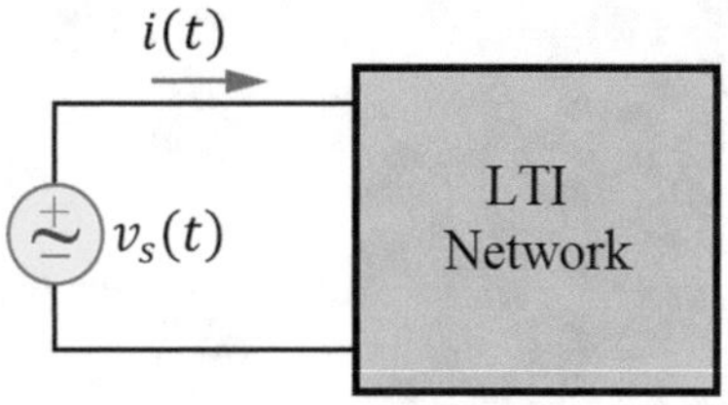

Figure 1.69 The circuit of problem 1.71

1.72. In the circuit of Figure 1.70, which of the relations below need to be held to transfer maximum average power to the right side of terminal a–b?

Difficulty level ○ Easy ○ Normal ● Hard

Calculation amount ○ Small ● Normal ○ Large

1) $\omega C = 0.01,\ \omega L = 50$
2) $\omega C = 0.01,\ \omega L = 200$
3) $\omega C = 100,\ \omega L = 2$
4) $\omega C = 100,\ \omega L = 0.5$

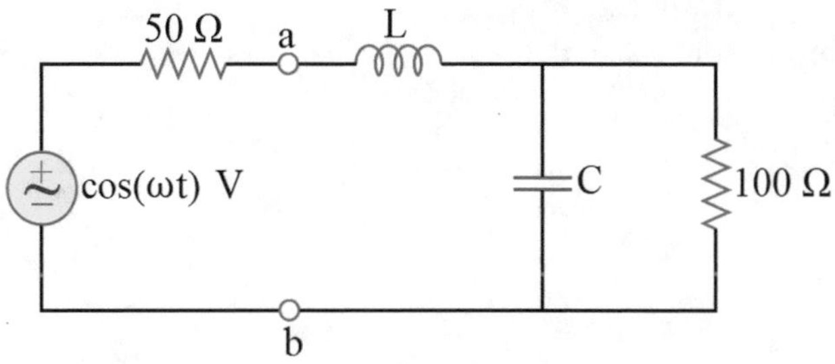

Figure 1.70 The circuit of problem 1.72

1.73. Calculate sinusoidal steady-state response of $v(t)$ in the circuit of Figure 1.71.

Difficulty level ○ Easy ○ Normal ● Hard

Calculation amount ○ Small ● Normal ○ Large

1) $\frac{8\sqrt{10}}{5}\cos(t)+4\cos(2t)\ V$
2) $\frac{8\sqrt{10}}{5}\cos(t)+4\cos\left(2t+30^{\circ}\right)\ V$
3) $\frac{8\sqrt{10}}{5}\cos\left(t-18.43^{\circ}\right)+4\cos(2t)\ V$
4) $\frac{8\sqrt{10}}{5}\cos\left(t-18.43^{\circ}\right)+4\cos\left(2t+30^{\circ}\right)\ V$

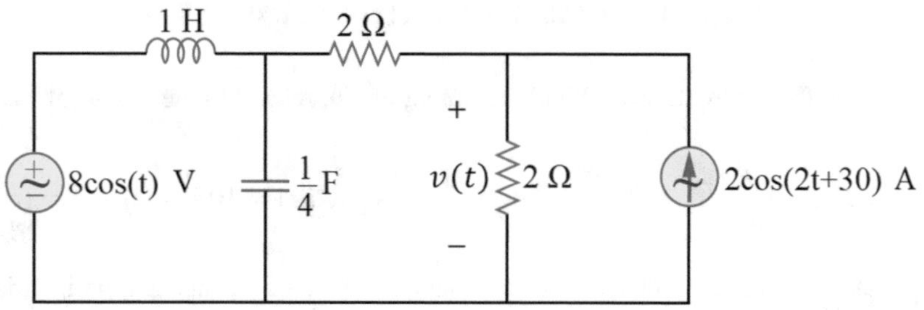

Figure 1.71 The circuit of problem 1.73

1.74. In the circuit of Figure 1.72.1, determine the value of R, L, and C so that the frequency response ($\mathbf{H}(j\omega)=\frac{\mathbf{I}(j\omega)}{\mathbf{V}(j\omega)}$) of Figure 1.72.2 is achieved.

Difficulty level ○ Easy ○ Normal ● Hard

Calculation amount ○ Small ● Normal ○ Large

1) $R=50,\ L=0.1\ mH,\ C=10\ mF$
2) $R=50,\ L=10\ mH,\ C=0.1\ mF$
3) $R=0.02,\ L=10\ mH,\ C=0.1\ mF$
4) $R=0.02,\ L=0.1\ mH,\ C=10\ mF$

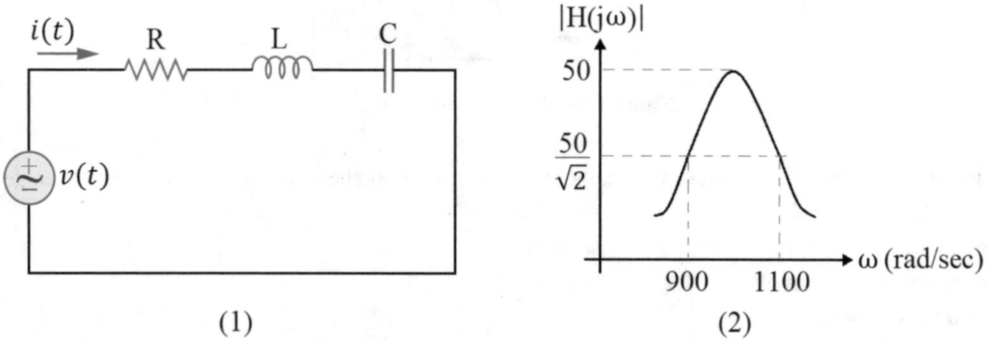

Figure 1.72 The circuit of problem 1.74

1.75. Calculate the sinusoidal steady-state response of the output voltage ($v_o(t)$) in the circuit of Figure 1.73.

Difficulty level ○ Easy ○ Normal ● Hard

Calculation amount ○ Small ● Normal ○ Large

1) $2+\sqrt{2}\cos\left(t+45^{\circ}\right)\ V$
2) $2\cos\left(t-45^{\circ}\right)\ V$
3) $2+\sqrt{2}\cos\left(t-45^{\circ}\right)\ V$
4) $2\cos\left(t+45^{\circ}\right)\ V$

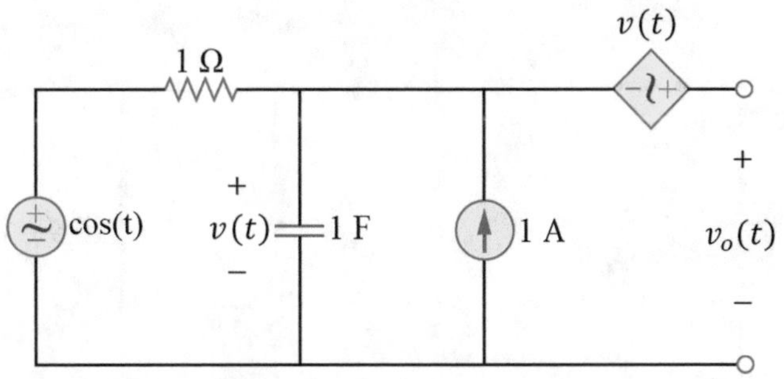

Figure 1.73 The circuit of problem 1.75

1.76. The circuit of Figure 1.74 shows three parallel impedances. For this circuit, we have the following information:

- The first impedance consumes 25 kW and 25 kVAr active and reactive powers, respectively.
- The second impedance absorbs 15 kVA apparent power with the leading power factor of 0.8.
- The third impedance consumes 11 kW active power with the unity power factor.

Determine the power factor of the circuit.

Difficulty level ○ Easy ○ Normal ● Hard

Calculation amount ○ Small ● Normal ○ Large

1) $\frac{2}{\sqrt{10}}$, lagging
2) $\frac{3}{\sqrt{10}}$, leading
3) $\frac{2}{\sqrt{10}}$, leading
4) $\frac{3}{\sqrt{10}}$, lagging

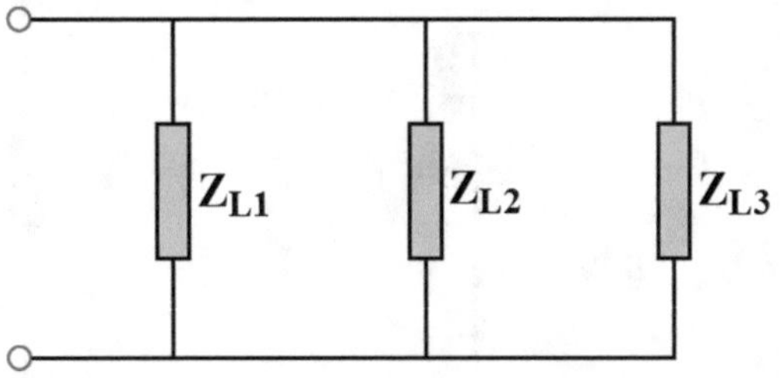

Figure 1.74 The circuit of problem 1.76

1.77. In the circuit of Figure 1.75, the maximum average power that can be transferred to the network is $\frac{1}{8}$ W. Determine the value of R and A.

Difficulty level ○ Easy ○ Normal ● Hard

Calculation amount ○ Small ● Normal ○ Large

1) $R = 1\ \Omega, A = 0.5\ V$
2) $R = 1\ \Omega, A = 1\ V$
3) $R = 3\ \Omega, A = 1\ V$
4) $R = 3\ \Omega, A = 0.5\ V$

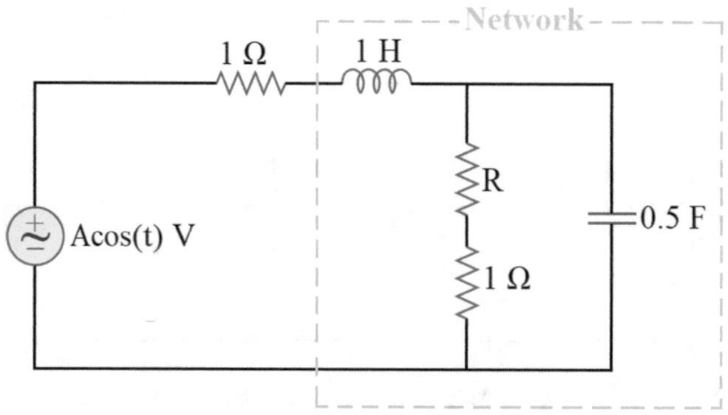

Figure 1.75 The circuit of problem 1.77

1.78. The circuit shown in Figure 1.76 is in sinusoidal steady state. Determine the Thevenin equivalent circuit seen by the load.

Difficulty level ○ Easy ○ Normal ● Hard

Calculation amount ○ Small ● Normal ○ Large

1) $\mathbf{V_{Th}} = j\ V, \mathbf{Z_{Th}} = 1\ \Omega$
2) $\mathbf{V_{Th}} = -j\ V, \mathbf{Z_{Th}} = 1\ \Omega$
3) $\mathbf{V_{Th}} = -j\ V, \mathbf{Z_{Th}} = (1 - j)\ \Omega$
4) $\mathbf{V_{Th}} = \left(\sqrt{2} + j\sqrt{2}\right)\ V, \mathbf{Z_{Th}} = (1 - j)\ \Omega$

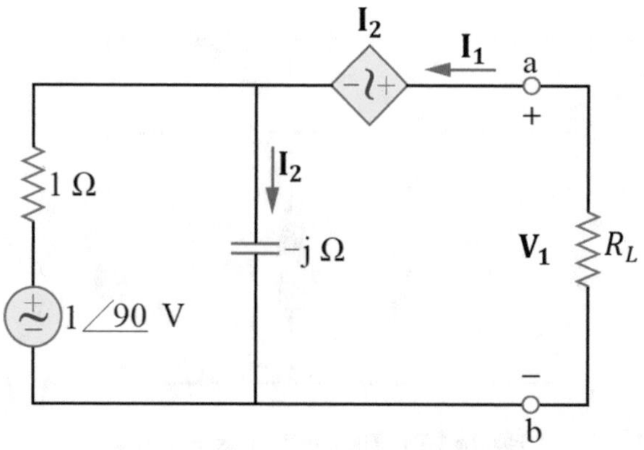

Figure 1.76 The circuit of problem 1.78

1.79. In the circuit of Figure 1.77, calculate the average power consumed in the indicated part of the circuit while its power factor is 1 in sinusoidal steady state.

Difficulty level ○ Easy ○ Normal ● Hard

Calculation amount ○ Small ● Normal ○ Large

1) 125 W
2) 250 W
3) 325 W
4) 500 W

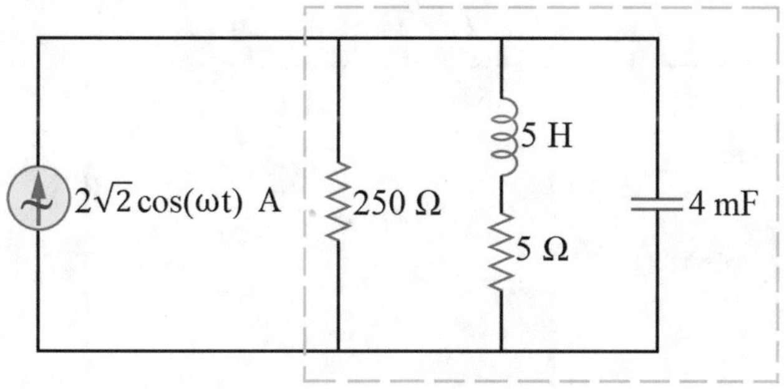

Figure 1.77 The circuit of problem 1.79

1.80. In the circuit of Figure 1.78, determine the ratio of the average power of the 1 Ω resistor to the average power of the voltage source in sinusoidal steady state.

Difficulty level ○ Easy ○ Normal ● Hard
Calculation amount ○ Small ● Normal ○ Large

1) $\frac{1}{2}$
2) $\frac{2}{3}$
3) $\frac{3}{4}$
4) 1

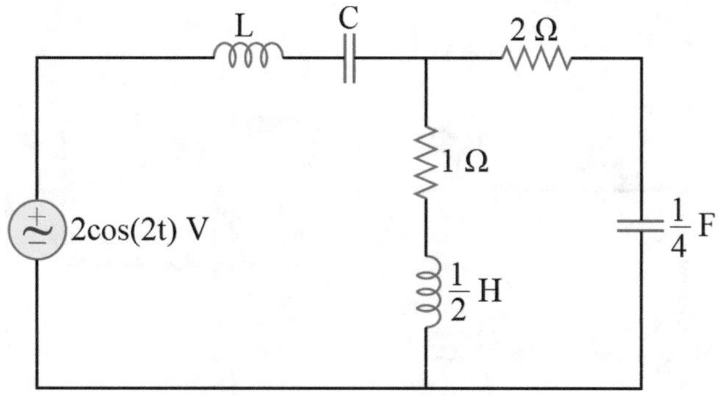

Figure 1.78 The circuit of problem 1.80

1.81. In the circuit of Figure 1.79, what percentage of total average power generated by the voltage source is consumed in the two 1 Ω resistors in sinusoidal steady state if the resistor of *R* is absorbing the maximum power?

Difficulty level ○ Easy ○ Normal ● Hard
Calculation amount ○ Small ● Normal ○ Large

1) 25%
2) 20%
3) 50%
4) 75%

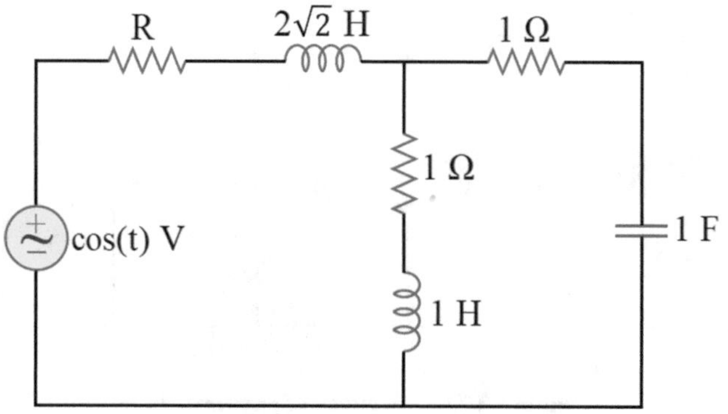

Figure 1.79 The circuit of problem 1.81

1.82. In the circuit of Figure 1.80, determine the inductance of the inductor so that the voltage and the current of the voltage source are in phase in sinusoidal steady state. Additionally, calculate the amplitude (peak value) of the current of the voltage source in this condition.

Difficulty level ○ Easy ○ Normal ● Hard

Calculation amount ○ Small ● Normal ○ Large

1) $2\ H$, $20\ A$
2) $2\ H$, $40\ A$
3) $1\ H$, $40\ A$
4) $1\ H$, $20\ A$

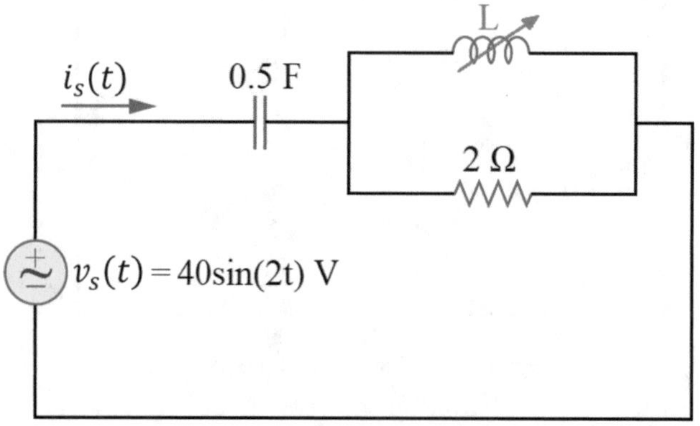

Figure 1.80 The circuit of problem 1.82

1.83. The circuit of Figure 1.81 is in sinusoidal steady state. The indicated network is in the resonance state and its maximum average power is 3 W. Calculate the reactive power of the current source.

Difficulty level ○ Easy ○ Normal ● Hard

Calculation amount ○ Small ○ Normal ● Large

1) $-4\ VAr$
2) $-4\sqrt{2}\ VAr$
3) $8\ VAr$
4) $4\sqrt{2}\ VAr$

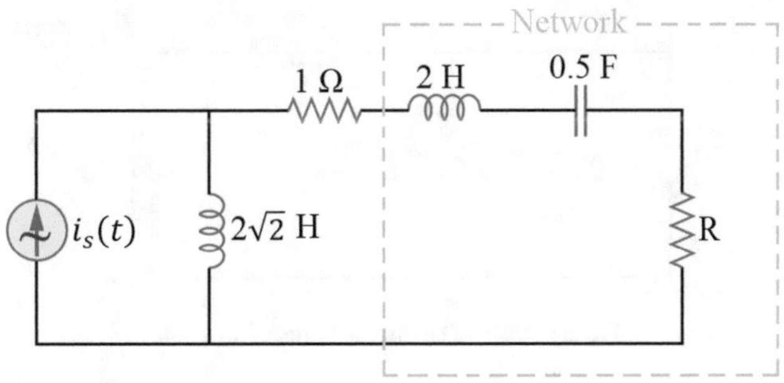

Figure 1.81 The circuit of problem 1.83

1.84. Calculate the average power of the dependent voltage source in the circuit of Figure 1.82.

Difficulty level ○ Easy ○ Normal ● Hard

Calculation amount ○ Small ○ Normal ● Large

1) 3 W generation
2) 6 W generation
3) 3 W consumption
4) 6 W consumption

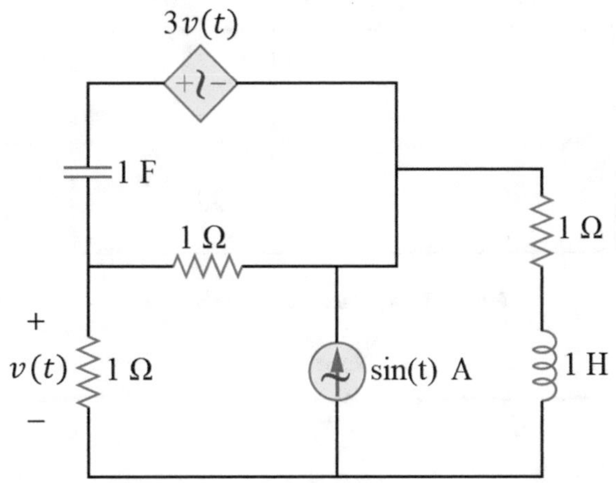

Figure 1.82 The circuit of problem 1.84

1.85. The circuit illustrated in Figure 1.83 is in sinusoidal steady state. Which of the following statements is incorrect if $\mathbf{Z_1} = (0.3 + j0.1)\ \Omega$, $\mathbf{Z_2} = (0.4 - j0.2)\ \Omega$, and $\mathbf{Z_3} = (0.2 + j0.4)\ \Omega$?

Difficulty level ○ Easy ○ Normal ● Hard

Calculation amount ○ Small ○ Normal ● Large

1) $|\mathbf{S_2}| = |\mathbf{S_3}|$
2) $P_2 = 2P_3$
3) $Q_1 = -2Q_2$
4) $Q_3 = 4Q_1$

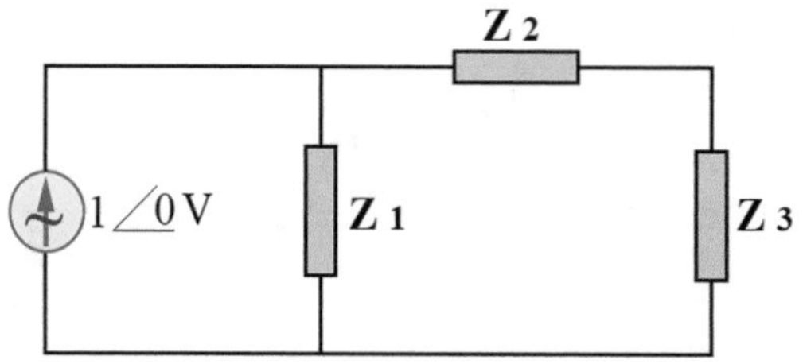

Figure 1.83 The circuit of problem 1.85

1.86. The circuit of Figure 1.84 is in sinusoidal steady state with the angular frequency of $\omega = 1\ rad/sec$. The maximum average power consumed by R is about $2\sqrt{5}\ W$. Calculate the total average power consumed in the circuit.

Difficulty level ○ Easy ○ Normal ● Hard

Calculation amount ○ Small ○ Normal ● Large

1) $2(\sqrt{5}+1)\ W$
2) $2\sqrt{5}\ W$
3) $(\sqrt{5}+1)\ W$
4) $\sqrt{5}\ W$

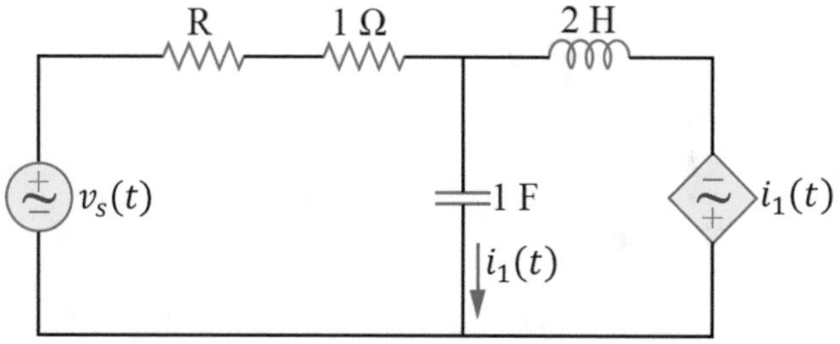

Figure 1.84 The circuit of problem 1.86

References

1. Rahmani-Andebili, M. (2020). DC Electrical circuit analysis: Practice problems, methods, and solutions, *Springer Nature*.

Solutions of Problems: Sinusoidal Steady-State Analysis

2

Abstract

In this chapter, the problems of the first chapter are fully solved, in detail, step-by-step, and with different methods.

2.1. Based on maximum average power transfer theorem, to transfer the maximum average power to a purely resistive load, the resistance of the load must be equal to the magnitude of the Thevenin impedance of the circuit seen by the load [1]. In other words:

$$R_L = |\mathbf{Z_{Th}}| \tag{1}$$

To calculate the Thevenin impedance ($\mathbf{Z_{Th}}$), we must turn off the voltage source (change it to a short circuit branch), as is shown in Figure 2.1.2 and in the following:

$$\mathbf{Z_{Th}} = (3 + j4)\ \Omega \Rightarrow |\mathbf{Z_{Th}}| = |3 + j4| = 5\ \Omega \tag{2}$$

Solving (1) and (2):

$$R_L = 5\ \Omega$$

Choice (1) is the answer.

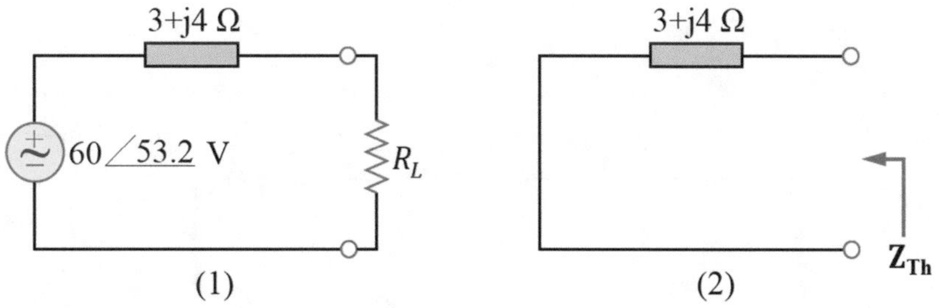

Figure 2.1 The circuit of solution of problem 2.1

2.2. As we know, to transfer maximum average power to a purely resistive component (R), the relation below must be held:

$$R = |\mathbf{Z_{Th}}| \tag{1}$$

where $|\mathbf{Z_{Th}}|$ is magnitude of the Thevenin impedance seen by the component.

M. Rahmani-Andebili, *AC Electrical Circuit Analysis*, https://doi.org/10.1007/978-3-030-60986-3_2

Herein, the voltage source needs to be turned off (changed to a short circuit branch), since we want to calculate the Thevenin impedance. Figure 2.2.2 shows the circuit in frequency domain. The impedances of the components (for $\omega = 1\ rad/sec$) can be calculated as follows:

$$\mathbf{Z}_{2\sqrt{2}\ \mathbf{H}} = j\omega L = j \times 1 \times 2\sqrt{2} = j2\sqrt{2}\ \Omega \tag{1}$$

$$\mathbf{Z}_{1\ \Omega} = 1\ \Omega \tag{2}$$

$$\mathbf{Z}_{1\ \mathbf{H}} = j\omega L = j \times 1 \times 1 = j\ \Omega \tag{3}$$

$$\mathbf{Z}_{1\ \mathbf{F}} = \frac{1}{j\omega C} = \frac{1}{j \times 1 \times 1} = -j\ \Omega \tag{4}$$

Thus:

$$\mathbf{Z}_{\mathbf{Th}} = j2\sqrt{2} + (1+j)||(1-j) = j2\sqrt{2} + \frac{(1+j)\times(1-j)}{(1+j)+(1-j)} = j2\sqrt{2} + \frac{2}{2} = \left(1 + j2\sqrt{2}\right)\ \Omega \tag{5}$$

By solving (1) and (5), we have:

$$R = \left|1 + j2\sqrt{2}\right| = 3\ \Omega$$

Choice (2) is the answer.

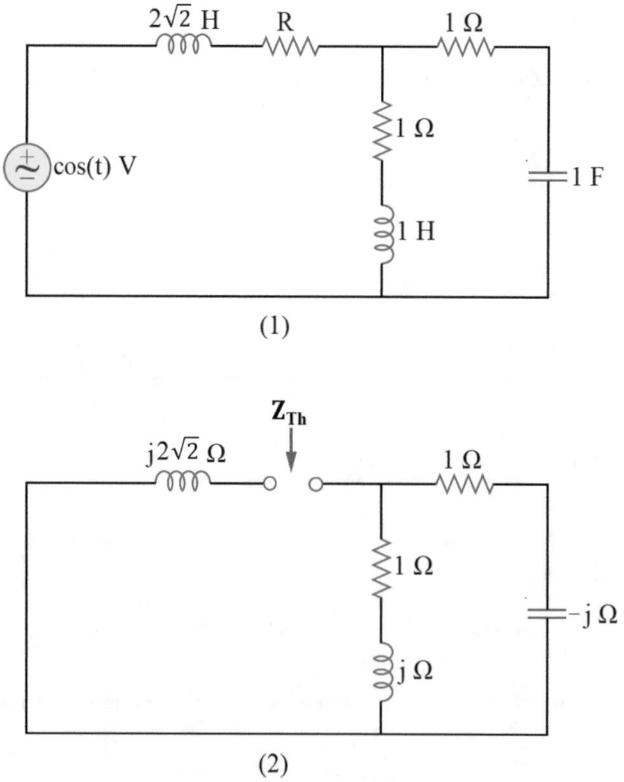

Figure 2.2 The circuit of solution of problem 2.2

2.3. The circuit of Figure 2.3.2 shows the main circuit in frequency domain. Herein, the phasor of cos(0.5t), that is, $1\angle 0^\circ$ is defined as the reference phasor, where "$\angle$" is the symbol of phase angle. Therefore, the phasor of the output voltage is $2\angle -30^\circ\ V$. The impedances of the components can be calculated as follows:

$$\mathbf{Z}_{1\,\Omega} = 1\,\Omega \tag{1}$$

$$\mathbf{Z}_{2\,\mathrm{F}} = \frac{1}{j\omega C} = \frac{1}{j \times 0.5 \times 2} = -j\,\Omega \tag{2}$$

Applying voltage division formula:

$$\mathbf{V_o} = \frac{-j}{-j+1} \times \mathbf{V_i} \Rightarrow (2\angle -30^\circ) = \frac{-j}{-j+1} \times \mathbf{V_i} \Rightarrow (2\angle -30^\circ) = \left(\frac{\sqrt{2}}{2}\angle -45^\circ\right)\mathbf{V_i}$$

$$\Rightarrow \mathbf{V_i} = \frac{(2\angle -30^\circ)}{\left(\frac{\sqrt{2}}{2}\angle -45^\circ\right)} = 2\sqrt{2}\angle 15^\circ\ V \tag{3}$$

The output voltage in time domain is:

$$v_i(t) = 2\sqrt{2}\cos\left(0.5t + 15^\circ\right)$$

Choice (2) is the answer.

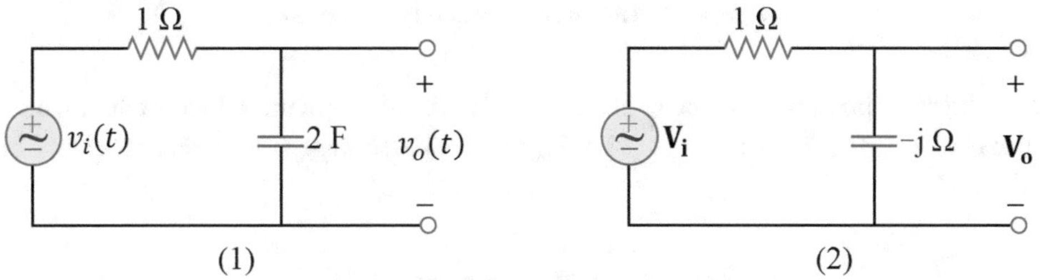

Figure 2.3 The circuit of solution of problem 2.3

2.4. Figure 2.4.2 shows the main circuit in frequency domain. The impedances of the components are as follows:

$$\mathbf{Z}_{0.5\,\Omega} = 0.5\,\Omega \tag{1}$$

$$\mathbf{Z}_{1\,F} = \frac{1}{j\omega C} = \frac{1}{j \times 2 \times 1} = -0.5j\,\Omega \tag{2}$$

The phasor of cos(2t), that is, $1\angle 0^\circ$ is defined as the reference phasor, where "$\angle$" is the symbol of phase angle. Therefore, the phasor of the current of the current source is $10\angle 0^\circ\ A$ or 10 A.

The requested voltage can be calculated by using Ohm's law as follows. In (3), $\mathbf{Z}$ is the impedance of the parallel connection of the resistor and capacitor:

$$\mathbf{Z} = 0.5 \| (-0.5j) = \frac{0.5 \times (-0.5j)}{0.5 + (-0.5j)} = \frac{-0.5j}{1-j} = \frac{\sqrt{2}}{4}\angle -45^\circ \tag{3}$$

$$\mathbf{V} = \mathbf{IZ} = (10\angle 0^\circ)\left(\frac{\sqrt{2}}{4}\angle -45^\circ\right) = \frac{5\sqrt{2}}{2}\angle -45^\circ\ V \tag{4}$$

Transferring to time domain:

$$v(t) = \frac{5\sqrt{2}}{2} \cos\left(2t - 45^{\circ}\right) V$$

Choice (4) is the answer.

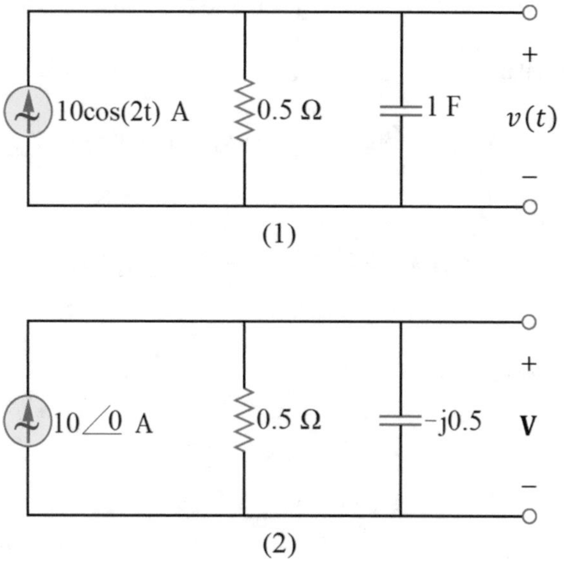

Figure 2.4 The circuit of solution of problem 2.4

2.5. Figure 2.5.2 illustrates the primary circuit in frequency domain. The phasor of the voltage of the voltage source is defined as the reference phasor ($1\angle 0^{\circ}$), where "$\angle$" is the symbol of phase angle. The impedances of the components are as follows:

$$\mathbf{Z}_{0.5\ \mathbf{F}} = \frac{1}{j\omega C} = \frac{1}{j \times 2 \times 0.5} = -j\ \Omega \tag{1}$$

$$\mathbf{Z}_{1\ \Omega} = 1\ \Omega \tag{2}$$

$$\mathbf{Z}_{1\ \mathbf{H}} = j\omega L = j \times 2 \times 1 = j2\ \Omega \tag{3}$$

By using the current division formula for the current of the inductor, we have:

$$\mathbf{I} = \frac{-j}{-j+1+j2} \times (1\angle 0^{\circ}) = \frac{-j}{1+j}\ A \tag{4}$$

Using Ohm's law for the inductor:

$$\mathbf{V} = \mathbf{I} \times (j2) \tag{5}$$

Solving (4) and (5):

$$\mathbf{V} = \frac{-j}{1+j} \times (2j) = \frac{2}{1+j} = (\sqrt{2}\angle -45^{\circ})\ V = \sqrt{2}e^{-j45}\ V$$

Choice (4) is the answer.

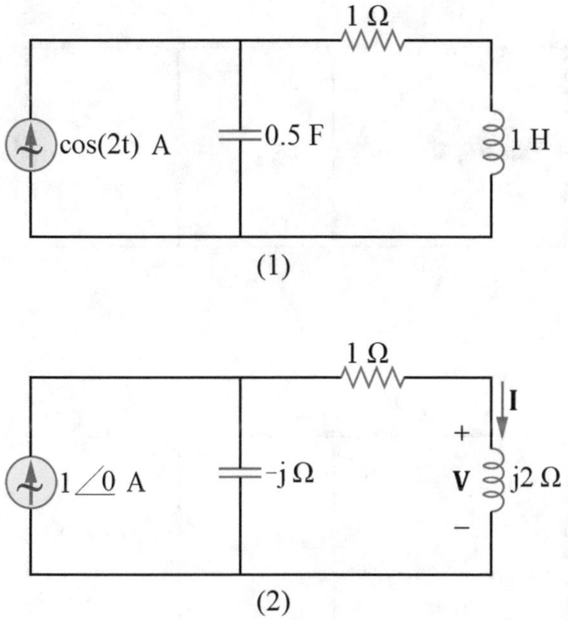

Figure 2.5 The circuit of solution of problem 2.5

2.6. Based on maximum average power transfer theorem, the relation below must be held to transfer the maximum average power to the load, where $\mathbf{Z_{Th}}^*$ is the complex conjugate of the Thevenin impedance seen by the load:

$$\mathbf{Z_L} = \mathbf{Z_{Th}}^* \tag{1}$$

The voltage source must be turned off (changed to a short circuit branch), since we need to calculate the Thevenin impedance. Figure 2.6.2 illustrates the circuit in frequency domain. The impedances of the components are as follows:

$$\mathbf{Z_{1\,H}} = j\omega L = j \times 1 \times 1 = j\,\Omega \tag{2}$$

$$\mathbf{Z_{0.5\,F}} = \frac{1}{j\omega C} = \frac{1}{j \times 1 \times 0.5} = -j2\,\Omega \tag{3}$$

$$\mathbf{Z_{1\,\Omega}} = 1\,\Omega \tag{4}$$

Therefore:

$$\mathbf{Z_{Th}} = 1 \,||\, (-j2) \,||\, (j) = \frac{1}{\frac{1}{1} + \frac{1}{-2j} + \frac{1}{j}} = \frac{2}{2-j} = \left(\frac{4}{5} + j\frac{2}{5}\right)\,\Omega \tag{5}$$

Solving (1) and (5):

$$\mathrm{Z_L} = \left(\frac{4}{5} + j\frac{2}{5}\right)^* = \left(\frac{4}{5} - j\frac{2}{5}\right)\,\Omega$$

Choice (2) is the answer.

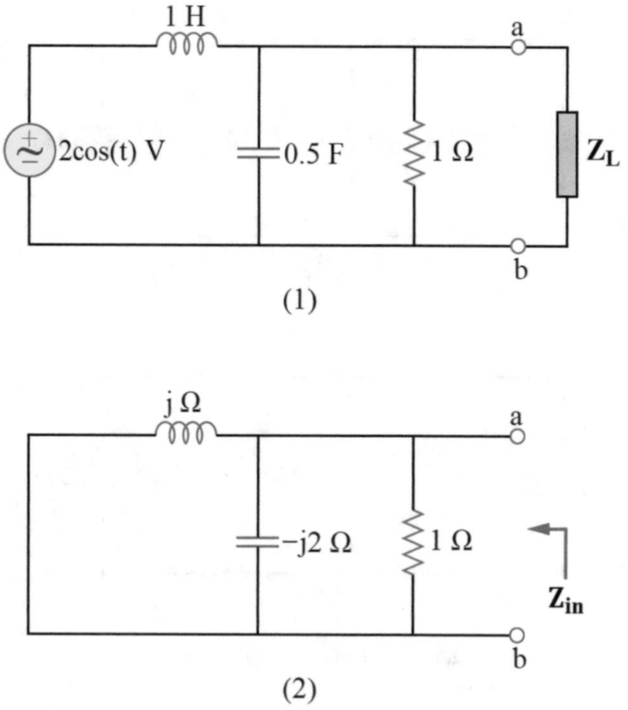

Figure 2.6 The circuit of solution of problem 2.6

2.7. Figure 2.7.2 shows the primary circuit in frequency domain. The phasor of cos(2*t*), that is, $1\angle 0^\circ$ is defined as the reference phasor, where "$\angle$" is the symbol of phase angle. Therefore, the phasor of the current of the current source is $4\angle 0^\circ\ A$ or 4 *A*. The impedances of the components are presented in the following.

$$\mathbf{Z}_{2\,\Omega} = 2\,\Omega \tag{1}$$

$$\mathbf{Z}_{0.5\,\mathrm{H}} = j\omega L = j \times 2 \times 0.5 = j\,\Omega \tag{2}$$

$$\mathbf{Z}_{0.5\,\mathrm{F}} = \frac{1}{j\omega C} = \frac{1}{j \times 2 \times 0.5} = -j\,\Omega \tag{3}$$

This problem can be solved by using the current division relation and Ohm's law as follows:

Applying the current division formula:

$$\mathbf{I}_\mathrm{L} = \frac{2}{2 + 2 + j + (-j)} \times 4 = 2\,A \tag{4}$$

Applying Ohm's law:

$$\mathbf{V}_\mathrm{L} = -\mathbf{I}_\mathrm{L}\mathbf{Z}_\mathrm{L} = -2 \times j = -j2\ V = 2\angle -90^\circ V \tag{5}$$

Transferring back to time domain:

$$v_L(t) = 2\cos\left(2t - 90^\circ\right)V = 2\sin(2t)\ V$$

Choice (1) is the answer.

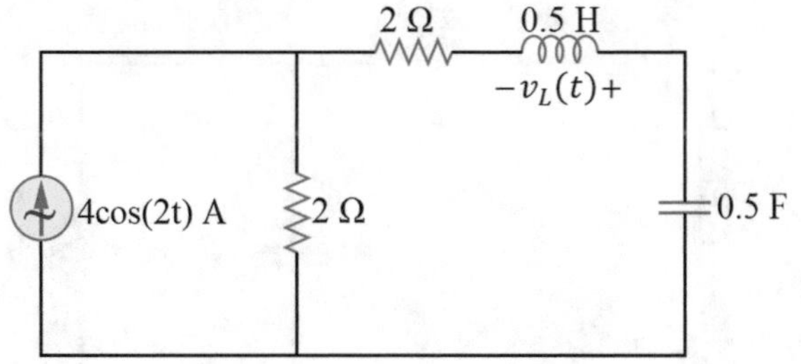

(1)

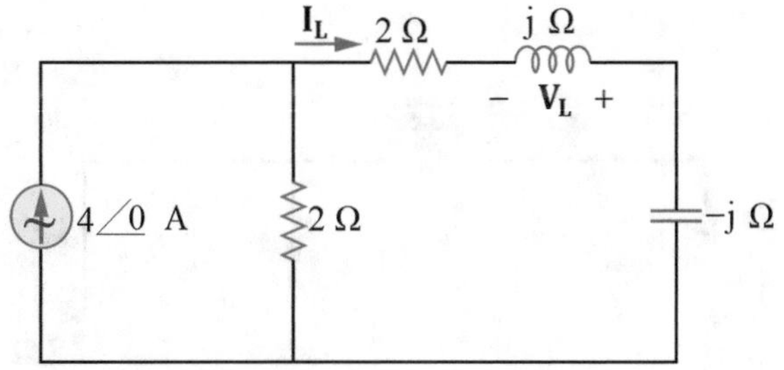

(2)

Figure 2.7 The circuit of solution of problem 2.7

2.8. The primary circuit is illustrated in frequency domain in Figure 2.8.2. The impedances of the components (for $\omega = \frac{1}{2}\ rad/sec$) are presented in the following:

$$\mathbf{Z_{4\ H}} = j\omega L = j \times \frac{1}{2} \times 4\ \Omega = j2\ \Omega \tag{1}$$

$$\mathbf{Z_{4\ mF}} = \frac{1}{j\omega C} = \frac{1}{j \times \frac{1}{2} \times \frac{1}{5}} = -j10\ \Omega \tag{2}$$

$$\mathbf{Z_{5\ \Omega}} = 5\ \Omega \tag{3}$$

Based on maximum average power transfer theorem, to transfer the maximum average power to $\mathbf{Z_s}$, the complex conjugate of the Thevenin impedance seen by $\mathbf{Z_s}$ must be equal to $\mathbf{Z_s}$. In other words, the relation below must be held:

$$\mathbf{Z_s} = \mathbf{Z_{Th}}^* \tag{4}$$

To determine the Thevenin impedance of the circuit, we need to turn off the voltage source (short circuit).

$$\mathbf{Z_{in}} = \mathbf{Z_{Th}} = j2 + (-j10)||(5) = j2 + \frac{(-j10) \times (5)}{(-j10) + (5)} = j2 + 4 - j2 = 4\ \Omega \tag{5}$$

Solving (4) and (5):

$$\mathbf{Z_s} = 4^* = 4\ \Omega$$

Choice (1) is the answer.

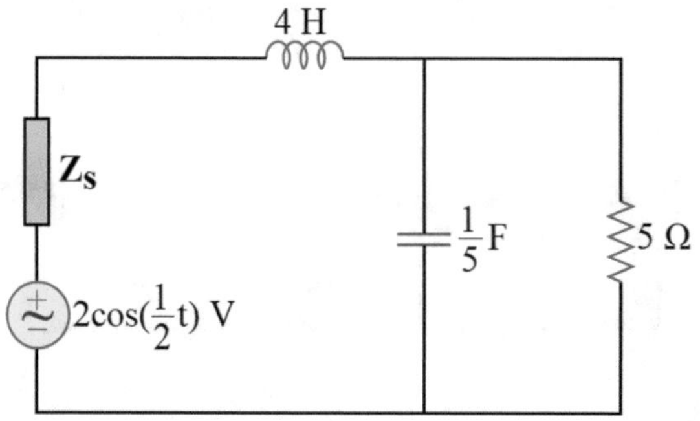

(1)

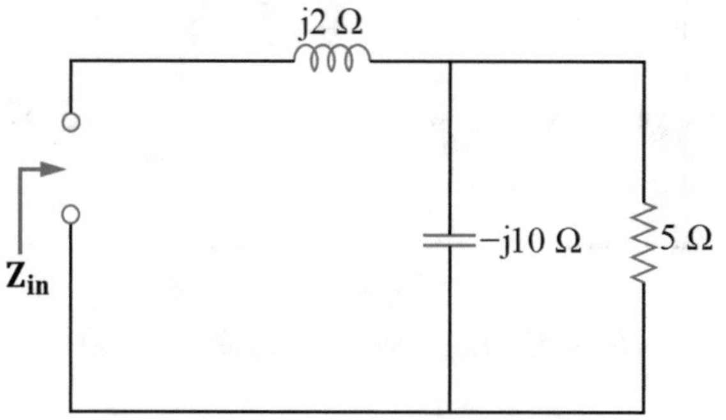

(2)

Figure 2.8 The circuit of solution of problem 2.8

2.9. The average power of a resistor can be calculated by using the following relation, where $|\mathbf{I}|$ and $|\mathbf{I}_{rms}|$ are the peak value and root mean square (rms) value of the amplitude of the current of the resistor, respectively.

$$P_R = \frac{1}{2}R|\mathbf{I}|^2 = R|\mathbf{I}_{rms}|^2 \tag{1}$$

The rms value of the amplitude of the current that includes different frequencies can be determined as follows:

$$|\mathbf{I}_{rms}| = \sqrt{\left(\frac{2}{\sqrt{2}}\right)^2 + \left(\frac{3}{\sqrt{2}}\right)^2} = \frac{\sqrt{26}}{2}\ A \tag{2}$$

Solving (1) and (2):

$$P_R = 4 \times \left(\frac{\sqrt{26}}{2}\right)^2 = 26\ W$$

Choice (1) is the answer.

2.10. The primary circuit is illustrated in frequency domain in Figure 2.9.2. The impedances of the components are presented in the following:

$$\mathbf{Z}_{0.5\ \mathbf{mH}} = j\omega L = j\omega \times 0.5 \times 10^{-3}\ \Omega \tag{1}$$

$$\mathbf{Z}_{1\ \Omega} = 1\ \Omega \tag{2}$$

$$\mathbf{Z}_{1\ \mathbf{mF}} = \frac{1}{j\omega C} = \frac{1}{j\omega \times 10^{-3}} = -j\frac{10^3}{\omega}\ \Omega \tag{3}$$

To determine the resonance frequency of the circuit, we need to equate the imaginary part of the input impedance of the circuit with zero and then solve it. In other words, we need to solve the following equation:

$$Im\{\mathbf{Z}_{\mathbf{in}}\} = 0 \tag{4}$$

From the circuit of Figure 2.9.2, we can write:

$$\mathbf{Z}_{\mathbf{in}} = j\omega \times 0.5 \times 10^{-3} + 1 || \left(-j\frac{10^3}{\omega}\right) + 1 = j\omega \times 0.5 \times 10^{-3} + \frac{-j\frac{10^3}{\omega}}{1 - j\frac{10^3}{\omega}} + 1$$

$$\Rightarrow \mathbf{Z}_{\mathbf{in}} = \left(1 + \frac{10^6}{\omega^2 + 10^6}\right) + j\omega\left(0.5 \times 10^{-3} - \frac{10^3}{\omega^2 + 10^6}\right) \tag{5}$$

$$Im\{\mathbf{Z}_{\mathbf{in}}\} = \omega\left(0.5 \times 10^{-3} - \frac{10^3}{\omega^2 + 10^6}\right) = \omega\left(\frac{0.5 \times 10^{-3}\omega^2 - 0.5 \times 10^3}{\omega^2 + 10^6}\right) \tag{6}$$

Solving (4) and (6):

$$0.5 \times 10^{-3}\omega^2 - 0.5 \times 10^3 = 0 \Rightarrow \omega^2 = 10^6 \Rightarrow \omega_0 = 10^3\ rad/sec$$

Choice (2) is the answer.

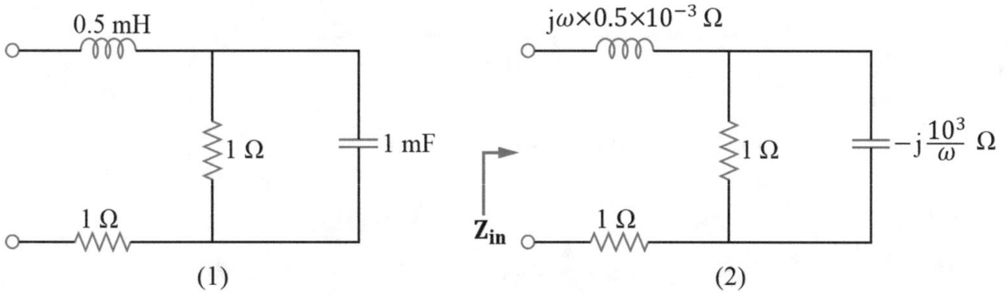

Figure 2.9 The circuit of solution of problem 2.10

2.11. Figure 2.10.2 shows the main circuit in frequency domain. The phasor of cos(100t), that is, $1\angle 0°$ is defined as the reference phasor, where "$\angle$" is the symbol of phase angle. Therefore, the phasor of the current of the current source is $10\angle 0°\ A$ or 10 A. The impedances of the components can be calculated as follows:

$$\mathbf{Z}_{10\ \Omega} = 10\ \Omega \tag{1}$$

$$\mathbf{Z}_{10\ \mathbf{mH}} = j\omega L = j \times 100 \times 10 \times 10^{-3} = j\ \Omega \tag{2}$$

$$\mathbf{Z}_{1\ \mathbf{mF}} = \frac{1}{j\omega C} = \frac{1}{j \times 100 \times 1 \times 10^{-3}} = -j10\ \Omega \tag{3}$$

This problem can be solved by using Ohm's law for the equivalent impedance of the circuit.

The equivalent impedance of the circuit, seen by the current source, is:

$$\mathbf{Z_{eq}} = (10) \parallel (10+j) \parallel (-10j) = \frac{1}{\frac{1}{10}+\frac{1}{10+j}+\frac{1}{-10j}} = \frac{1}{\frac{j(10+j)+(j10)-(10+j)}{j10(10+j)}}$$

$$\Rightarrow \mathbf{Z_{eq}} = \frac{1}{\frac{-11+19j}{j10\,(10+j)}} = \frac{-10+100j}{-11+j19} = (4.58\angle -24.3°)\Omega \tag{4}$$

Using Ohm's law:

$$\mathbf{V} = \mathbf{IZ_{eq}} = (10\angle 0°)\,(4.58\angle -24.3°) = (4.58\angle -24.3°)V \tag{5}$$

The voltage of the capacitor in time domain is:

$$v_c(t) = v(t) = 45.8\cos\left(100t - 24.3^{\circ}\right)\,V$$

Choice (1) is the answer.

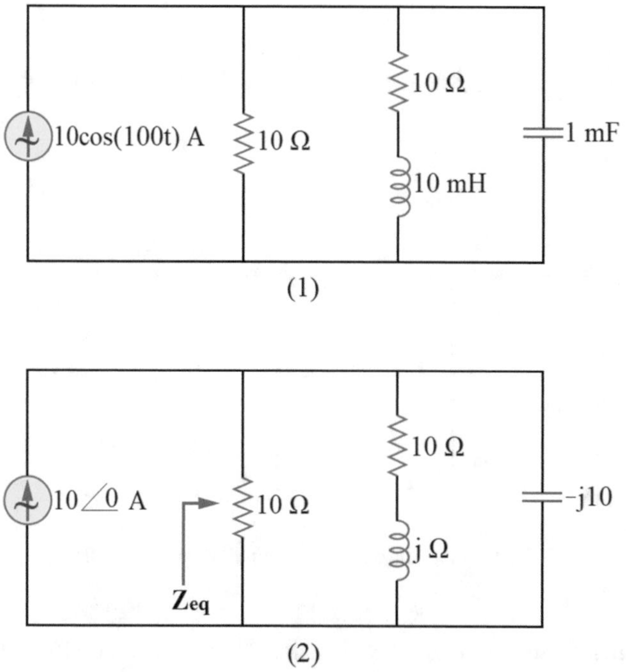

Figure 2.10 The circuit of solution of problem 2.11

2.12. Figure 2.11.2 shows the main circuit in frequency domain. The phasor of cos(t), that is, $1\angle 0°$ is defined as the reference phasor, where "$\angle$" is the symbol of phase angle. Therefore, the phasor of the voltage of the source is $A\angle 0°\ V$ or A *Volt*. The impedances of the components are presented in the following:

$$\mathbf{Z_{1\,F}} = \frac{1}{j\omega C} = \frac{1}{j\times 1\times 1} = -j\,\Omega \tag{1}$$

$$\mathbf{Z_{1\,\Omega}} = 1\,\Omega \tag{2}$$

$$\mathbf{Z_{1\,H}} = j\omega L = j\times 1\times 1 = j\,\Omega \tag{3}$$

$$\mathbf{Z}_{2\,\mathbf{H}} = j\omega L = j \times 1 \times 2 = j2\ \Omega \tag{4}$$

The current can be calculated by using Ohm's law, but, first, we need to calculate the impedance of the indicated part of the circuit (see Figure 2.11.2) as follows:

$$\mathbf{Z}_{\mathbf{Total}} = (-j)||(1+j)||(j2) = \frac{1}{\frac{1}{-j} + \frac{1}{1+j} + \frac{1}{j2}} = \frac{1}{j + \frac{1}{2}(1-j) - \frac{1}{2}j} = \frac{1}{\frac{1}{2}} = 2\ \Omega \tag{5}$$

Using Ohm's law for the indicated part of the circuit:

$$\mathbf{I} = \frac{A\angle 0^\circ}{\mathbf{Z}_{\mathbf{Total}}} = \frac{A}{2}\ Ampere \tag{6}$$

By transferring to time domain, we have:

$$i(t) = \frac{1}{2}A\cos(t)A$$

Choice (3) is the answer.

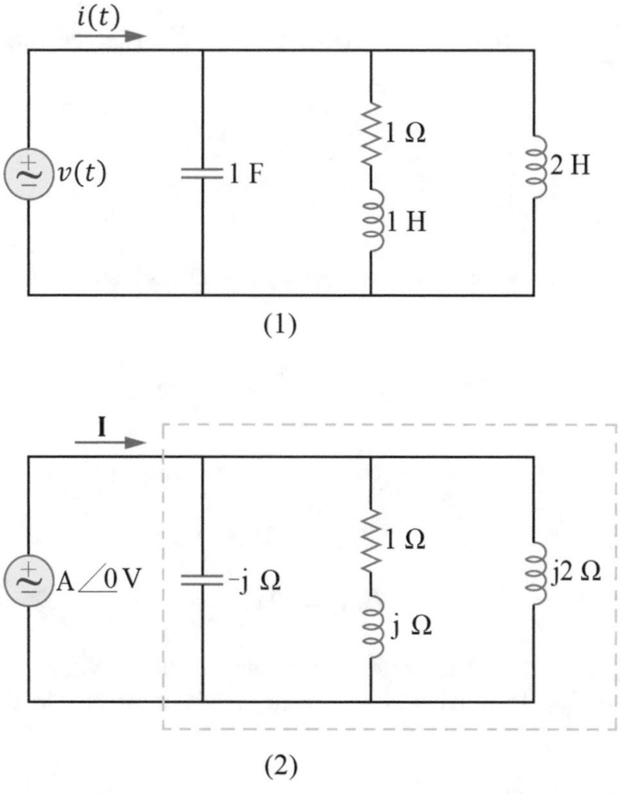

Figure 2.11 The circuit of solution of problem 2.12

2.13. The impedance of the series connection of a 2 Ω resistor with a capacitor with the capacitance of C is:

$$\mathbf{Z}_{\mathbf{Series}} = 2 + \frac{1}{j\omega C} = \left(2 - j\frac{1}{\omega C}\right)\Omega \tag{1}$$

In addition, the impedance of the parallel connection of a 4 Ω resistor with a capacitor with the capacitance of $\frac{1}{20}$ F is:

$$\mathbf{Z}_{\mathbf{Parallel}} = 4||\left(\frac{1}{j\omega\frac{1}{20}}\right) = 4||\left(-j\frac{20}{\omega}\right) = \frac{4 \times \left(-j\frac{20}{\omega}\right)}{4 + \left(-j\frac{20}{\omega}\right)} = \frac{-j80}{4\omega - j20}\ \Omega$$

$$\Rightarrow \mathbf{Z}_{\mathbf{Parallel}} = \frac{-1600 - j320\omega}{16\omega^2 + 400} = \left(\frac{1600}{16\omega^2 + 400} - j\frac{320\omega}{16\omega^2 + 400}\right)\ \Omega \qquad (2)$$

Based on the information given in the problem:

$$\mathbf{Z}_{\mathbf{Series}} = \mathbf{Z}_{\mathbf{Parallel}} \qquad (3)$$

Solving (1), (2), and (3):

$$2 - j\frac{1}{\omega C} = \frac{1600}{16\omega^2 + 400} - j\frac{320\omega}{16\omega^2 + 400} \Rightarrow \begin{cases} 2 = \dfrac{1600}{16\omega^2 + 400} & (4) \\ -j\dfrac{1}{\omega C} = -j\dfrac{320\omega}{16\omega^2 + 400} & (5) \end{cases}$$

$$\xrightarrow{(4)} 32\omega^2 + 800 = 1600 \Rightarrow \omega = 5\ rad/sec \qquad (6)$$

$$\overset{(5)}{\Rightarrow} \frac{1}{\omega C} = \frac{320\omega}{16\omega^2 + 400} \Rightarrow 16\omega^2 + 400 = 320\omega^2 C \xrightarrow{Using\ (6)} 16 \times 5^2 + 400 = 320 \times 5^2 \times C$$

$$\Rightarrow C = \frac{800}{8000} = 0.1\ F$$

Choice (3) is the answer.

2.14. First, we need to determine and draw the circuit in frequency domain, as is shown in Figure 2.12.2. The impedances of the components are as follows:

$$\mathbf{Z}_{\mathbf{1\ F}} = \frac{1}{j\omega C} = \frac{1}{j\omega \times 1} = \frac{1}{j\omega}\ \Omega \qquad (1)$$

$$\mathbf{Z}_{\mathbf{1\ H}} = j\omega L = j\omega \times 1 = j\omega\ \Omega \qquad (2)$$

$$\mathbf{Z}_{\mathbf{1\ \Omega}} = 1\ \Omega \qquad (3)$$

$$\mathbf{Z}_{\mathbf{10\ \Omega}} = 10\ \Omega \qquad (4)$$

First Method: By looking at Figure 2.12.2, we can quickly figure it out that $\mathbf{V_{ab}}$ is zero, since the right-side part of the circuit is a Wheatstone bridge. In a Wheatstone bridge circuit, illustrated in Figure 2.12.3, the condition of $\mathbf{Z_1 \times Z_3 = Z_2 \times Z_4}$ holds.

As can be noticed from Figure 2.12.2, we have:

$$\frac{1}{j\omega} \times j\omega = 1 \times 1 \Rightarrow 1 = 1 \qquad (5)$$

which satisfies the condition. Therefore, $\mathbf{V_{ab}} = 0$ and consequently $|\mathbf{V_{ab}}| = 0$. Hence, the value of $|\mathbf{V_{ab}}|$ does not depend on the frequency of current source. Choice (3) is the answer.

Second Method: We can directly calculate $|\mathbf{V_{ab}}|$ in the circuit of Figure 2.12.2 as follows:

$$\mathbf{V_{ab}} = \mathbf{V_a} - \mathbf{V_b} = 1 \times \mathbf{I_a} - j\omega \times \mathbf{I_b} = \frac{1 + j\omega}{1 + j\omega + 1 + \frac{1}{j\omega}}\mathbf{I} - j\omega \times \frac{1 + \frac{1}{j\omega}}{1 + j\omega + 1 + \frac{1}{j\omega}}\mathbf{I}$$

$$\mathbf{V_{ab}} = \frac{1 + j\omega - (j\omega + 1)}{1 + j\omega + 1 + \frac{1}{j\omega}}\mathbf{I} = 0 \quad (6)$$

$$\Rightarrow |\mathbf{V_{ab}}| = 0 \quad (7)$$

Choice (3) is the answer.

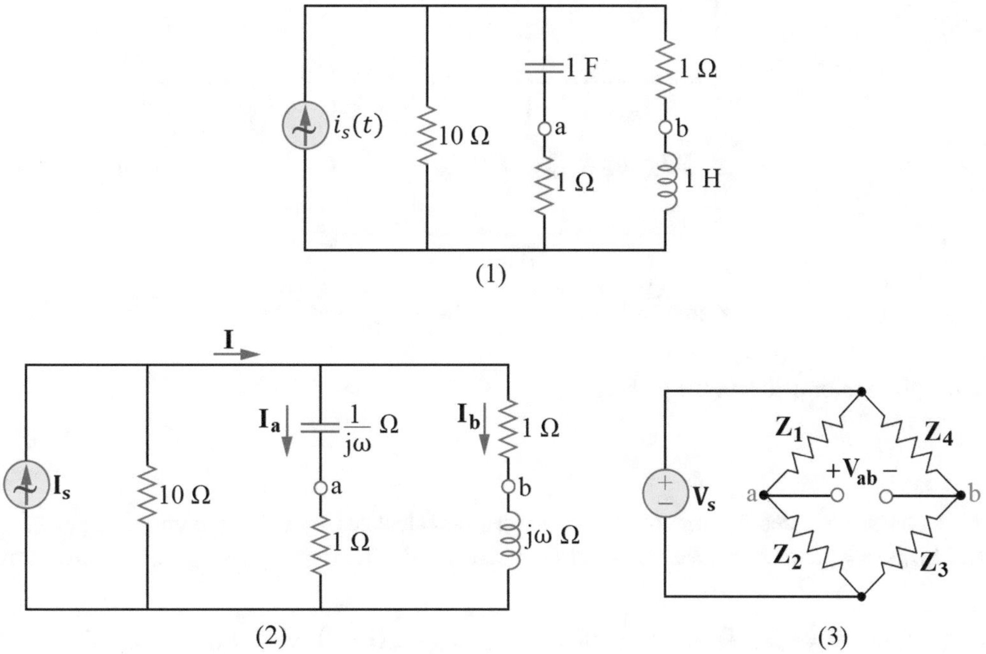

Figure 2.12 The circuit of solution of problem 2.14

2.15. Nodal analysis is the best method to be applied to solve the problem.

Applying Ohm's law for the series connection of the inductor and the 3 Ω resistor in the circuit of Figure 2.13:

$$\mathbf{I_L} = \frac{20 - \mathbf{V}}{3 + j4} \quad (1)$$

Applying KCL in the top node:

$$-5 + \frac{\mathbf{V} - 20}{3 + j4} + \frac{\mathbf{V}}{2 - j4} = 0 \Rightarrow \mathbf{V}\left(\frac{1}{3 + j4} + \frac{1}{2 - j4}\right) = 5 + \frac{20}{3 + j4} \Rightarrow \mathbf{V} = \frac{5 + \frac{20}{3+j4}}{\frac{1}{3+j4} + \frac{1}{2-j4}}$$

$$\Rightarrow \mathbf{V} = \frac{\frac{35+j20}{3+j4}}{\frac{5}{22-j4}} = \frac{850 + j300}{15 + j20} \quad (2)$$

Solving (1) and (2):

$$\mathbf{I_L} = \frac{20 - \frac{850+j300}{15+j20}}{3+j4} = \frac{\frac{-550+j100}{15+j20}}{3+j4} = \frac{-550+j100}{-35+j120} = (2+j4)\ A$$

Choice (4) is the answer.

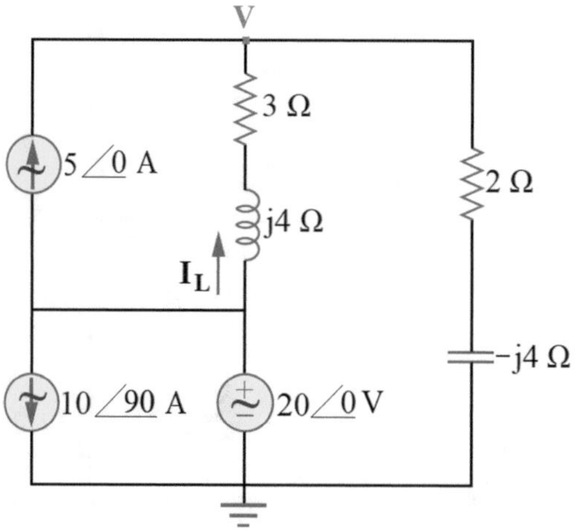

Figure 2.13 The circuit of solution of problem 2.15

2.16. Based on the given information in the problem:

$$P_{R_1} = 1.5\ W \tag{1}$$

The average power generated by the voltage source can be calculated by using the complex power relation, where $\mathbf{V_{source}}$ and $\mathbf{I_{source}}$ are the phasor (peak value) of the voltage and current of the voltage source, respectively.

$$\mathbf{S_{source}} = \frac{1}{2}\mathbf{V_{source}}\mathbf{I_{source}}^* = \frac{1}{2}(10)\left(\frac{1}{2}(1+j)\right)^* = \frac{5}{2}(1-j)\ VA \Rightarrow P_{source} = 2.5\ W \tag{2}$$

The average power generated by the voltage source is consumed in the resistors, thus:

$$P_{source} = P_{R_1} + P_{2\ \Omega}$$

Solving (1) and (2):

$$2.5\ W = 1.5\ W + P_{2\ \Omega} \Rightarrow P_{2\ \Omega} = 1\ W \tag{3}$$

The average power relation for a resistor is as follows, where $|\mathbf{I}|$ is the amplitude (peak value) of the current of the resistor.

$$P = \frac{1}{2}R|\mathbf{I}|^2 \xrightarrow{Using\ (3)} 1\ W = \frac{1}{2} \times 2 \times |\mathbf{I_{2\ \Omega}}|^2 \Rightarrow |\mathbf{I_{2\ \Omega}}|^2 = 1 \Rightarrow |\mathbf{I_{2\ \Omega}}| = 1\ A$$

Choice (2) is the answer.

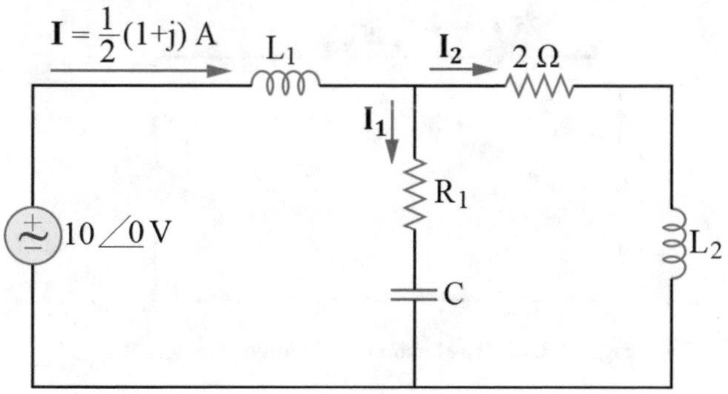

Figure 2.14 The circuit of solution of problem 2.16

2.17. Based on the information given in the problem, the resonance frequency and the $-3\ dB$ bandwidth are:

$$f_0 = 50000\ Hz \tag{1}$$

$$BW = 500\ Hz \tag{2}$$

In a series RLC circuit, the resonance frequency and the $-3\ dB$ bandwidth of the circuit can be calculated by using the following relations:

$$\omega_0 = \frac{1}{\sqrt{LC}}\ rad/sec \Rightarrow f_0 = \frac{1}{2\pi\sqrt{LC}}\ Hz \tag{3}$$

$$BW = \frac{R}{L}\ rad/sec = \frac{R}{2\pi L}\ Hz \tag{4}$$

Solving (1) and (3):

$$50000 = \frac{1}{2\pi\sqrt{LC}} \tag{5}$$

Solving (2) and (4):

$$500 = \frac{R}{2\pi L} \Rightarrow 500 = \frac{25+25}{2\pi L} \Rightarrow L = \frac{1}{20\pi} = 0.0159 = 15.9\ mH \tag{6}$$

Solving (5) and (6):

$$50000 = \frac{1}{2\pi\sqrt{0.0159 \times C}} \Rightarrow C = 0.64\ nF$$

Choice (4) is the answer.

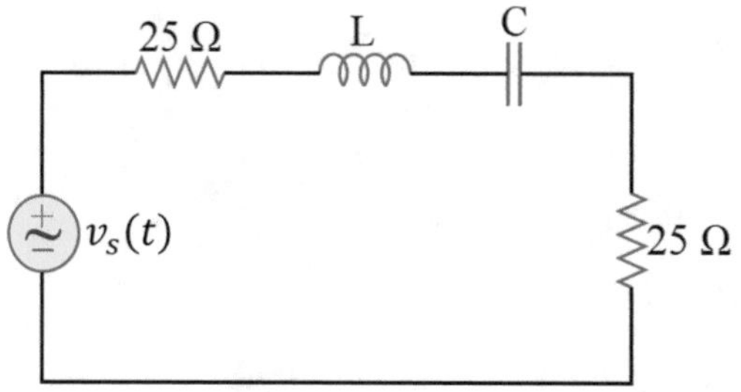

Figure 2.15 The circuit of solution of problem 2.17

2.18. The circuit of Figure 2.16.2 illustrates the primary circuit in frequency domain. The phasor of the voltage of the voltage source is defined as the reference phasor ($1\angle 0^\circ$), where "$\angle$" is the symbol of phase angle. The impedances of the components are as follows:

$$\mathbf{Z_L} = j\omega L\ \Omega \tag{1}$$

$$\mathbf{Z_C} = \frac{1}{j\omega C}\ \Omega \tag{2}$$

Using voltage division formula for the left-side capacitor:

$$\mathbf{V_1} = \frac{\frac{1}{j\omega C}}{\frac{1}{j\omega C} + j\omega L} \times (1\angle 0^\circ) = \frac{1}{1 - \omega^2 LC}\ V \tag{3}$$

Using voltage division formula for the right-side inductor:

$$\mathbf{V_2} = \frac{j\omega L}{j\omega L + \frac{1}{j\omega C}} \times (1\angle 0^\circ) = \frac{-\omega^2 LC}{-\omega^2 LC + 1}\ V \tag{4}$$

Therefore, the requested voltage is:

$$\mathbf{V} = \mathbf{V_1} - \mathbf{V_2} = \frac{1}{1 - \omega^2 LC} - \frac{-\omega^2 LC}{-\omega^2 LC + 1} = \frac{1 + \omega^2 LC}{1 - \omega^2 LC}$$

Choice (1) is the answer.

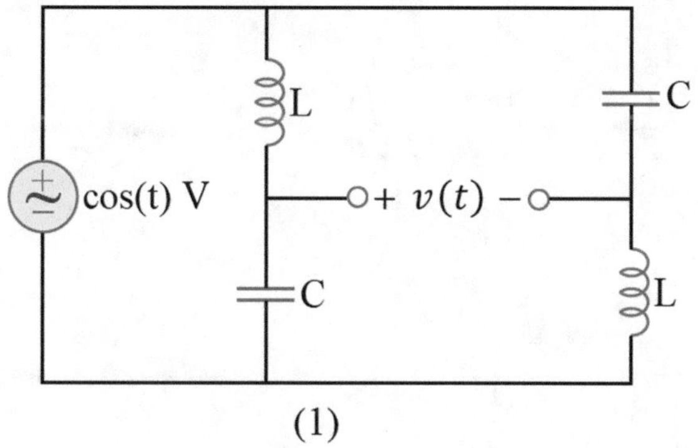

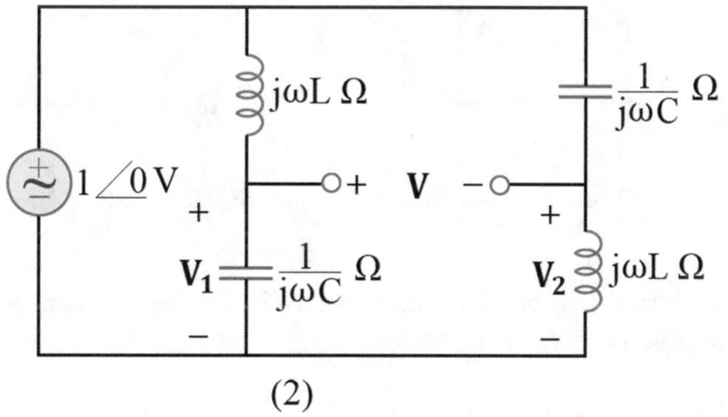

Figure 2.16 The circuit of solution of problem 2.18

2.19. In a parallel RLC circuit, the $-3\ dB$ bandwidth and the quality factor of the circuit can be calculated by using the following relations:

$$BW_{Parallel} = \frac{1}{RC}\ rad/\sec \Rightarrow BW_{Parallel} = \frac{1}{R \times 0.5}\ rad/\sec \tag{1}$$

$$Q_{Parallel} = R\sqrt{\frac{C}{L}} \Rightarrow Q_{Parallel} = R\sqrt{\frac{0.5}{2}} = \frac{R}{2} \tag{2}$$

In addition, the quality factor of the series RLC circuit can be calculated as follows:

$$Q_{Series} = \frac{1}{R}\sqrt{\frac{L}{C}} = \frac{1}{R}\sqrt{\frac{1}{C'}} \tag{3}$$

Based on the given information:

$$BW_{Parallel} = 1\ rad/\sec \tag{4}$$

$$Q_{Parallel} = Q_{Series} \tag{5}$$

Solving (1) and (4):

$$\frac{1}{R \times 0.5} = 1 \Rightarrow R = 2\ \Omega \tag{6}$$

Solving (2) and (3):

$$\frac{R}{2}=\frac{1}{R}\sqrt{\frac{1}{C'}}\xrightarrow{Using\ (6)}\frac{2}{2}=\frac{1}{2}\sqrt{\frac{1}{C'}}\Rightarrow\sqrt{\frac{1}{C'}}=2\Rightarrow C'=\frac{1}{4}\ F$$

Choice (4) is the answer.

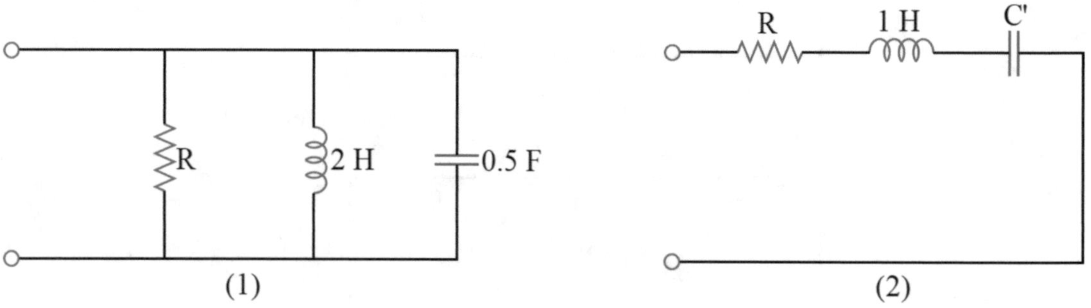

Figure 2.17 The circuit of solution of problem 2.19

2.20. Although the circuit is an AC circuit, the problem can be solved in time domain because the circuit is a resistive circuit. Herein, nodal analysis is applied to solve the problem.

By applying KCL in the left-side node, we can write:

$$-10\sin(4t)+\frac{v_1(t)}{1}+\frac{v_1(t)-v_3(t)}{1}=0\Rightarrow -10\sin(4t)+2v_1(t)-v_3(t)=0 \quad (1)$$

Applying KCL in the right-side node:

$$\frac{v_3(t)-v_1(t)}{1}+\frac{v_3(t)}{1}+2v_1(t)=0\Rightarrow v_1(t)=-2v_3(t) \quad (2)$$

Solving (1) and (2):

$$-10\sin(4t)+2(-2v_3(t))-v_3(t)=0\Rightarrow v_3(t)=-2\sin(4t)$$

Choice (3) is the answer.

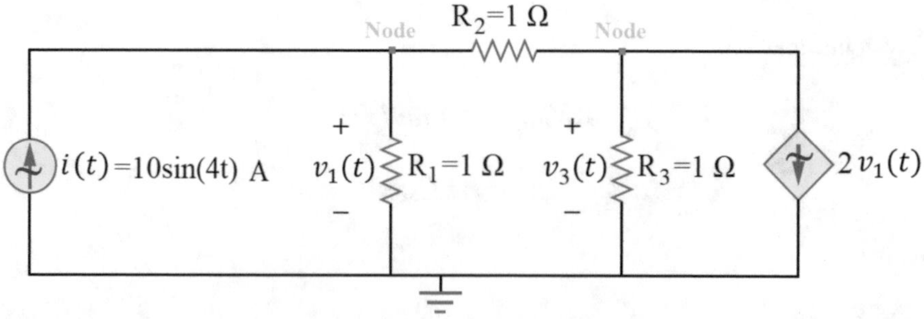

Figure 2.18 The circuit of solution of problem 2.20

2.21. Based on the information given in the problem:

$$V_{rms} = 100\ V \tag{1}$$

$$\mathbf{S}_A = 100e^{j30}\ kVA = (100\angle 30^\circ)\ kVA \tag{2}$$

$$\mathbf{S}_B = 100e^{j-30}\ kVA = (100\angle -30^\circ)\ kVA \tag{3}$$

The total complex power of the circuit can be calculated as follows:

$$\mathbf{S} = \mathbf{S}_A + \mathbf{S}_B = 100\angle 30^\circ + 100\angle -30^\circ = (100\sqrt{3}\angle 0^\circ)\ kVA = 100\sqrt{3}\ kVA \tag{4}$$

As we know, the apparent power of a component can be calculated by using the following relation, where V_{rms} and I_{rms} are the magnitude (rams value) of the voltage and the current of the component, respectively.

$$S = |\mathbf{S}| = V_{rms}\, I_{rms} \tag{5}$$

Solving (1), (4), and (5):

$$I_{rms} = \frac{100\sqrt{3}\ kVA}{100\ V} = \sqrt{3}\ kA = 1000\sqrt{3}\ A$$

Choice (4) is the answer.

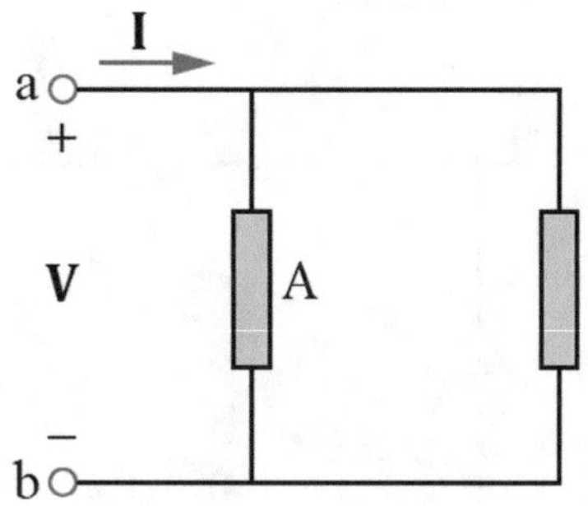

Figure 2.19 The circuit of solution of problem 2.21

2.22. Resonance frequency of a circuit is a frequency, in which zero value is achieved for the imaginary part of the input admittance (or input impedance) of the circuit. In this problem, we can determine the input admittance of the circuit and equate its imaginary part with zero and then solve it. In other words, the equation below must be solved:

$$Im\{\mathbf{Y_{in}}\} = 0 \tag{1}$$

The primary circuit is illustrated in frequency domain in Figure 2.20.2. The impedances of the components are presented in the following:

$$\mathbf{Z_{R_1}} = R_1 \tag{2}$$

$$\mathbf{Z_{R_2}} = R_2 \tag{3}$$

$$\mathbf{Z_L} = j\omega L \tag{4}$$

$$\mathbf{Z_C} = \frac{1}{j\omega C} \tag{5}$$

$$\mathbf{Y_{in}} = \frac{1}{R_1} + \frac{1}{R_2 + j\omega L} + \frac{1}{\frac{1}{j\omega C}} = \frac{1}{R_1} + \frac{R_2 - j\omega L}{{R_2}^2 + \omega^2 L^2} + j\omega C$$

$$\Rightarrow \mathbf{Y_{in}} = \left(\frac{1}{R_1} + \frac{R_2}{{R_2}^2 + \omega^2 L^2}\right) + j\omega\left(C - \frac{L}{{R_2}^2 + \omega^2 L^2}\right) \tag{6}$$

Solving (1) and (6):

$$\omega\left(C - \frac{L}{{R_2}^2 + \omega^2 L^2}\right) = 0 \Rightarrow C - \frac{L}{{R_2}^2 + \omega^2 L^2} \Rightarrow {R_2}^2 + \omega^2 L^2 = \frac{L}{C} \Rightarrow \omega^2 = \frac{1}{LC} - \frac{{R_2}^2}{L^2} \Rightarrow \omega = \sqrt{\frac{1}{LC} - \left(\frac{R_2}{L}\right)^2}$$

Choice (1) is the answer.

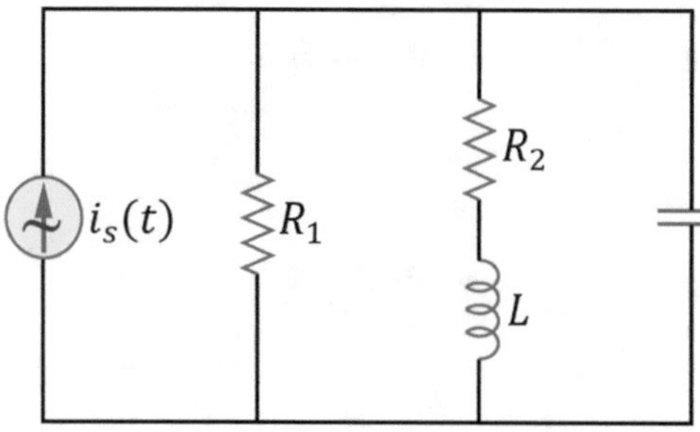

(1)

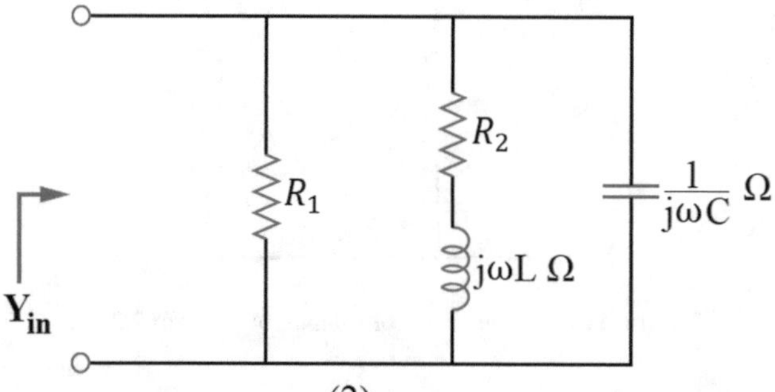

(2)

Figure 2.20 The circuit of solution of problem 2.22

2.23. Based on maximum average power transfer theorem, the relation below must be held to transfer the maximum average power to the load, where $\mathbf{Z_{Th}}^*$ is the complex conjugate of the Thevenin impedance seen by the load:

$$\mathbf{Z_x} = \mathbf{Z_{Th}}^* \tag{1}$$

Figure 2.21.2 shows the primary circuit in frequency domain. The impedances of the components are as follows:

$$\mathbf{Z_{1\,\Omega}} = 1 \tag{2}$$

$$\mathbf{Z}_{\frac{1}{2}\,\mathbf{H}} = j\omega L = j \times 1 \times \frac{1}{2} = j\frac{1}{2}\ \Omega \tag{3}$$

$$\mathbf{Z}_{\frac{1}{4}\,F} = \frac{1}{j\omega C} = \frac{1}{j \times 1 \times \frac{1}{4}} = -j4\ \Omega \tag{4}$$

To determine the Thevenin impedance, we need to turn off all the power sources. Thus, the voltage source is replaced by a short circuit branch, as is shown in Figure 2.21.2.

$$\mathbf{Z_{Th}} = \mathbf{Z_{in}} = \left(1 + j\frac{1}{2}\right)||(1 + (-j4)||(-j4)) = \left(1 + j\frac{1}{2}\right)||(1 - j2)$$

$$\Rightarrow \mathbf{Z_{in}} = \frac{\left(1 + j\frac{1}{2}\right) \times (1 - j2)}{\left(1 + j\frac{1}{2}\right) + (1 - j2)} = \frac{2 - j\frac{3}{2}}{2 - j\frac{3}{2}} = 1\ \Omega \tag{5}$$

Solving (1) and (5):

$$\mathrm{Z_x} = 1^* = 1\ \Omega$$

Choice (2) is the answer.

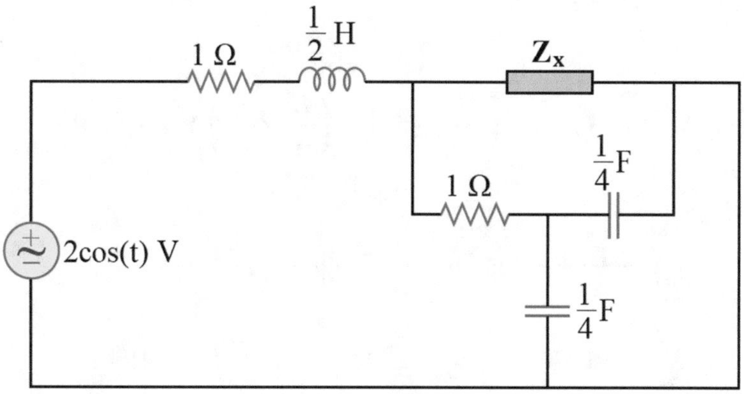

(1)

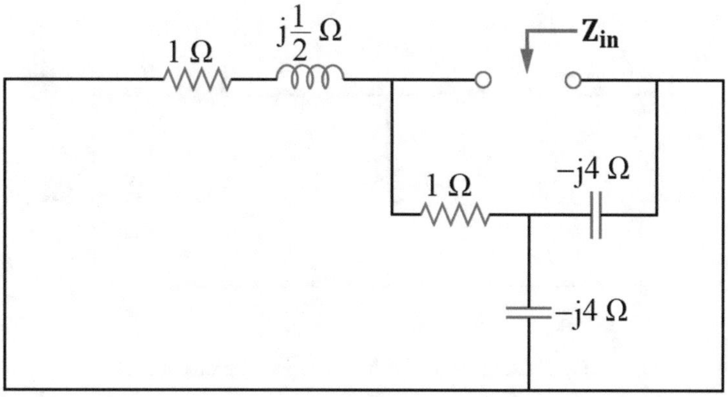

(2)

Figure 2.21 The circuit of solution of problem 2.23

2.24. The primary circuit is illustrated in frequency domain in Figure 2.22.2. The phasor of cos(t), that is, $1\angle 0°$ is defined as the reference phasor, where "$\angle$" is the symbol of phase angle. Therefore, the phasor of the voltage of the voltage source is $(1\angle 0°)V$ or 1 V. The impedances of the components are presented in the following:

$$\mathbf{Z}_{1\ \mathbf{H}} = j\omega L = j \times 1 \times 1 = j\,\Omega \quad (1)$$

$$\mathbf{Z}_{2\ \Omega} = 2\,\Omega \quad (2)$$

$$\mathbf{Z}_{1\ \mathbf{F}} = \frac{1}{j\omega C} = \frac{1}{j \times 1 \times 1} = -j\,\Omega \quad (3)$$

The current of the voltage source can be calculated as follows:

$$\mathbf{I} = \frac{1}{j + 2||(-j + 2 + j)} = \frac{1}{j + 2||2} = \frac{1}{1 + j} = \left(\frac{\sqrt{2}}{2}\angle{-45°}\right) A \quad (4)$$

Using current division relation:

$$\mathbf{I_o} = \frac{2}{2 + (-j + 2 + j)} \times \left(\frac{\sqrt{2}}{2}\angle{-45°}\right) = \left(\frac{\sqrt{2}}{4}\angle{-45°}\right) A \quad (5)$$

Using Ohm's law:

$$\mathbf{V_o} = j \times \mathbf{I_o} = j \times \left(\frac{\sqrt{2}}{4}\angle{-45°}\right) = \left(\frac{\sqrt{2}}{4}\angle 45°\right) V = \frac{\sqrt{2}}{4} e^{j45}\ V$$

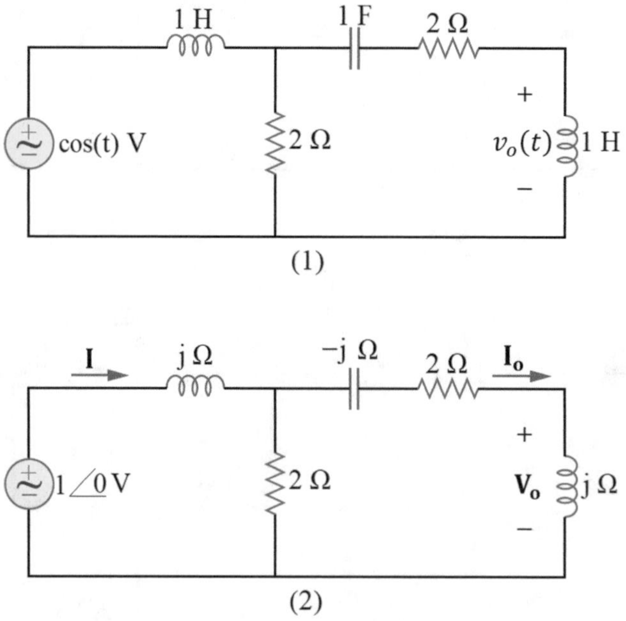

Figure 2.22 The circuit of solution of problem 2.24

Choice (3) is the answer.

2.25. The primary circuit is illustrated in frequency domain in Figure 2.23.2. The phasor of cos(100t), that is, $1\angle 0°$ is defined as the reference phasor, where "$\angle$" is the symbol of phase angle. Therefore, the phasor of the voltage of the voltage source is $(100\angle 0°)V$ or 100 V. The impedances of the components are presented in the following:

$$\mathbf{Z}_{100\ \Omega} = 100\,\Omega \quad (1)$$

$$\mathbf{Z}_{0.1\ \mathbf{mF}} = \frac{1}{j\omega C} = \frac{1}{j \times 100 \times 0.1 \times 10^{-3}} = -j100\ \Omega \tag{2}$$

$$\mathbf{Z}_{1\ \mathbf{H}} = j\omega L = j \times 100 \times 1 = j100\ \Omega \tag{3}$$

It is suggested to simplify the circuit, by combining the indicated part of the circuit shown in Figure 2.23.2.

$$\mathbf{Z} = (-j100)||(100 + j100) = \frac{(-j100) \times (100 + j100)}{(-j100) + (100 + j100)} = (100 - j100)\ \Omega \tag{4}$$

Using voltage division formula in the circuit of Figure 2.23.3:

$$\mathbf{V_C} = \frac{100 - j100}{100 = j100 + 100} \times 100 = (0.63\angle{-18.4°})\ V \tag{5}$$

Transferring back to time domain:

$$v_c(t) = 0.63\cos\left(100t - 18.4^{\circ}\right)V$$

Choice (1) is the answer.

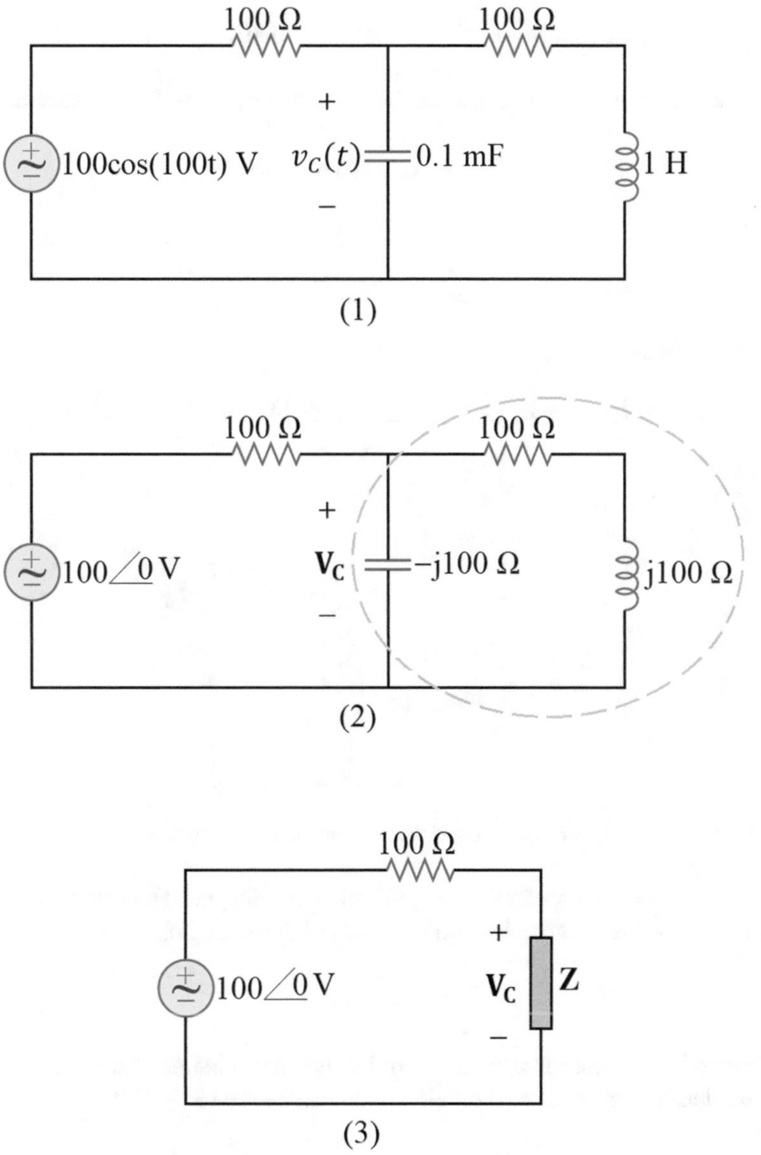

Figure 2.23 The circuit of solution of problem 2.25

2.26. In this problem, the phase angle of the voltage of the power source is defined as the reference phase angle. Therefore:

$$\mathbf{V} = (1000\angle 0^\circ)\ V\ rms = 1000\ V\ rms \tag{1}$$

where, " $\angle$ " is the symbol of phase angle. Based on the information given in the problem:

$$P_R = 1000\ W \tag{2}$$

$$S_Z = 1000\ VA \tag{3}$$

$$PF_Z = 0.6\ Lagging \tag{4}$$

Using average power relation for the resistor:

$$P_R = V_{R,rms} I_{R,rms} \xrightarrow{Using\ (1),(2)} 1000 = 1000 I_{R,rms} \Rightarrow I_{R,rms} = 1\ A \Rightarrow \mathbf{I_{R,rms}} = 1\angle 0^\circ \tag{5}$$

In (5), we consider the fact that the voltage and current of a resistor are in phase.

Using apparent power relation for the impedance:

$$S_Z = V_{Z,rms} I_{Z,rms} \xrightarrow{Using\ (1),(3)} 1000 = 1000 I_{Z,rms} \Rightarrow I_{Z,rms} = 1\ A \tag{6}$$

The phase angle of the current of the impedance can be calculated by using (4) as follows:

$$\angle \mathbf{I_Z} = -\cos^{-1}(0.6) = -53^\circ \tag{7}$$

Solving (6) and (7):

$$\mathbf{I_{Z,rms}} = (1\angle -53^\circ)\ A \tag{8}$$

Applying KCL in the top node:

$$\mathbf{I_{rms}} = \mathbf{I_{R,rms}} + \mathbf{I_{Z,rms}} = 1\angle 0^\circ + 1\angle -53^\circ = (1.79\angle -26.5^\circ)\ A$$

Choice (1) is the answer.

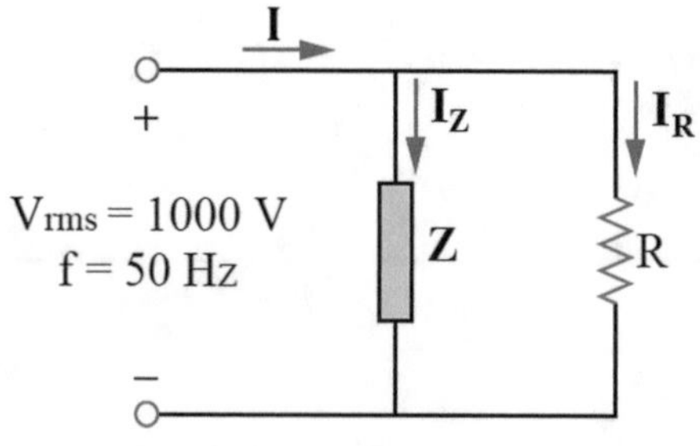

Figure 2.24 The circuit of solution of problem 2.26

2.27. Based on maximum average power transfer theorem, the relation below must be held to transfer the maximum average power to the load, where $\mathbf{Z_{Th}}^*$ is the complex conjugate of the Thevenin impedance seen by the load:

$$\mathbf{Z_L} = \mathbf{Z_{Th}}^* \tag{1}$$

Herein, a heuristic method is the best approach to determine the Thevenin impedance seen by the load. Herein, we need to turn off the independent voltage source (short circuit), as can be seen in 2.25.2.

Applying KVL in the left-side mesh of the circuit of Figure 2.25.2:

$$(5 + j10)\mathbf{I} + 5\mathbf{I} = 0 \Rightarrow (10 + j10)\mathbf{I} = 0 \Rightarrow \mathbf{I} = 0 \qquad (2)$$

Equation (2) implies that the voltage of the dependent voltage source is zero. Therefore, it is replaced by a short circuit branch. Consequently, the branch parallel to the dependent voltage source is short-circuited and eliminated. Figure 2.25.3 shows the updated circuit.

$$\mathbf{Z_{Th}} = \mathbf{Z_{in}} = 1||(j3) = \frac{1 \times (j3)}{1 + (j3)} = (0.9 + j0.3)\ \Omega \qquad (3)$$

Solving (1) and (3):

$$Z_L = \mathbf{Z_{Th}}^* = (0.9 + j0.3)^* = (0.9 - j0.3)\ \Omega$$

Choice (3) is the answer.

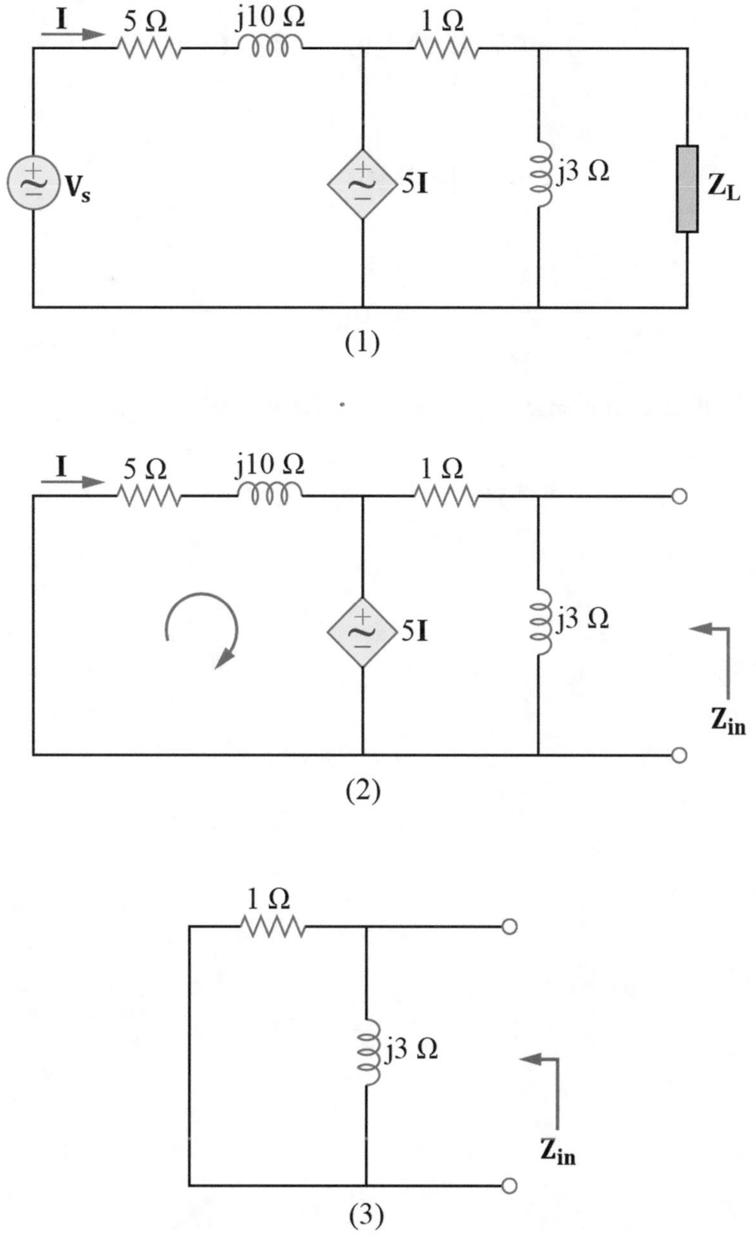

Figure 2.25 The circuit of solution of problem 2.27

2.28. The primary circuit is illustrated in frequency domain in Figure 2.26.2. Herein, the DC current source is turned off (open circuit), since we want to determine the average power of the dependent voltage source supplied just by the AC current source.

In this problem, the phase angle of the current of the current source is defined as the reference phase angle. Therefore, its phasor is $1\angle 0^\circ$ or 1 A. Herein, "$\angle$" is the symbol of phase angle.

To simplify the circuit, source transformation theorem is applied for the parallel connection of the current source and the 2 Ω resistor, as is shown in Figure 2.26.3.

Since no current is flowing through the 3 Ω resistor, **V** is the voltage of the 1 Ω resistor.

Using Ohm's law for the 1 Ω resistor:

$$\mathbf{V} = 1 \times \mathbf{I} = \mathbf{I} \tag{1}$$

KVL in the left-side mesh:

$$-2 + 2\mathbf{I} + 2\mathbf{V} + 2\mathbf{I} + \mathbf{I} = 0 \Rightarrow -2 + 5\mathbf{I} + 2\mathbf{V} = 0 \tag{2}$$

Solving (1) and (2):

$$-2 + 5\mathbf{I} + 2\mathbf{I} = 0 \Rightarrow \mathbf{I} = \frac{2}{7}\ A \tag{3}$$

Solving (1) and (3):

$$\mathbf{V} = \frac{2}{7}\ V \tag{4}$$

Now, the power of the dependent voltage source can be calculated as follows:

$$P = \frac{1}{2}|2\mathbf{V}||\mathbf{I}| = \frac{1}{2} \times \frac{4}{7} \times \frac{2}{7} = \frac{4}{49}\ W$$

Choice (1) is the answer.

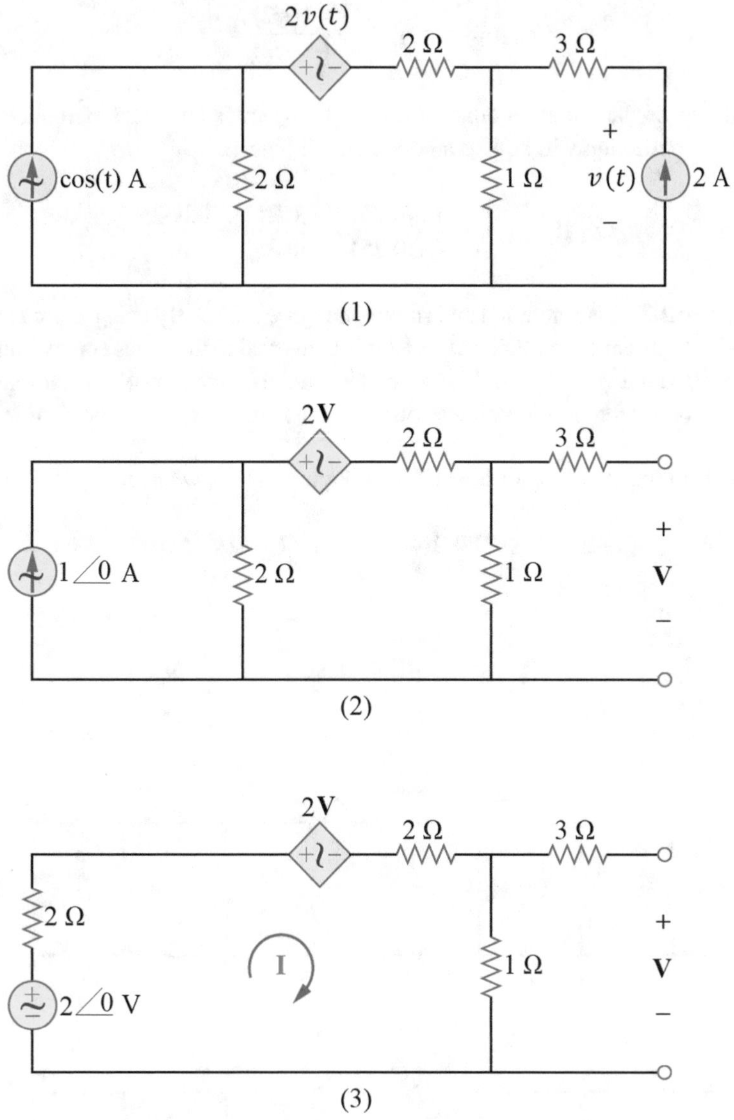

Figure 2.26 The circuit of solution of problem 2.28

2.29. The circuit of Figure 2.27.2 shows the primary circuit in frequency domain. The phasor of cos(t), that is, $1\angle 0^\circ$ is defined as the reference phasor, where "$\angle$" is the symbol of phase angle. Therefore, the phasors of the currents of the current sources are 1 A and 2 A. The impedances of the components are as follows:

$$\mathbf{Z_{2\ \Omega}} = 2\ \Omega \tag{1}$$

$$\mathbf{Z_{2\ H}} = j\omega L = j \times 1 \times 2 = j2\ \Omega \tag{2}$$

$$\mathbf{Z_{0.5\ F}} = \frac{1}{j\omega C} = \frac{1}{j \times 1 \times 0.5} = -j2\ \Omega \tag{3}$$

$$\mathbf{Z_{4\ F}} = \frac{1}{j\omega C} = \frac{1}{j \times 1 \times 4} = -j0.25\ \Omega \tag{4}$$

$$\mathbf{Z_{0.25\ H}} = j\omega L = j \times 1 \times 0.25 = j0.25\ \Omega \tag{5}$$

$$\mathbf{Z_{1\ \Omega}} = 1\ \Omega \tag{6}$$

As can be noticed from the circuit of Figure 2.27.2, the impedance of the parallel connection of 2 H inductor and 0.5 F capacitor is infinite (theoretically undefined). Thus, this connection can be considered as an open circuit branch:

$$\mathbf{Z}_{2\ \mathbf{H}}||\mathbf{Z}_{0.5\ \mathbf{F}} = \frac{(j2) \times (-j2)}{(j2) + (-j2)} = \frac{4}{0} \tag{7}$$

Likewise, the impedance of the parallel connection of 4 F capacitor and 0.25 H inductor is infinite (theoretically undefined). Hence, this connection will be like an open circuit branch.

$$\mathbf{Z}_{4\ \mathbf{F}}||\mathbf{Z}_{0.25\ \mathbf{H}} = \frac{(-j0.25) \times (j0.25)}{(-j0.25) + (j0.25)} = \frac{0.0625}{0} \tag{8}$$

Now, the circuit of Figure 2.27.2 is simplified and shown in Figure 2.27.3. By using source transformation theorem, the parallel connection of 1 A current source and 2 Ω resistor is converted to the series connection of 2 V voltage source and 2 Ω resistor, as is illustrated in Figure 2.27.4. Likewise, the parallel connection of 2 A current source and 1 Ω resistor is converted to the series connection of 2 V voltage source and 1 Ω resistor, as can be seen in Figure 2.27.4.

Now, by applying KVL in the only loop of the circuit of Figure 2.27.4, we have:

$$-2 + 2\mathbf{I} + 2\mathbf{I} + \mathbf{I} + 2 = 0 \Rightarrow 5\mathbf{I} = 0 \Rightarrow \mathbf{I} = 0 \tag{9}$$

Transferring to time domain:

$$i(t) = 0\ A$$

Choice (4) is the answer.

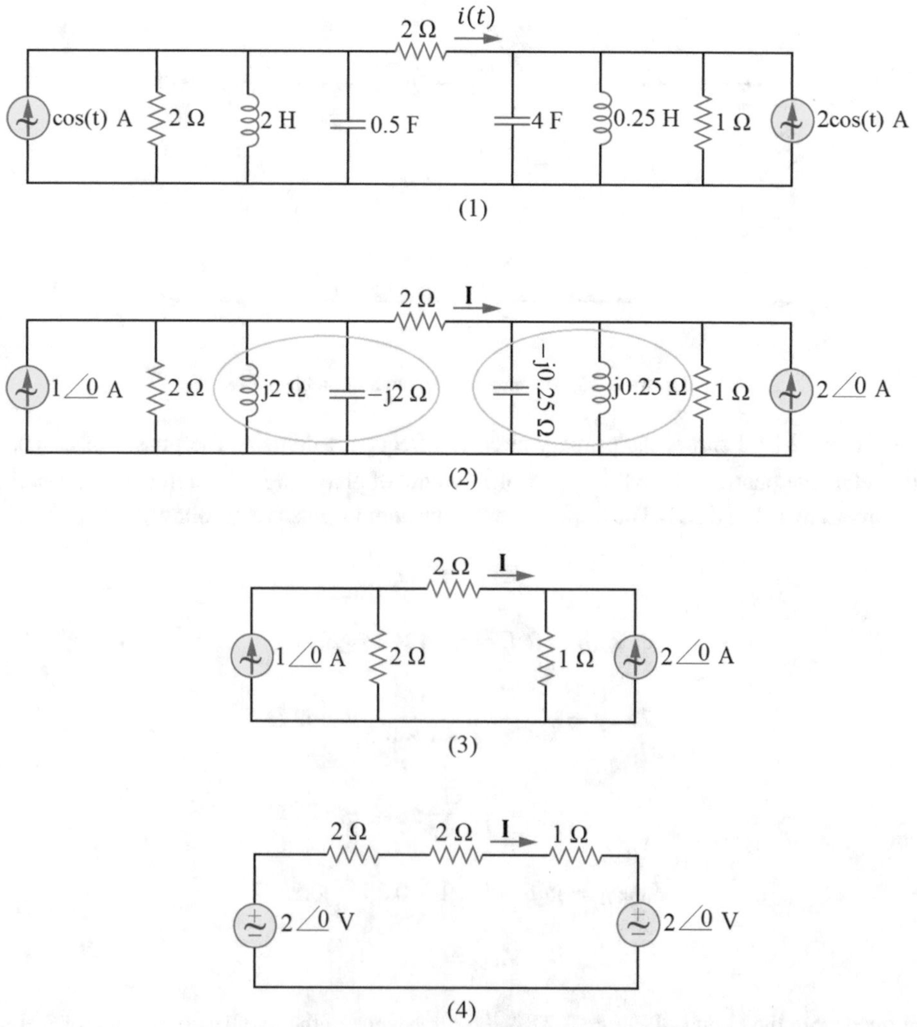

Figure 2.27 The circuit of solution of problem 2.29

2.30. The circuit includes two independent power sources with different frequencies. One of them is a DC source (DC current source); thus, its frequency is zero, and the other one (AC voltage source) has the angular frequency of 1 rad/sec. Therefore, we must use superposition theorem to solve the problem.

Part 1: The circuit of Figure 2.28.2 shows the main circuit, which only includes the DC current source. Hence, the voltage source is shut down (short circuit), and the inductor is replaced by a short circuit branch.

Since the resistor is short-circuited, its voltage is zero.

$$v(t) = 0 \tag{1}$$

By applying KVL in the indicated loop (see Figure 2.28.2), we have:

$$-v_{o1}(t) + v(t) = 0 \Rightarrow v_{o1}(t) = v(t) \tag{2}$$

Solving (1) and (2):

$$v_{o1}(t) = 0 \tag{3}$$

Part 2: The circuit of Figure 2.28.3 illustrates the circuit in frequency domain, while the current source is turned off (open circuit). The phasor of cos(t), that is, $1\angle 0°$ is defined as the reference phasor, where "$\angle$" is the symbol of phase angle. The impedances of the components are determined as follows:

$$\mathbf{Z}_{1\,\mathbf{H}} = j\omega L = j \times 1 \times 1 = j\,\Omega \tag{4}$$

$$\mathbf{Z}_{1\,\Omega} = 1\,\Omega \tag{5}$$

Now, by applying voltage division formula for the resistor, we have:

$$\mathbf{V} = \frac{1}{1+j} \times 1 = \frac{1}{\sqrt{2}\angle 45°} = \frac{1}{\sqrt{2}}\angle -45°)\ V \tag{6}$$

Applying KVL in the indicated loop (see Figure 2.28.3):

$$-\mathbf{V}_{o2} + \mathbf{V} + \mathbf{V} = 0 \Rightarrow \mathbf{V}_{o2} = 2\mathbf{V} \tag{7}$$

By solving (6) and (7):

$$\mathbf{V}_{o2} = (\sqrt{2}\angle -45°)\ V \tag{8}$$

Transferring to time domain:

$$v_{o2}(t) = \sqrt{2}\cos\left(t - 45^{\circ}\right)V \tag{9}$$

Based on superposition theorem:

$$v_o(t) = v_{o1}(t) + v_{o2}(t) = 0 + \sqrt{2}\cos\left(t - 45\right) = \sqrt{2}\cos\left(t - 45^{\circ}\right)\ V$$

Choice (4) is the answer.

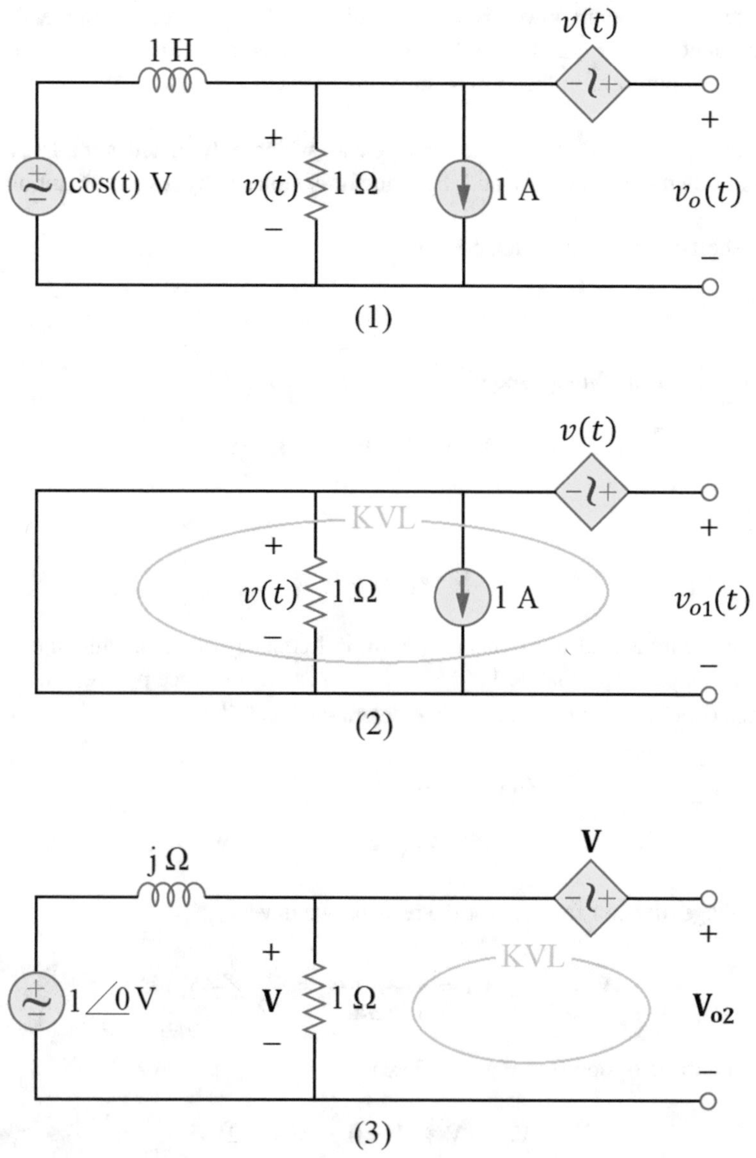

Figure 2.28 The circuit of solution of problem 2.30

2.31. The voltage of the voltage source includes three terms with different frequencies. Therefore, the problem must be solved by using superposition theorem.

Part 1: The circuit has been redrawn in Figure 2.29.2 to exhibit it in frequency domain. For the unknown angular frequency of ω, the impedance of each component is as follows:

$$\mathbf{Z}_{1\,\Omega} = 1\ \Omega \tag{1}$$

$$\mathbf{Z}_{1\,\mathrm{F}} = \frac{1}{j\omega C} = \frac{1}{j\omega \times 1} = -j\frac{1}{\omega}\ \Omega \tag{2}$$

By using voltage division formula, we can calculate the output voltage as follows:

$$\mathbf{V_o} = \frac{-j\frac{1}{\omega}}{1+\left(-j\frac{1}{\omega}\right)}\mathbf{V_s} \Rightarrow |\mathbf{V_o}| = \left|\frac{-j\frac{1}{\omega}}{1+\left(-j\frac{1}{\omega}\right)}\right||\mathbf{V_s}| = \frac{\frac{1}{\omega}}{\sqrt{1+\left(\frac{1}{\omega}\right)^2}}|\mathbf{V_s}| = \frac{1}{\sqrt{\omega^2+1}}|\mathbf{V_s}| \tag{3}$$

Now, for the voltage of $4\cos\left(\frac{1}{2}t\right)\ V$, the peak value and the root mean square (rms) value of the output voltage are:

$$|\mathbf{V_{o1}}| = \frac{1}{\sqrt{\left(\frac{1}{2}\right)^2+1}} \times 4 = \frac{8}{\sqrt{5}}\ V \Rightarrow \mathbf{V}_{1rms} = \frac{1}{\sqrt{2}} \times \frac{8}{\sqrt{5}} = \frac{8}{\sqrt{10}} \tag{4}$$

Part 2: Likewise, for the voltage of $-\frac{4}{3}\cos\left(\frac{3}{2}t\right)\ V$, the peak value and the rms value of the output voltage are:

$$|\mathbf{V_{o2}}| = \frac{1}{\sqrt{\left(\frac{3}{2}\right)^2+1}} \times \frac{4}{3} = \frac{8}{3\sqrt{13}}\ V \Rightarrow \mathbf{V}_{2rms} = \frac{1}{\sqrt{2}} \times \frac{8}{3\sqrt{13}} = \frac{8}{3\sqrt{26}} \tag{5}$$

Part 3: Similarly, for the voltage of $\frac{4}{5}\cos\left(\frac{5}{2}t\right)\ V$, the peak value and the rms value of the output voltage are:

$$|\mathbf{V_{o3}}| = \frac{1}{\sqrt{\left(\frac{5}{2}\right)^2+1}} \times \frac{4}{5} = \frac{8}{5\sqrt{29}}\ V \Rightarrow \mathbf{V}_{3rms} = \frac{1}{\sqrt{2}} \times \frac{8}{5\sqrt{29}} = \frac{8}{5\sqrt{58}} \tag{6}$$

Now, by using the relation below, we can calculate the rms value of the output voltage:

$$\mathbf{V}_{rms} = \sqrt{\mathbf{V}_{1rms}^{\ 2} + \mathbf{V}_{2rms}^{\ 2} + \mathbf{V}_{3rms}^{\ 2}} = \sqrt{\left(\frac{8}{\sqrt{10}}\right)^2 + \left(\frac{8}{3\sqrt{26}}\right)^2 + \left(\frac{8}{5\sqrt{58}}\right)^2}$$

$$\Rightarrow \mathbf{V}_{rms} = \sqrt{\frac{64}{10} + \frac{64}{234} + \frac{64}{1450}} = 2.59\ V$$

Choice (3) is the answer.

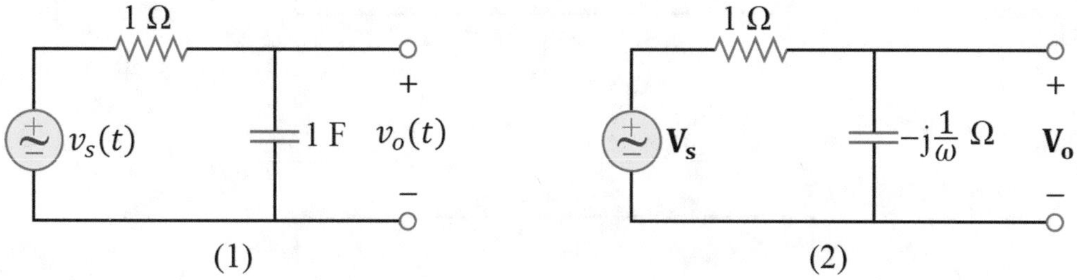

Figure 2.29 The circuit of solution of problem 2.31

2.32. Based on the problem, we need to determine the relation between the amplitudes of the input and output voltages in frequency domain. The circuit of Figure 2.30.2 shows the primary circuit in frequency domain. The impedances of the components are determined as follows:

$$\mathbf{Z_{1\,F}} = \frac{1}{j\omega C} = \frac{1}{j\omega}\ \Omega \tag{1}$$

$$\mathbf{Z_{2\,\Omega}} = 2\ \Omega \tag{2}$$

Applying KCL in the indicated node:

$$\frac{\mathbf{V}_o - \mathbf{V}_i}{\frac{1}{j\omega}} + g_m\mathbf{V}_i + \frac{\mathbf{V}_o}{2} = 0 \Rightarrow \left(\frac{1}{2} + j\omega\right)\mathbf{V}_o + (g_m - j\omega)\mathbf{V}_i = 0 \Rightarrow \mathbf{V}_o = \frac{-g_m + j\omega}{\frac{1}{2} + j\omega}\mathbf{V}_i \tag{3}$$

The amplitude of the output voltage is:

$$\Rightarrow |\mathbf{V}_o| = \left|\frac{-g_m + j\omega}{\frac{1}{2} + j\omega}\right| |\mathbf{V}_i| = \frac{\sqrt{g_m{}^2 + \omega^2}}{\sqrt{\left(\frac{1}{2}\right)^2 + \omega^2}} |\mathbf{V}_i| \tag{4}$$

Based on the given information:

$$|\mathbf{V}_o| \geq |\mathbf{V}_i| \tag{5}$$

Solving (4) and (5):

$$\frac{\sqrt{g_m{}^2 + \omega^2}}{\sqrt{\left(\frac{1}{2}\right)^2 + \omega^2}} |\mathbf{V}_i| \geq |\mathbf{V}_i| \Rightarrow \sqrt{g_m{}^2 + \omega^2} \geq \sqrt{\left(\frac{1}{2}\right)^2 + \omega^2}$$

$$\Rightarrow g_m{}^2 + \omega^2 \geq \left(\frac{1}{2}\right)^2 + \omega^2 \Rightarrow g_m{}^2 \geq \left(\frac{1}{2}\right)^2 \Rightarrow |g_m| \geq \frac{1}{2}$$

Choice (4) is the answer.

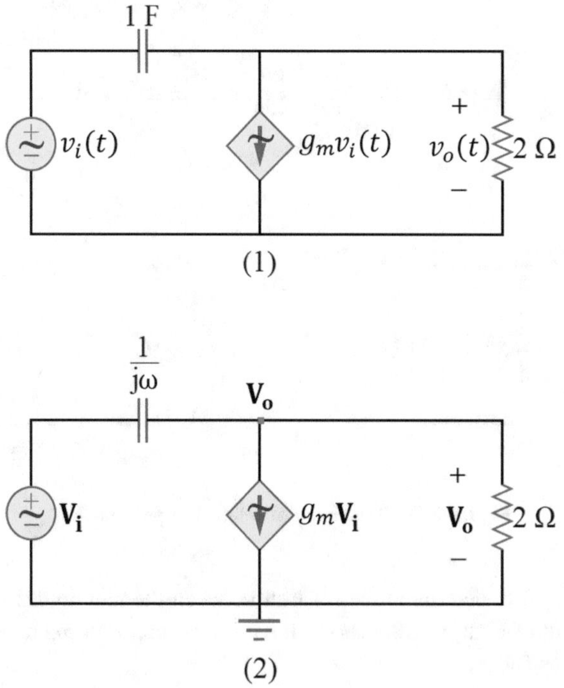

Figure 2.30 The circuit of solution of problem 2.32

2.33. Based on maximum average power transfer theorem, the relation below must be held to transfer the maximum average power to the load, where $\mathbf{Z_{Th}}^*$ is the complex conjugate of the Thevenin impedance seen by the load:

$$\mathbf{Z_L} = \mathbf{Z_{Th}}^* \tag{1}$$

As is shown in Figure 2.31.2, to calculate the Thevenin impedance seen by the load, we need to turn off all the independent sources (herein, independent voltage source), apply a test source (herein, a test voltage source) to the terminal, and determine the value of $\frac{\mathbf{V_t}}{\mathbf{I_t}}$. Nodal analysis can be applied to solve the problem.

Applying KCL in node "a":

$$\mathbf{I} + \mathbf{I_t} = 0 \Rightarrow \mathbf{I} = -\mathbf{I_t} \quad (2)$$

Using Ohm's law for the 1 Ω resistor:

$$\mathbf{I} = \frac{\mathbf{V} - \mathbf{V_t}}{1} \Rightarrow \mathbf{I} = \mathbf{V} - \mathbf{V_t} \xrightarrow{Using\ (1)} \mathbf{V} = \mathbf{V_t} - \mathbf{I_t} \quad (3)$$

Applying KCL in the indicated node:

$$\frac{\mathbf{V}}{j2} - 3\mathbf{I} + \mathbf{I} = 0 \xrightarrow{Using\ (2),(3)} \frac{\mathbf{V_t} - \mathbf{I_t}}{j2} - 2(-\mathbf{I_t}) = 0$$

$$\Rightarrow \left(\frac{1}{j2}\right)\mathbf{V_t} + \left(-\frac{1}{j2} + 2\right)\mathbf{I_t} = 0 \Rightarrow \frac{\mathbf{V_t}}{\mathbf{I_t}} = \frac{\frac{1}{j2} - 2}{\frac{1}{j2}} = 1 - j4 \Rightarrow \mathbf{Z_{Th}} = (1 - j4)\ \Omega \quad (4)$$

Solving (1) and (4):

$$\mathbf{Z_L} = (1 - j4)^* = (1 + j4)\ \Omega$$

Choice (4) is the answer.

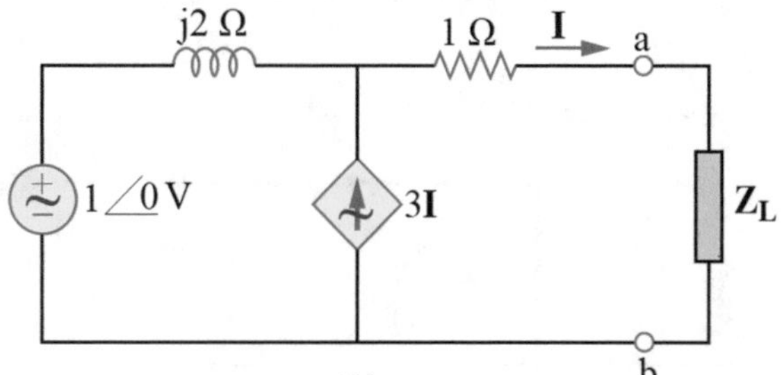

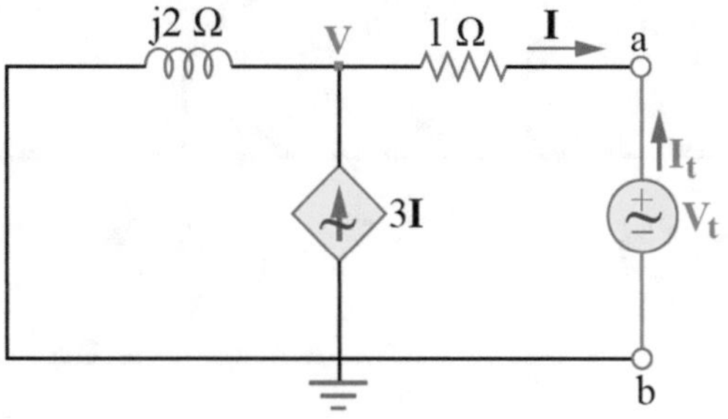

Figure 2.31 The circuit of solution of problem 2.33

2.34. The circuit of Figure 2.32.2 shows the primary circuit in frequency domain while the terminal is short-circuited and node "b" is grounded. The problem has two power sources; however, both are operated in the same angular frequency ($\omega = 3$ *rad/sec*). Thus, we do not need to apply superposition theorem. The impedances of the components are determined as follows:

$$\mathbf{Z}_{\frac{1}{3}\,\mathbf{F}} = \frac{1}{j\omega C} = \frac{1}{j \times 3 \times \frac{1}{3}} = -j\ \Omega \quad (1)$$

$$\mathbf{Z}_{\frac{1}{6}H} = j\omega L = j \times 3 \times \frac{1}{6} = j0.5\ \Omega \tag{2}$$

In this problem, the phasor of cos(3*t*) is defined as the reference phasor ($1\angle 0°$, where "$\angle$" is the symbol of phase angle). Therefore, the voltage of the voltage source and the current of the current source are $(2\angle 0°)V$ and $(1\angle 0°)A$, respectively. Nodal analysis can be applied to solve the problem.

By applying Ohm's law for the capacitor, we can write:

$$\mathbf{I_C} = \frac{2-0}{-j} = j2 \tag{3}$$

Applying KCL in the indicated node:

$$\frac{0-2}{-j} - (1\angle -90°) + \frac{0-\frac{1}{4}\mathbf{I_C}}{j0.5} + \mathbf{I_{sc}} = 0 \Rightarrow -j + j\frac{1}{2}\mathbf{I_C} + \mathbf{I_{sc}} = 0 \tag{4}$$

Solving (3) and (4):

$$-j + j\frac{1}{2} \times j2 + \mathbf{I_{sc}} = 0 \Rightarrow \mathbf{I_{sc}} = (1+j)\ A = (\sqrt{2}\angle 45°)\ A \tag{5}$$

Transferring from frequency domain to time domain:

$$i_{sc}(t) = \sqrt{2}\cos(3t + 45^\circ)\ A$$

Choice (1) is the answer.

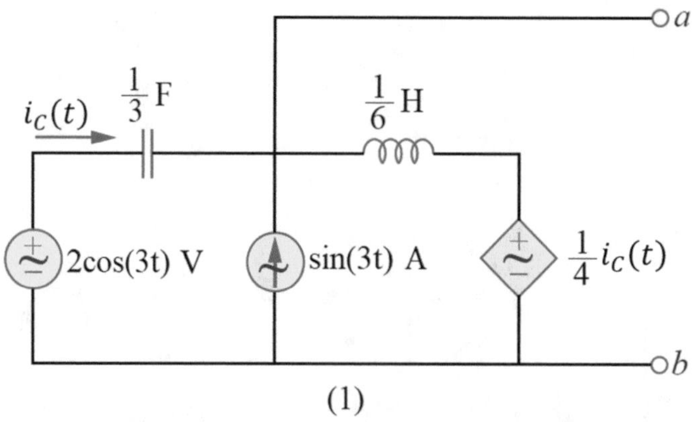

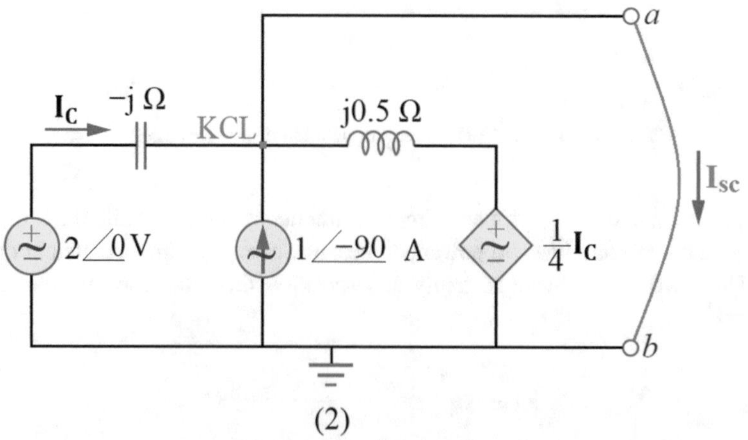

Figure 2.32 The circuit of solution of problem 2.34

2.35. The main circuit, shown in Figure 2.33.1, includes two current sources with different angular frequencies. Therefore, superposition theorem must be applied in this problem, as can be seen in Figs. 2.33.2–3.

Part 1: Figure 2.33.2 illustrates the primary circuit in frequency domain for the left-side current source with the angular frequency of 1 rad/sec. Another current source is turned off (open circuit). Herein, the phasor of cos(t), that is, $1\angle 0^\circ$ is defined as the reference phasor, where "$\angle$" is the symbol of phase angle. Hence, the phasor of the current of the left-side current source is $10\angle 0^\circ\ A$. The impedances are determined as follows:

$$\mathbf{Z}_{1\,\mathbf{F}} = \frac{1}{j\omega C} = \frac{1}{j \times 1 \times 1} = -j\,\Omega \tag{1}$$

$$\mathbf{Z}_{1\,\Omega} = 1\,\Omega \tag{2}$$

Applying current division relation:

$$\mathbf{I}_{o1} = \frac{1}{1 + (-j) + 1 + 1} \times 10 = \frac{10}{3-j}\,A \tag{3}$$

$$\mathbf{I}_{o1} = 1 \times \mathbf{I}_{o1} = \frac{10}{3-j}\,V = (3.16\angle 18.43^\circ)\,V \tag{4}$$

The output voltage in time domain is:

$$v_{o1}(t) = 3.16\cos\left(t + 18.43^\circ\right) \tag{5}$$

Part 2: Figure 2.33.3 exhibits the main circuit in frequency domain for the bottom current source with the angular frequency of 2 rad/sec. Another current source is turned off (open circuit). In this part, the phasor of cos($2t$), that is, $1\angle 0^\circ$ is defined as the reference phasor, where "$\angle$" is the symbol of phase angle. Thus, the phasor of the current of the bottom current source is $2\angle -30^\circ$. The impedances can be calculated as follows:

$$\mathbf{Z}_{1\,\mathbf{F}} = \frac{1}{j\omega C} = \frac{1}{j \times 2 \times 1} = -j0.5\,\Omega \tag{6}$$

$$\mathbf{Z}_{1\,\Omega} = 1\,\Omega \tag{7}$$

Applying current division formula:

$$\mathbf{I}_{o2} = \frac{1}{1 + (-j0.5) + 1 + 1} \times (2\angle -30^\circ) = \frac{2\angle -30^\circ}{3 - j0.5}\,A = (0.66\angle -20.54^\circ)\,A \tag{8}$$

$$\mathbf{V}_{o2} = 1 \times \mathbf{I}_{o2} = (0.66\angle -20.54^\circ)\,V \tag{9}$$

The output voltage in time is:

$$v_{o2}(t) = 0.66\cos\left(2t - 20.54^\circ\right) \tag{10}$$

By using superposition theorem and (5) and (10), we can determine the total output voltage in sinusoidal steady state:

$$v_o(t) = v_{o1}(t) + v_{o2}(t) = 3.16\cos(t + 18.43) + 0.66\cos(2t - 20.54)$$

Choice (4) is the answer.

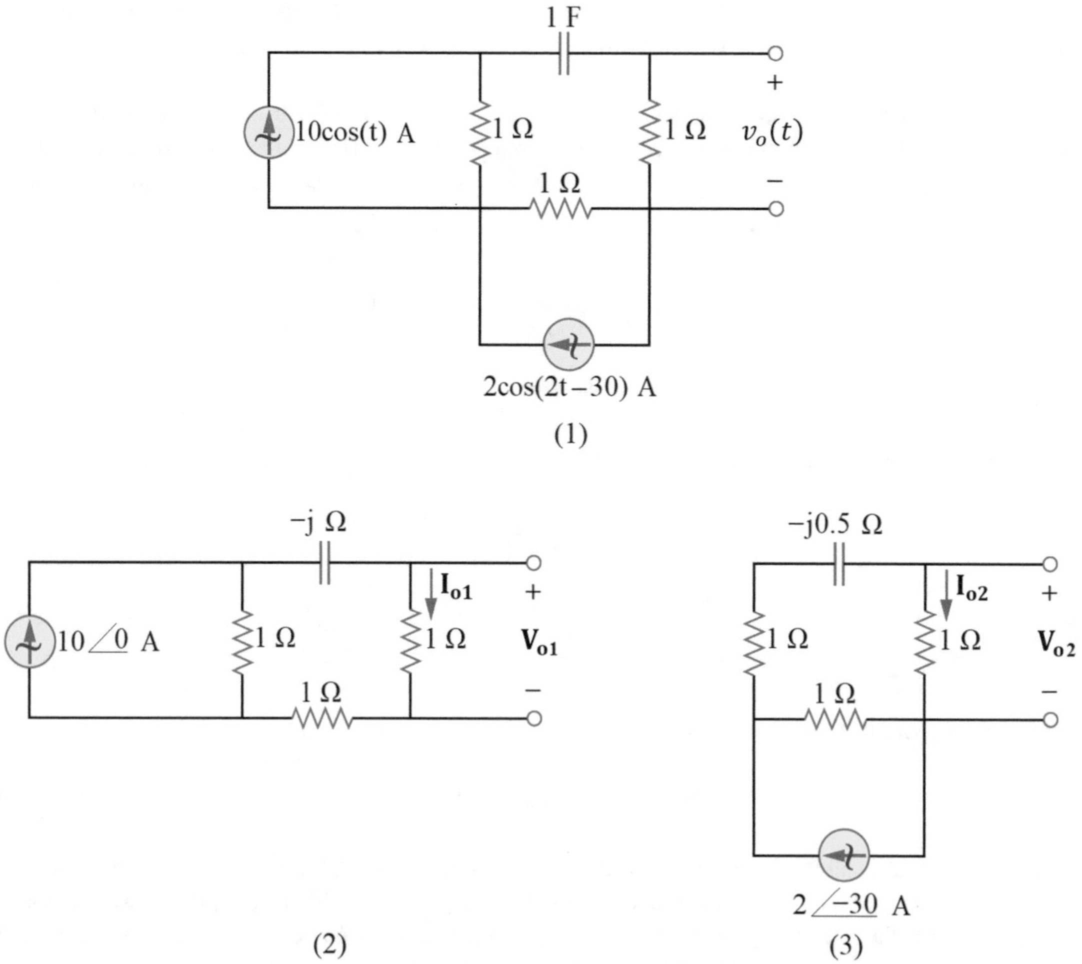

Figure 2.33 The circuit of solution of problem 2.35

2.36. To transfer the maximum average power to the load, the impedance of the load must be equal to the complex conjugate of the Thevenin impedance of the circuit seen by the load, that is:

$$\mathbf{Z_L} = \mathbf{Z_{Th}}^* \tag{1}$$

The circuit of Figure 2.34.2 shows the main circuit in frequency domain, where a test voltage source has been connected to the terminal and the independent voltage source has been shut down (short circuited). The value of $\frac{\mathbf{V_t}}{\mathbf{I_t}}$ will give us the Thevenin impedance of the circuit seen by the load. The impedances of the components are as follows:

$$\mathbf{Z_{25\,\Omega}} = 25\ \Omega \tag{2}$$

$$\mathbf{Z_{20\,H}} = j\omega L = j \times 1 \times 20 = j20\ \Omega \tag{3}$$

$$\mathbf{Z_{1\,\Omega}} = 1\ \Omega \tag{4}$$

$$\mathbf{Z_{3\,H}} = j\omega L = j \times 1 \times 3 = j3\ \Omega \tag{5}$$

The circuit of Figure 2.34.2 can be analyzed by using nodal analysis.

Applying KCL in the supernode:

$$-\mathbf{I_t} + \frac{\mathbf{V_t}}{j3} + \frac{\mathbf{V_t} - 5\mathbf{I_x}}{1} = 0 \Rightarrow -\mathbf{I_t} + \left(1 - j\frac{1}{3}\right)\mathbf{V_t} - 5\mathbf{I_x} = 0 \tag{6}$$

Defining $\mathbf{I_x}$ based on the node voltages, by using Ohm's law:

$$\mathbf{I_x} = \frac{0 - 5\mathbf{I_x}}{25 + j20} \Rightarrow (25 + j20)\mathbf{I_x} = -5\mathbf{I_x} \Rightarrow (30 + j20)\mathbf{I_x} = 0 \Rightarrow \mathbf{I_x} = 0 \tag{7}$$

Solving (6) and (7):

$$-\mathbf{I_t} + \left(1 - j\frac{1}{3}\right)\mathbf{V_t} - 5 \times 0 = 0 \Rightarrow \left(1 - j\frac{1}{3}\right)\mathbf{V_t} = \mathbf{I_t} \Rightarrow \frac{\mathbf{V_t}}{\mathbf{I_t}} = \frac{1}{\left(1 - j\frac{1}{3}\right)} = 0.9 + j0.3$$

$$\Rightarrow \mathbf{Z_{Th}} = (0.9 + j0.3)\ \Omega \tag{8}$$

Solving (1) and (8):

$$\Rightarrow \mathbf{Z_L} = (0.9 + j0.3)^* = (0.9 - j0.3)\ \Omega$$

Choice (2) is the answer.

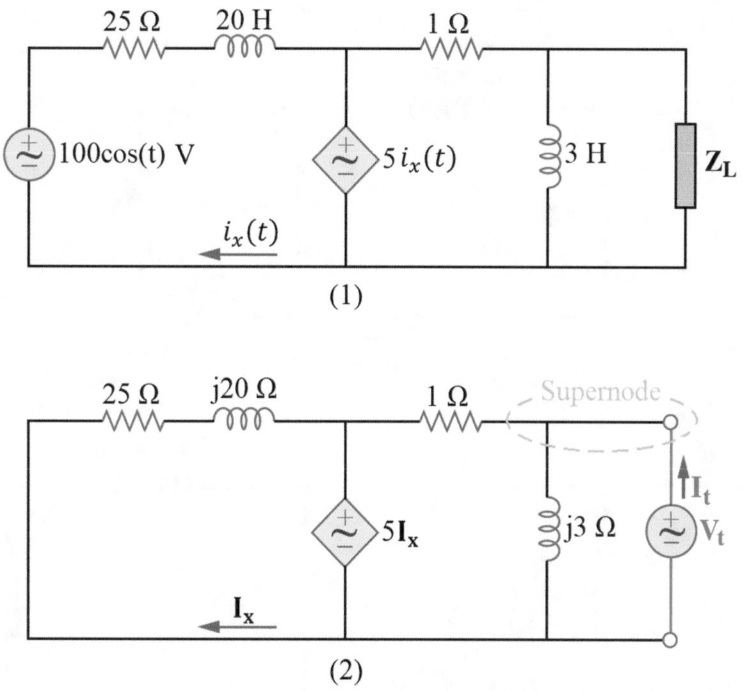

Figure 2.34 The circuit of solution of problem 2.36

2.37. Based on maximum average power transfer theorem, the maximum average power can be transferred to the load if:

$$\mathbf{Z_L} = \mathbf{Z_{Th}}^* \tag{1}$$

In this regard, the maximum transferrable average power can be calculated by using the following relation, where $\mathbf{V_{Th}}$ and R_{Th} are the phasor (peak value) of the Thevenin voltage and the resistance of the Thevenin impedance seen by the load, respectively.

$$P_{L,max} = \frac{1}{8}\frac{|\mathbf{V_{Th}}|^2}{R_{Th}} \tag{2}$$

Therefore, we need to determine the Thevenin equivalent circuit seen by the load. To determine the Thevenin impedance, the independent voltage source is turned off (short-circuited) and the input impedance is calculated, as can be seen in Figure 2.35.2.

$$\mathbf{Z_{Th}} = \mathbf{Z_{in}} = (15 + j20)||(-j20) = \frac{(15 + j20) \times (-j20)}{(15 + j20) + (-j20)} = (26.66 - j20)\ \Omega \tag{3}$$

Solving (1) and (3):

$$Z_L = (26.66 - j20)^* = (26.66 + j20)\ \Omega \tag{4}$$

Moreover, to determine the Thevenin voltage of the circuit, the open circuit voltage of the load needs to be calculated, as can be seen in Figure 2.35.3.

$$\mathbf{V_{Th}} = \mathbf{V_{oc}} = \frac{-j20}{-j20 + (15 + j20)} \times (10\angle 0^\circ) = \frac{-j200}{15} = -j13.33\ V \tag{5}$$

Solving (2), (3), and (5):

$$P_{L,max} = \frac{1}{8}\frac{|-j13.33|^2}{26.66} = 0.83\ W$$

Choice (1) is the answer.

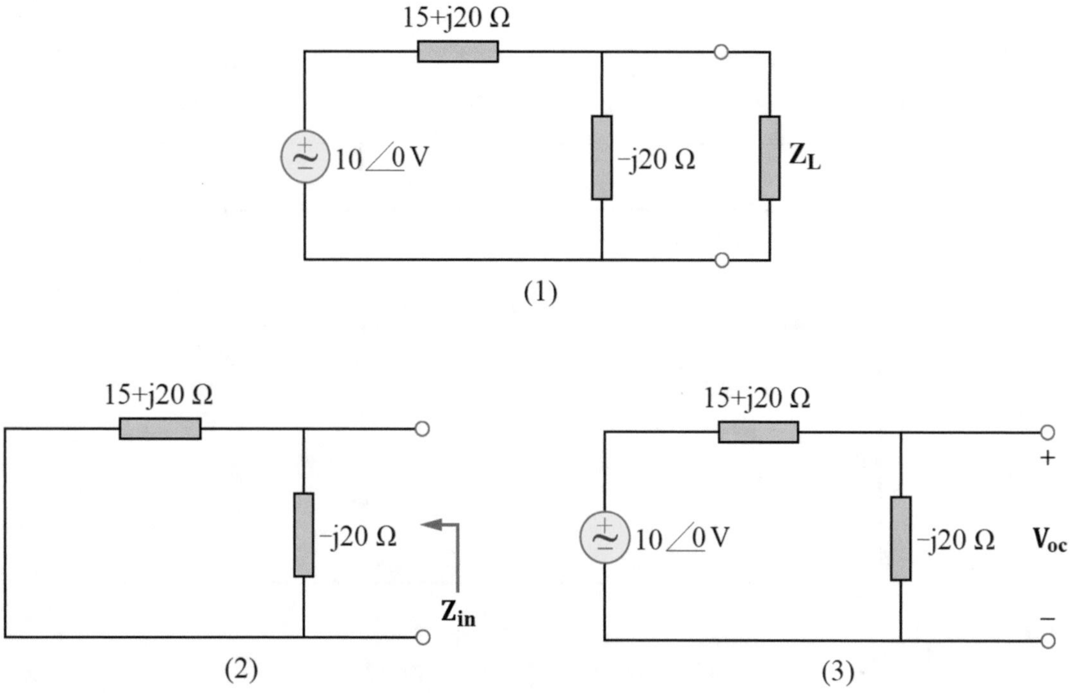

Figure 2.35 The circuit of solution of problem 2.37

2.38. Figure 2.36.2 illustrates the primary circuit in frequency domain. The phasor of cos(1000t), that is, $1\angle 0^\circ$ is defined as the reference phasor, where "$\angle$" is the symbol of phase angle. Therefore, the phasor of the voltage of the independent voltage source is $(30\angle 0^\circ)$ V or 30 V. The impedances of the components are as follows:

$$\mathbf{Z_{3\,\Omega}} = 3\ \Omega \tag{1}$$

$$\mathbf{Z_{1.5\,\Omega}} = 1.5\ \Omega \tag{2}$$

$$\mathbf{Z_{1\,H}} = j\omega L = j1000 \times 4 \times 10^{-3} = j4\ \Omega \tag{3}$$

$$\mathbf{Z}_{0.5\ \mathrm{F}} = \frac{1}{j\omega C} = \frac{1}{j1000 \times 500 \times 10^{-6}} = -2j\ \Omega \tag{4}$$

The problem can be simply solved by using nodal analysis as follows:

Applying KCL in the left-side node:

$$\frac{\mathbf{V}-30}{3} + \frac{\mathbf{V}}{1.5} + \mathbf{I} = 0 \Rightarrow \mathbf{V} + \mathbf{I} = 10 \Rightarrow \mathbf{V} = 10 - \mathbf{I} \tag{5}$$

Applying KCL in the right-side node:

$$-\mathbf{I} + \frac{\mathbf{V}}{j4} + \frac{\mathbf{V}-2\mathbf{I}}{-j2} = 0 \Rightarrow \mathbf{V}\left(\frac{1}{j4} + \frac{1}{-j2}\right) - (1+j)\mathbf{I} = 0$$

$$\Rightarrow \left(j\frac{1}{4}\right)\mathbf{V} - (1+j)\mathbf{I} = 0 \tag{6}$$

Solving (5) and (6):

$$\left(j\frac{1}{4}\right)(10-\mathbf{I}) - (1+j)\mathbf{I} = 0 \Rightarrow j\frac{5}{2} - \left(1 + \frac{5}{4}\right)\mathbf{I} = 0 \Rightarrow \mathbf{I} = \frac{j\frac{5}{2}}{1+j\frac{5}{4}} = (1.5\angle 39^\circ)\ A \tag{7}$$

The current in time domain is:

$$i(t) = 1.5\cos\left(1000t + 39^\circ\right) A$$

Choice (2) is the answer.

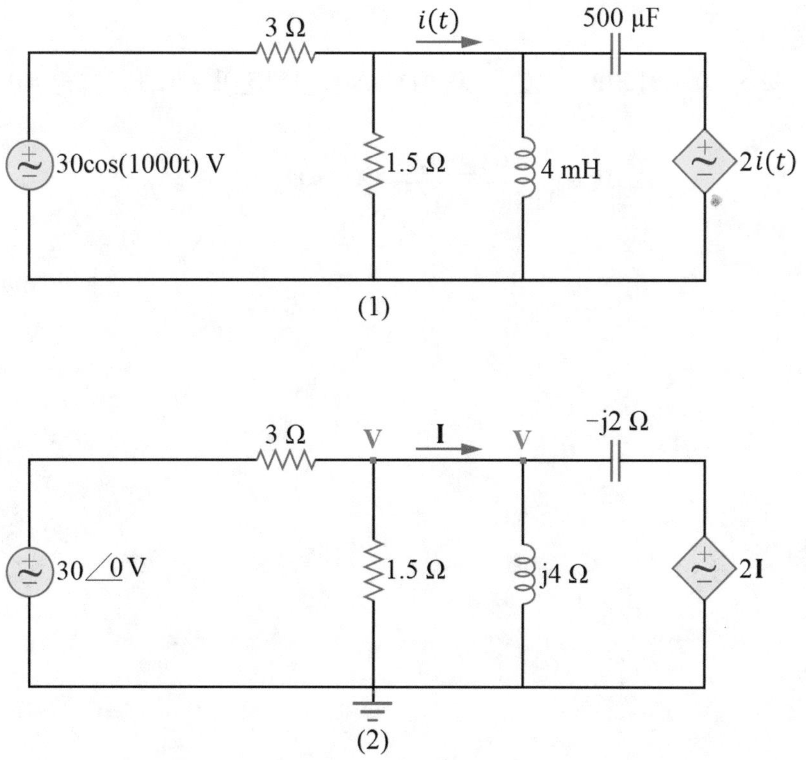

Figure 2.36 The circuit of solution of problem 2.38

2.39. Figure 2.37.2 illustrates the main circuit in frequency domain. The phasor of $\cos(2t)$, that is, $1\angle 0^\circ$ is defined as the reference phasor, where "$\angle$" is the symbol of phase angle. Therefore, the phasor of the current of the current source is $(10\angle 0^\circ)\ A$ or $10\ A$. The impedances of the components are as follows:

$$\mathbf{Z}_{3\ \Omega} = 3\ \Omega \tag{1}$$

$$\mathbf{Z}_{2\ \mathrm{H}} = j\omega L = j \times 2 \times 2 = j4\ \Omega \tag{2}$$

$$\mathbf{Z}_{6\ \Omega} = 6\ \Omega \tag{3}$$

$$\mathbf{Z}_{0.1\ \mathrm{F}} = \frac{1}{j\omega C} = \frac{1}{j \times 2 \times 0.1} = -j5\ \Omega \tag{4}$$

$$\mathbf{Z}_{8\ \Omega} = 8\ \Omega \tag{5}$$

$$\mathbf{Z}_{4\ \mathrm{H}} = j\omega L = j \times 2 \times 4 = j8\ \Omega \tag{6}$$

Now, we should simplify the circuit by calculating the impedance of the indicated part of the circuit of Figure 2.37.2 as follows:

$$\mathbf{z} = (j4) \parallel (6 - j5) = \frac{(j4)\text{–}(6 - j5)}{(j4)+(6 - j5)} = \frac{20 - j24}{6 - \mathrm{j}} = (5.13\angle 59.6^\circ)\ \Omega \tag{7}$$

By applying KCL in the left-side node, we can write:

$$-10 + \mathbf{I} + \mathbf{I}' = 0 \Rightarrow \mathbf{I}' = 10 - \boldsymbol{I} \tag{8}$$

Applying KVL in the indicated loop:

$$-3\mathbf{I} + \mathbf{I}'\mathbf{Z} - 5\mathbf{I} = 0 \xrightarrow{Using\ (7)} -8\mathbf{I} + \mathbf{I}'(5.13\angle 59.6^\circ) = 0 \tag{9}$$

Solving (8) and (9):

$$-8\mathbf{I} + (10 - \mathbf{I})(5.13\angle 59.6^\circ) = 0 \Rightarrow (11.48\angle -157.3^\circ)\mathbf{I} + 51.3\angle 59.6^\circ = 0$$

$$\Rightarrow \mathbf{I} = \frac{-51.3\angle 59.6^\circ}{11.48\angle -157.3^\circ} = (4.47\angle 36.9^\circ)\ A$$

Choice (1) is the answer.

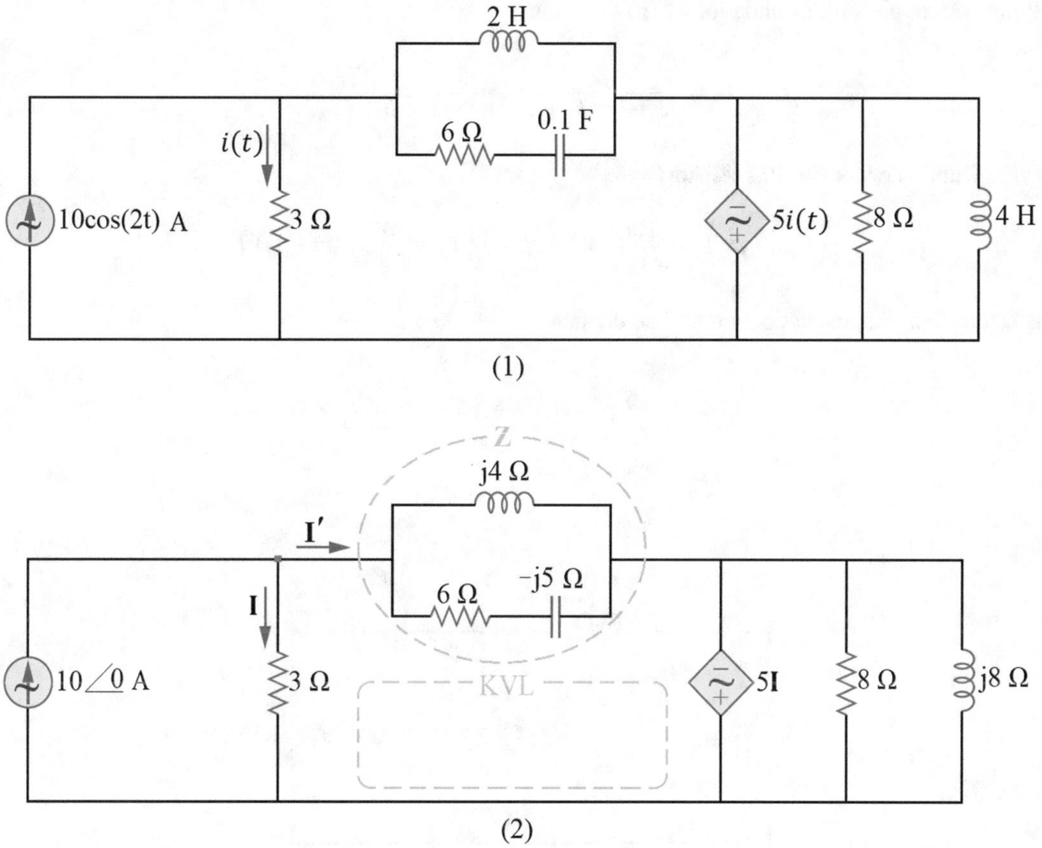

Figure 2.37 The circuit of solution of problem 2.39

2.40. Figure 2.38.2 shows the circuit in frequency domain. The phasor of sin(4*t*), that is, $1\angle 0°$ is defined as the reference phasor, where "$\angle$" is the symbol of phase angle. Therefore, the phasor of the voltage of the voltage source is $5\angle 0°$ *V* or 5 *V*. The impedances of the components are as follows:

$$\mathbf{Z}_{2\ \Omega} = 2\ \Omega \tag{1}$$

$$\mathbf{Z}_{0.5\ \mathbf{H}} = j\omega L = j \times 4 \times 0.5 = j2\ \Omega \tag{2}$$

$$\mathbf{Z}_{\frac{1}{8}\ \mathbf{F}} = \frac{1}{j\omega C} = \frac{1}{j \times 4 \times \frac{1}{8}} = -j2\ \Omega \tag{3}$$

$$\mathbf{Z}_{1\ \Omega} = 1\ \Omega \tag{4}$$

We can simplify the circuit by calculating the impedance seen by the voltage source, as is presented in the following.

$$\mathbf{Z}_{\mathbf{in}} = 2 + (j2)||(1 - j2) = 2 + \frac{(j2)(1 - j2)}{(j2) + (1 - j2)} = 2 + \frac{4 + j2}{1} = 6 + j2 \tag{5}$$

Applying Ohm's law for the input impedance of the circuit:

$$\mathbf{I} = \frac{\mathbf{V}}{\mathbf{Z}_{\mathbf{in}}} = \frac{5}{6 + j2}\ A \tag{6}$$

Applying current division formula for the 1 Ω resistor:

$$\mathbf{I_o} = \frac{j2}{j2+(-j2)+1} \times \frac{5}{6+j2} = \frac{10j}{6+j2}\ \text{A} \tag{7}$$

Applying Ohm's law for the 1 Ω resistor:

$$\mathbf{V_o} = \mathbf{I_o} \times 1 = \frac{10j}{6+j2} \times 1 = \left(\frac{1}{2} + j\frac{3}{2}\right)\text{V} = \left(\frac{\sqrt{10}}{2}\angle(tan^{-1}(3))^{\circ}\right)V \tag{8}$$

Transferring from frequency domain to time domain:

$$v_o(t) = \frac{\sqrt{10}}{2}\sin\left(4t + \left(tan^{-1}(3)\right)^{\circ}\right)V$$

Choice (3) is the answer.

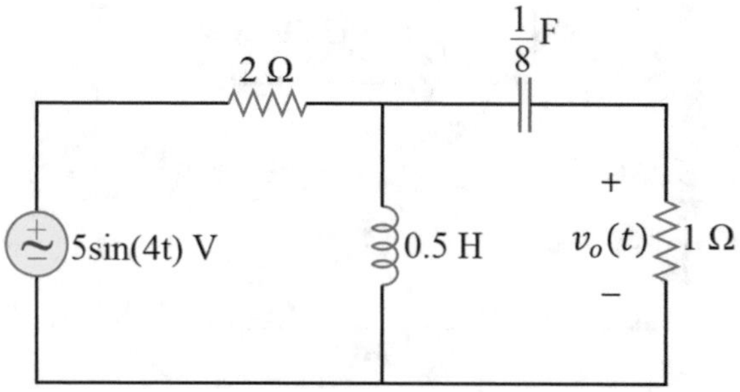

(1)

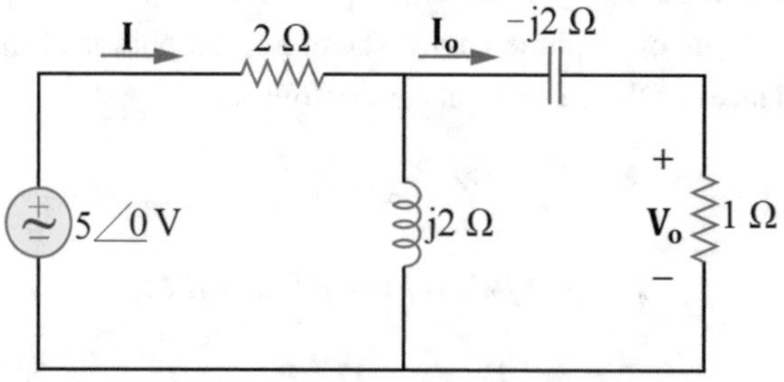

(2)

Figure 2.38 The circuit of solution of problem 2.40

2.41. Figure 2.39.2 shows the primary circuit in frequency domain. The phasor of cos(100*t*), that is, $1\angle 0^{\circ}$ is defined as the reference phasor, where "$\angle$" is the symbol of phase angle. Thus, the phasor of the current of the current source is $10\angle 0^{\circ}\ A$ or 10 *A*. The impedances of the components are presented in the following:

$$\mathbf{Z}_{\frac{2}{5}\,\mathbf{mF}} = \frac{1}{j\omega C} = \frac{1}{j \times 100 \times \frac{2}{5} \times 10^{-3}} = -j25\ \Omega \tag{1}$$

$$\mathbf{Z}_{50\ \Omega} = 50\ \Omega \tag{2}$$

$$\mathbf{Z}_{\frac{5}{6}\ \mathrm{mF}} = \frac{1}{j\omega C} = \frac{1}{j \times 100 \times \frac{5}{6} \times 10^{-3}} = -j12\ \Omega \tag{3}$$

$$\mathbf{Z}_{\frac{3}{5}\ \mathrm{H}} = j\omega L = j \times 100 \times \frac{3}{5} = j60\ \Omega \tag{4}$$

To solve the problem, it is suggested to simplify the circuit by using source transformation theorem for the parallel connection of the independent current source and the left-side impedance as well as determining the equivalent impedance of the right-side parallel branches, as is shown in Figure 2.39.3.

Based on source transformation theorem, the voltage of the voltage source is:

$$\mathbf{V} = (10\angle 0^\circ) \times (-j25) = -j250 \tag{5}$$

In addition, the equivalent impedance of the right-side parallel branches is:

$$(-j12)||(j60) = \frac{(-j12) \times (j60)}{(-j12) + (j60)} = -j15\ \Omega \tag{6}$$

Now, by applying KVL in the loop (see Figure 2.39.3), we can write:

$$j250 + (-j25)\mathbf{I} + \mathbf{V_x} - 0.4\mathbf{V_x} + (-j15)\mathbf{I} = 0$$

$$\Rightarrow j250 + (-j40)\mathbf{I} + 0.6\mathbf{V_x} = 0 \tag{7}$$

Applying Ohm's law for the resistor:

$$\mathbf{V_x} = 50\mathbf{I} \tag{8}$$

Solving (7) and (8):

$$j250 + (-j40)\mathbf{I} + 0.6 \times (50\mathbf{I}) = 0 \Rightarrow j250 + (30 - j40)\mathbf{I} = 0$$

$$\Rightarrow \mathbf{I} = \frac{-j250}{30 - j40} \tag{9}$$

Solving (8) and (9):

$$\mathbf{V_x} = 50 \times \left(\frac{-j250}{30 - j40}\right) = (200 - j150)\ V$$

Choice (4) is the answer.

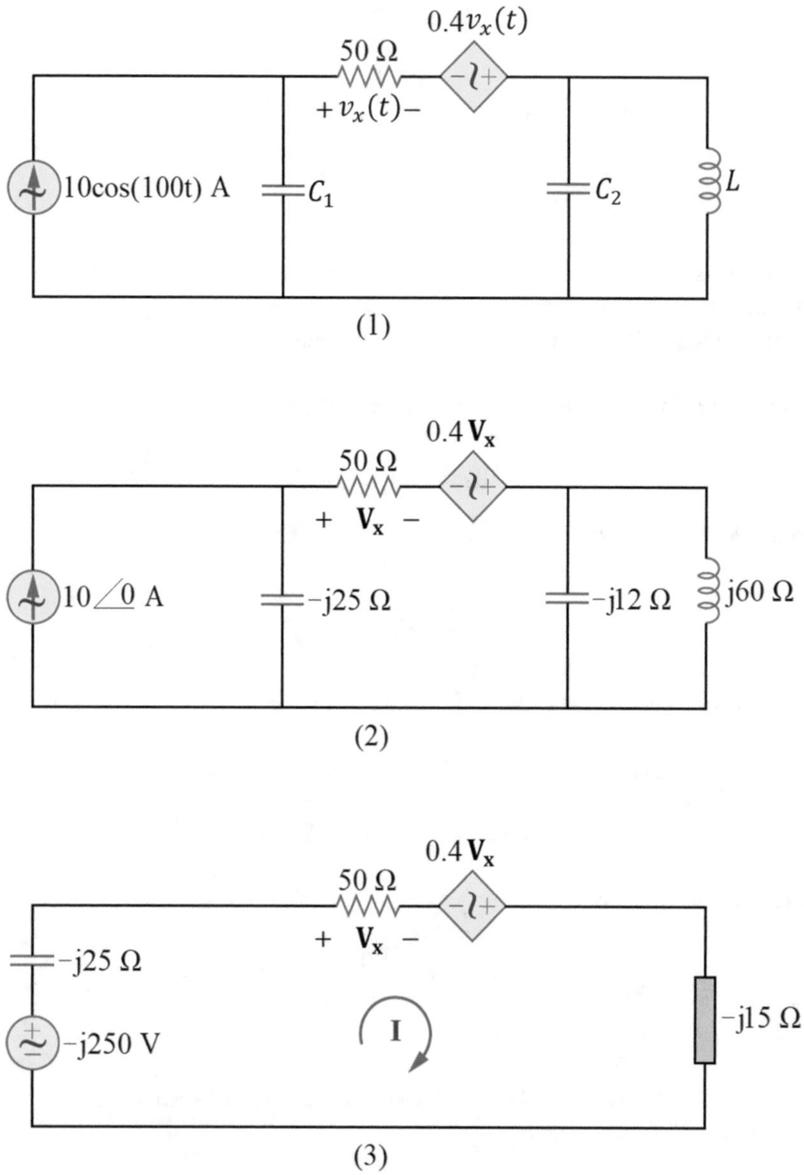

Figure 2.39 The circuit of solution of problem 2.41

2.42. Although the circuit includes two power sources, it is operated with a single frequency. Therefore, we do not need to apply superposition theorem. Figure 2.40.2 illustrates the primary circuit in frequency domain. The phasor of cos(10*t*), that is, $1\angle 0°$ is defined as the reference phasor, where "$\angle$" is the symbol of phase angle. Thus, the phasor of the voltages of the left-side and right-side voltage sources are $1\angle 0°\ V$ and $4\angle 0°\ V$, respectively. The impedances of the components are presented in the following:

$$\mathbf{Z}_{\mathbf{10}\ \Omega} = 10\ \Omega \tag{1}$$

$$\mathbf{Z}_{\mathbf{5\ H}} = j\omega L = j \times 10 \times 5 = j50\ \Omega \tag{2}$$

$$\mathbf{Z}_{\mathbf{2\ mF}} = \frac{1}{j\omega C} = \frac{1}{j \times 10 \times 2 \times 10^{-3}} = -j50\ \Omega \tag{3}$$

$$\mathbf{Z}_{\mathbf{1}\ \Omega} = 1\ \Omega \tag{4}$$

$$\mathbf{Z}_{\mathbf{0.2\ H}} = j\omega L = j \times 10 \times 0.2 = j2\ \Omega \tag{5}$$

The frequency of the power sources is equal to the resonance frequency of the parallel connection of the 5 H inductor and 2 mF capacitor, as can be seen in the following.

$$\omega_0 = \frac{1}{\sqrt{LC}} = \frac{1}{\sqrt{5 \times 2 \times 10^{-3}}} = \frac{1}{\sqrt{10^{-2}}} = 10\ rad/sec \quad (6)$$

Thus, this part of the circuit will behave like an open circuit branch.

Or, we can calculate the impedance of the parallel connection of the 5 H inductor and 2 mF capacitor in the circuit of Figure 2.40.2 as follows:

$$(j50)||(-j50) = \frac{(j50) \times (-j50)}{(j50) + (-j50)} = \frac{2500}{0} = \infty \quad (7)$$

Therefore, this part of the circuit has an infinite impedance (theoretically undefined). Hence, no current will flow through it.

By applying KVL in the right-side mesh (see Figure 2.40.2), we can write:

$$-4 + (j2)\mathbf{I} + \mathbf{I} = 0 \Rightarrow \mathbf{I} = \frac{4}{2 + j2} = (2\angle{-45°})\ A \quad (8)$$

The requested current in time domain is:

$$i(t) = \sqrt{2}\cos\left(10t - 45^{\circ}\right)\ A$$

Choice (3) is the answer.

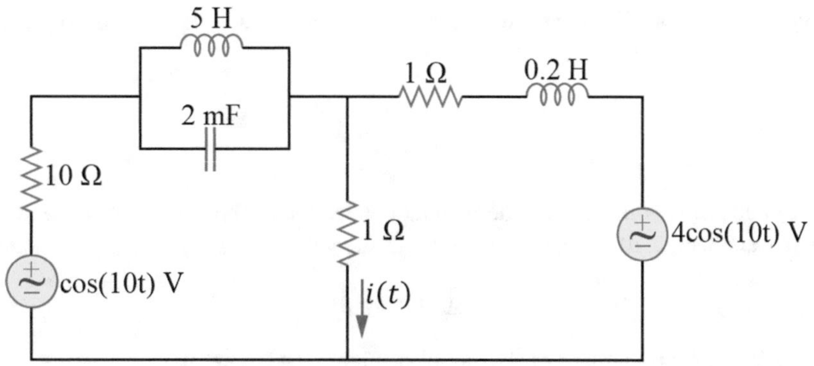

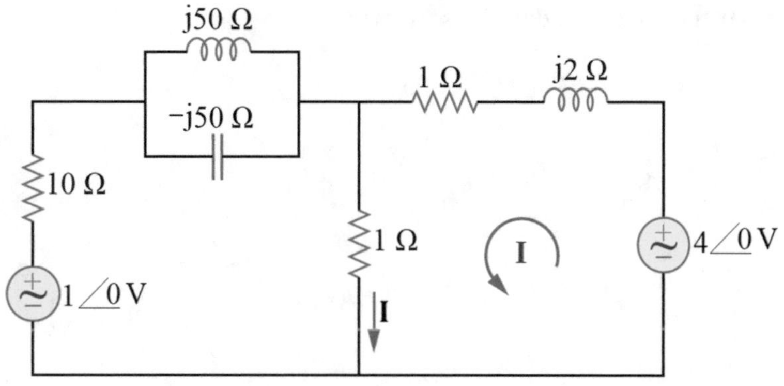

Figure 2.40 The circuit of solution of problem 2.42

2.43. Figure 2.41.2 exhibits the primary circuit in frequency domain. The phasor of sin(1000t), that is, $1\angle 0°$ is defined as the reference phasor, where "$\angle$" is the symbol of phase angle. Thus, the phasor of the voltage of the voltage source is $30\angle 0°$ V or 30 V. The impedances of the components are presented in the following.

$$\mathbf{Z}_{6\ \Omega} = 6\ \Omega \tag{1}$$

$$\mathbf{Z}_{3\ \Omega} = 3\ \Omega \tag{2}$$

$$\mathbf{Z}_{1\ \mathrm{mH}} = j\omega L = j \times 1000 \times 1 \times 10^{-3} = j\ \Omega \tag{3}$$

$$\mathbf{Z}_{1\ \mathrm{mF}} = \frac{1}{j\omega C} = \frac{1}{j \times 1000 \times 1 \times 10^{-3}} = -j\ \Omega \tag{4}$$

$$\mathbf{Z}_{1\ \Omega} = 1\ \Omega \tag{5}$$

This problem can be simply solved by using source transformation theorem, as is shown in the following. Based on this theorem, the series connection of the independent voltage source and the 6 Ω resistor can be replaced by the parallel connection of a current source (with the current of 5 A) and the same resistor (see Figure 2.41.3).

$$\mathbf{I} = \frac{30\angle 0°}{6} = 5\angle 0° \tag{6}$$

In addition, the indicated part of the circuit of Figure 2.41.2 can be simultaneously simplified as follows (see Figure 2.41.3):

$$(j)||(1-j) = \frac{(j) \times (1-j)}{(j) + (1-j)} = (1+j)\ \Omega \tag{7}$$

Now, the impedance of the parallel connection of the 6 Ω and 3 Ω resistors can be calculated as follows (see Figure 2.41.4):

$$6||3 = \frac{6 \times 3}{6+3} = 2\ \Omega \tag{8}$$

By applying source transformation theorem, the parallel connection of the current source and the 2 Ω resistor can be changed to the series connection of a voltage source (with the voltage of 10 V) and 2 Ω resistor (see Figure 2.41.5).

$$\mathbf{V} = (5\angle 0°) \times 2 = 10\angle 0° \tag{9}$$

Now, by applying KVL in the only mesh of the circuit of Figure 2.41.5, we have:

$$-10 + 2\mathbf{I} - 2\mathbf{I} + (1+j)\mathbf{I} = 0 \Rightarrow \mathbf{I} = \frac{10}{1+j} = (5\sqrt{2}\angle -45°)\ A \tag{10}$$

The current can be transferred back to time domain as follows:

$$i(t) = 5\sqrt{2}\sin\left(1000t - 45^{\circ}\right) A$$

Choice (2) is the answer.

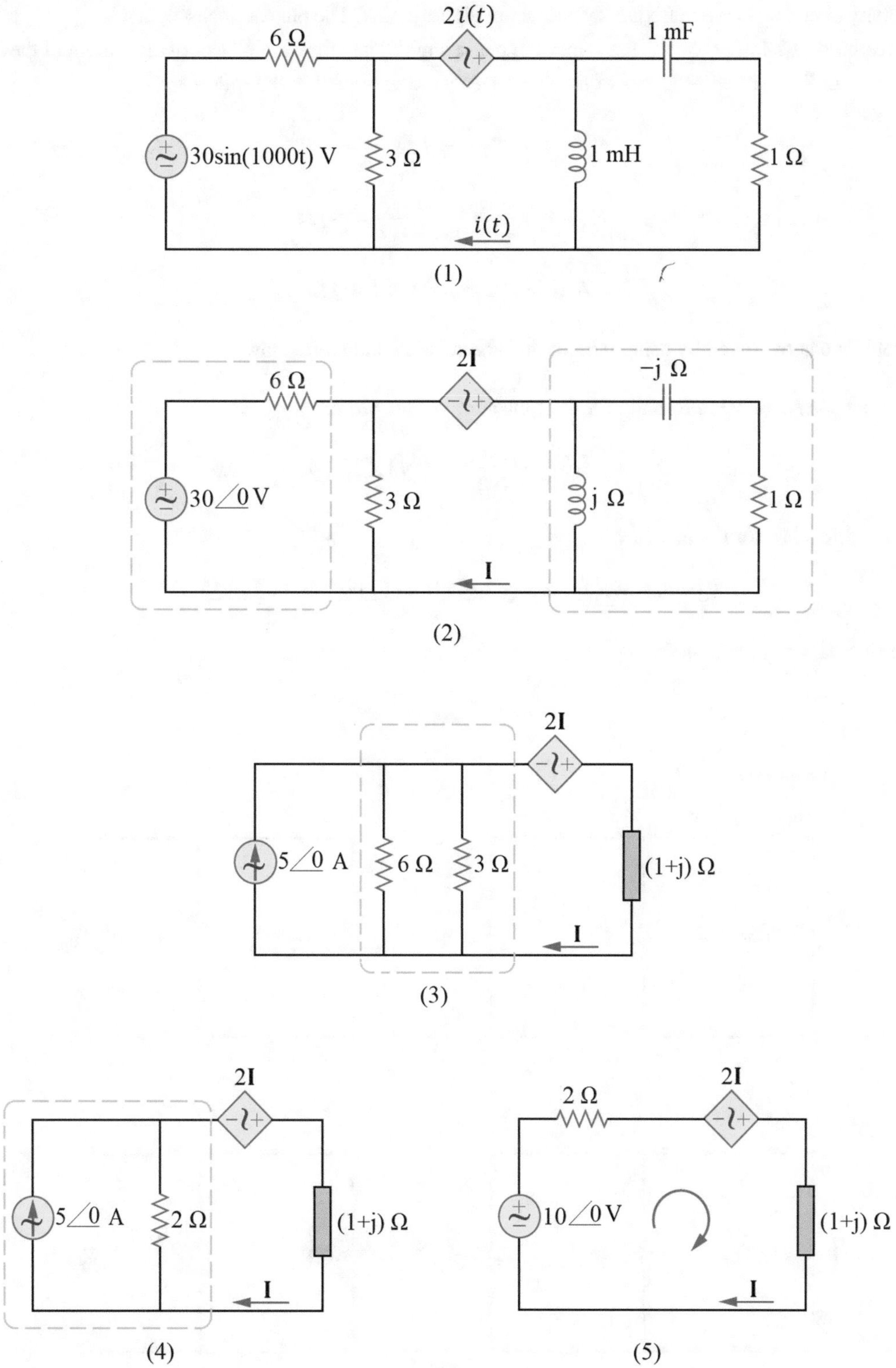

Figure 2.41 The circuit of solution of problem 2.43

2.44. The primary circuit is shown in frequency domain in Figure 2.42.2. The phasor of cos(t), that is, $1\angle 0^\circ$ is defined as the reference phasor, where "$\angle$" is the symbol of phase angle. Therefore, the phasor of the current of the current source is $2\angle 0^\circ$ A or 2 A. The impedances of the components are presented in the following:

$$\mathbf{Z}_{1\,\Omega} = 1\,\Omega \quad (1)$$

$$\mathbf{Z}_{1\,\mathrm{F}} = \frac{1}{j\omega C} = \frac{1}{j \times 1 \times 1} = -j\,\Omega \quad (2)$$

$$\mathbf{Z}_{1\,\mathrm{H}} = j\omega L = j \times 1 \times 1 = j\,\Omega \quad (3)$$

This problem can be solved by using current division relation and Ohm's law.

By applying current division formula for the capacitor, we can write:

$$\mathbf{I}_C = \frac{1}{1 + (-j)} \times 2 = \sqrt{2}\angle 45^\circ \ A \quad (4)$$

Applying Ohm's law for the inductor:

$$\mathbf{V_o} = -2\mathbf{I}_C\mathbf{Z}_{1\,\mathrm{H}} = -2 \times (\sqrt{2}\angle 45^\circ) \times (1\angle 90^\circ) = 2\sqrt{2}\angle -45^\circ V \quad (5)$$

Transferring back to time domain:

$$v_o(t) = 2\sqrt{2}\cos\left(t - 45^\circ\right) V$$

Choice (1) is the answer.

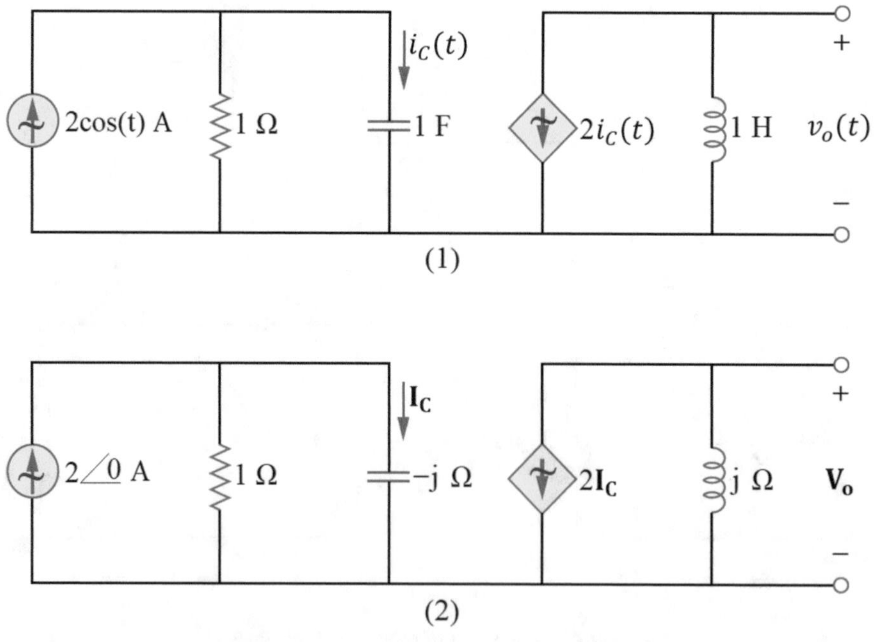

Figure 2.42 The circuit of solution of problem 2.44

2.45. Figure 2.43.2 shows the primary circuit in frequency domain. The impedances of the components are presented in the following:

$$\mathbf{Z}_{4\,\Omega} = 4\,\Omega \tag{1}$$

$$\mathbf{Z}_{4\,\mathrm{H}} = j\omega L = j4\omega\,\Omega \tag{2}$$

$$\mathbf{Z}_{1\,\mathrm{F}} = \frac{1}{j\omega C} = \frac{1}{j\omega}\,\Omega \tag{3}$$

$$\mathbf{Z}_{1\,\Omega} = 1\,\Omega \tag{4}$$

To determine the resonance frequency of the circuit, we need to find the input impedance (or the input admittance) of the circuit and equate its imaginary part with zero and then solve it. In other words:

$$Im\{\mathbf{Z}_{\mathbf{in}}(j\omega)\} = 0 \Rightarrow \omega = \omega_0 \tag{5}$$

$$Im\{\mathbf{Y}_{\mathbf{in}}(j\omega)\} = 0 \Rightarrow \omega = \omega_0 \tag{6}$$

The input impedance of the circuit can be calculated as follows:

$$\mathbf{Z}_{\mathbf{in}} = 4 + j4\omega + \left(\frac{1}{j\omega}\right)||1 = 4 + j4\omega + \frac{\left(\frac{1}{j\omega}\right) \times 1}{\left(\frac{1}{j\omega}\right) + 1} = 4 + j4\omega + \frac{-j}{\omega - j}$$

$$\Rightarrow \mathbf{Z}_{\mathbf{in}} = 4 + j4\omega + \frac{1 - j\omega}{\omega^2 + 1} = 4 + \frac{1}{\omega^2 + 1} + j\left(4\omega - \frac{\omega}{\omega^2 + 1}\right) \tag{7}$$

Solving (5) and (7):

$$4\omega - \frac{\omega}{\omega^2 + 1} = 0 \Rightarrow \omega^2 + 1 = \frac{1}{4} \Rightarrow \omega^2 = -\frac{3}{4} \Rightarrow \omega = \pm j\frac{\sqrt{3}}{2}$$

Therefore, the circuit does not have any resonance frequency because no real non-zero value was found for ω.

Choice (4) is the answer.

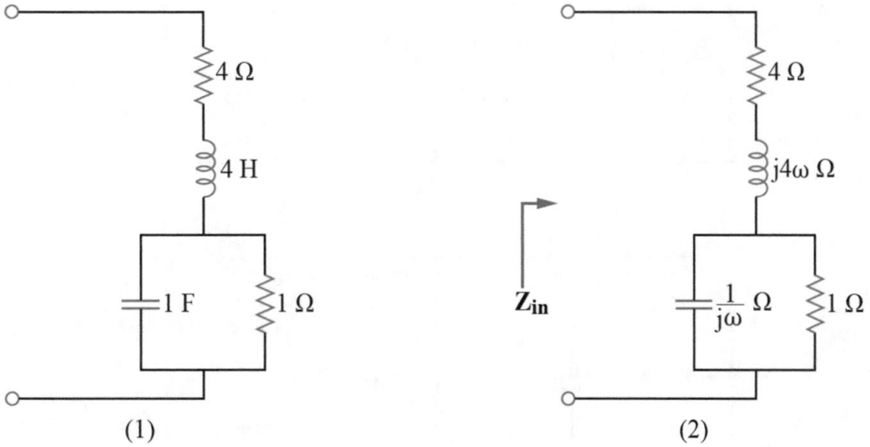

Figure 2.43 The circuit of solution of problem 2.45

2.46. The primary circuit is shown in frequency domain in Figure 2.44.2. The phasor of cos(2*t*), that is, $1\angle 0°$ is defined as the reference phasor, where "$\angle$" is the symbol of phase angle. Therefore, the phasor of the current of the current source is $2\angle 0°$ *A* or 2 *A*. The impedances of the components are presented in the following:

$$\mathbf{Z}_{1\,\mathbf{H}} = j\omega L = j \times 2 \times 1 = j2\ \Omega \quad (1)$$

$$\mathbf{Z}_{\frac{1}{4}\,\mathbf{F}} = \frac{1}{j\omega C} = \frac{1}{j \times 2 \times \frac{1}{4}} = -j2\ \Omega \quad (2)$$

$$\mathbf{Z}_{1\,\Omega} = 1\ \Omega \quad (3)$$

$$\mathbf{Z}_{\frac{1}{2}\,\mathbf{F}} = \frac{1}{j\omega C} = \frac{1}{j \times 2 \times \frac{1}{2}} = -j\ \Omega \quad (4)$$

$$\mathbf{Z}_{\frac{1}{2}\,\mathbf{H}} = j\omega L = j \times 2 \times \frac{1}{2} = j\ \Omega \quad (5)$$

The problem can be solved by using nodal analysis. By applying KCL in the indicated supernode, we can write:

$$-2 + \frac{\mathbf{V_x}}{j2} + \frac{\mathbf{V_x}}{-j2} + \frac{\mathbf{V_x}}{1-j} + \frac{\mathbf{V_x}}{1+j} = 0 \Rightarrow\ -2 + \mathbf{V_x}\left(\frac{1}{j2} + \frac{1}{-j2} + \frac{1}{1-j} + \frac{1}{1+j}\right) = 0$$

$$\Rightarrow\ -2 + \mathbf{V_x}\left(0 + \frac{1+j+1-j}{1+1}\right) = 0 \Rightarrow\ -2 + \mathbf{V_x}\left(\frac{2}{2}\right) = 0 \Rightarrow \mathbf{V_x} = 2 \quad (6)$$

Now, by applying voltage division formula, we have:

$$\mathbf{V} = \frac{1}{1+(-j)} \times 2 = \frac{2}{1-j} = \sqrt{2}\angle 45°\ V \quad (7)$$

By transferring to time domain, we can write:

$$v(t) = \sqrt{2}\cos\left(2t + 45^{\circ}\right)$$

Choice (3) is the answer.

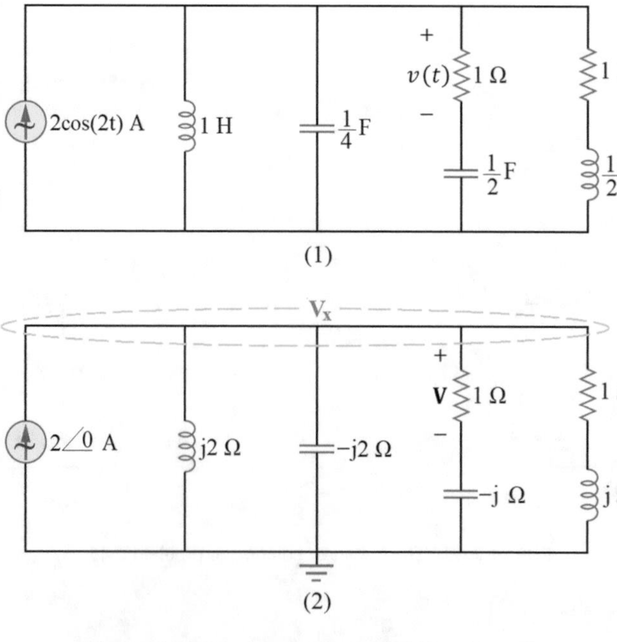

Figure 2.44 The circuit of solution of problem 2.46

2.47. Based on the information given in the problem:

$$\mathbf{V_1} = (10 + j2)\ V \quad (1)$$

$$\mathbf{V_2} = (12 + j12)\ V \quad (2)$$

Applying KCL in node 1:

$$\frac{\mathbf{V_1} - \mathbf{V_{s1}}}{2 + j2} + \frac{\mathbf{V_1}}{-j2} + \frac{\mathbf{V_1}}{\frac{1}{2}} + \frac{\mathbf{V_1} - \mathbf{V_2}}{j4} = 0 \quad (3)$$

Solving (1), (2), and (3):

$$\frac{(10 + j2) - \mathbf{V_{s1}}}{2 + j2} + \frac{(10 + j2)}{-j2} + \frac{(10 + j2)}{\frac{1}{2}} + \frac{(10 + j2) - (12 + j12)}{j4} = 0$$

$$\Rightarrow \frac{(10 + j2) - \mathbf{V_{s1}}}{2 + j2} - 1 + j5 + 20 + j4 - 2.5 + j0.5 = 0$$

$$\Rightarrow \frac{(10 + j2) - \mathbf{V_{s1}}}{2 + j2} = -16.5 - j9.5 \Rightarrow (10 + j2) - \mathbf{V_{s1}} = -14 - j52$$

$$\Rightarrow \mathbf{V_{s1}} = (24 + j54)\ V$$

Choice (4) is the answer.

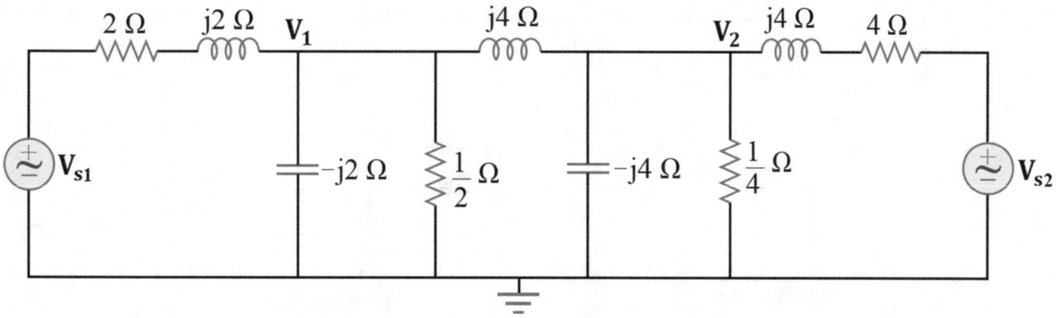

Figure 2.45 The circuit of solution of problem 2.47

2.48. The average power of the resistor can be calculated by using the relation below:

$$P_R = \frac{1}{2}\frac{|\mathbf{V}|^2}{R} \quad (1)$$

The primary circuit is illustrated in frequency domain in Figure 2.46.2. The phasor of $\cos(2t)$, that is, $1\angle 0^\circ$ is defined as the reference phasor, where "$\angle$" is the symbol of phase angle. Therefore, the phasors of the voltage of the left-side and right-side voltage sources are $1\angle 0^\circ\ V$ and $1\angle -90^\circ\ V$, respectively. The impedances of the components are presented in the following:

$$\mathbf{Z_{1\ F}} = \frac{1}{j\omega C} = \frac{1}{j \times 2 \times 1} = -j0.5\ \Omega \quad (2)$$

$$\mathbf{Z}_{1\,\Omega} = 1\ \Omega \tag{3}$$

$$\mathbf{Z}_{1\,\mathrm{H}} = j\omega L = j \times 2 \times 1 = j2\ \Omega \tag{4}$$

Applying KCL in the indicated node:

$$\frac{\mathbf{V}_x - (-1)}{-j0.5} + \frac{\mathbf{V}_x}{1} + \frac{\mathbf{V}_x - (1\angle -90^\circ)}{j2} = 0 \Rightarrow \mathbf{V}_x\left(\frac{1}{-j0.5} + 1 + \frac{1}{j2}\right) + \frac{1}{-j0.5} - \frac{(1\angle -90^\circ)}{j2} = 0$$

$$\Rightarrow \mathbf{V_X}\,(1 + j1.5) + j2 + 0.5 = 0 \Rightarrow \mathbf{V_X} = \frac{-0.5 - j2}{1 + j1.5} = (1.14\angle -160.34^\circ)\ V \tag{5}$$

Solving (1) and (5):

$$P_R = \frac{1}{2}\frac{|\mathbf{V}|^2}{R} = \frac{1}{2}\frac{(1.14)^2}{1} = 0.65\ W \tag{6}$$

Choice (1) is the answer.

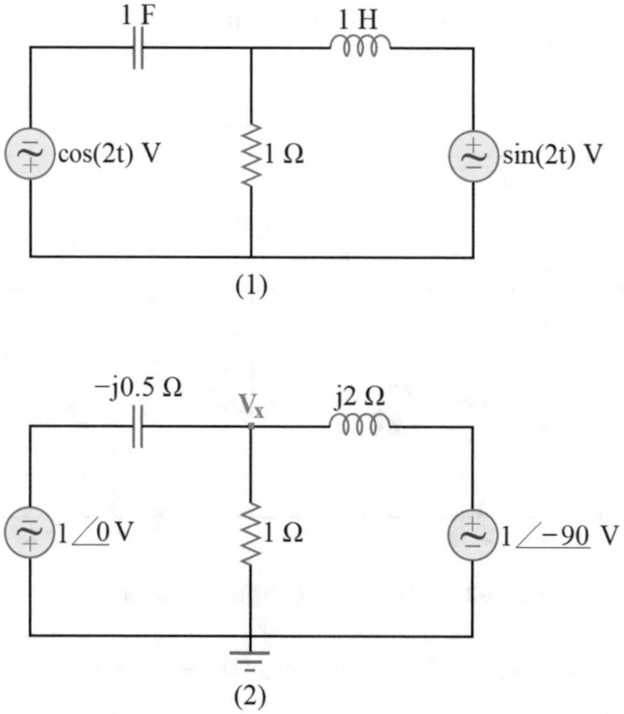

Figure 2.46 The circuit of solution of problem 2.48

2.49. Figure 2.47.2 shows the primary circuit in frequency domain. The impedances of the components are as follows:

$$\mathbf{Z}_{1\,\Omega} = 1 \tag{1}$$

$$\mathbf{Z_L} = j\omega L = j\omega \times 1 = j\omega \tag{2}$$

$$\mathbf{Z_C} = \frac{1}{j\omega C} = -j\frac{1}{\omega C} \tag{3}$$

Based on the information given in the problem:

$$\mathbf{Z_{in}} = \mathbf{Y_{in}} \quad (4)$$

On the other hand, based on the definition of the impedance and the admittance, we know that:

$$\mathbf{Z_{in}} = \frac{1}{\mathbf{Y_{in}}} \quad (5)$$

By solving (4) and (5), we have:

$$\mathbf{Z_{in}} = \frac{1}{\mathbf{Z_{in}}} \Rightarrow \mathbf{Z_{in}}^2 = 1 \Rightarrow \begin{cases} \mathbf{Z_{in}} = 1 \\ \mathbf{Z_{in}} = -1 \end{cases} \Rightarrow \begin{cases} |\mathbf{Z_{in}}| = 1 \ (6), \angle \mathbf{Z_{in}} = 0° \quad (7) \\ |\mathbf{Z_{in}}| = 1 \ (8), \angle \mathbf{Z_{in}} = 180° \ (9) \end{cases}$$

Herein, "$\angle$ " is the symbol of phase angle.

$$\mathbf{Z_{in}} = (1 + j\omega)||\left(1 - j\frac{1}{\omega C}\right) = \frac{(1 + j\omega) \times \left(1 - j\frac{1}{\omega C}\right)}{(1 + j\omega) + \left(1 - j\frac{1}{\omega C}\right)} = \frac{\left(1 + \frac{1}{C}\right) + j\left(\omega - \frac{1}{\omega C}\right)}{(2) + j\left(\omega - \frac{1}{\omega C}\right)} \quad (10)$$

Solving (6) and (10):

$$\frac{\sqrt{\left(1 + \frac{1}{C}\right)^2 + \left(\omega - \frac{1}{\omega C}\right)^2}}{\sqrt{(2)^2 + \left(\omega - \frac{1}{\omega C}\right)^2}} = 1 \Rightarrow \left(1 + \frac{1}{C}\right)^2 + \left(\omega - \frac{1}{\omega C}\right)^2 = (2)^2 + \left(\omega - \frac{1}{\omega C}\right)^2$$

$$\Rightarrow \left(1 + \frac{1}{C}\right)^2 = (2)^2 \Rightarrow 1 + \frac{1}{C} = 2 \Rightarrow C = 1\ F$$

Choice (3) is the answer.

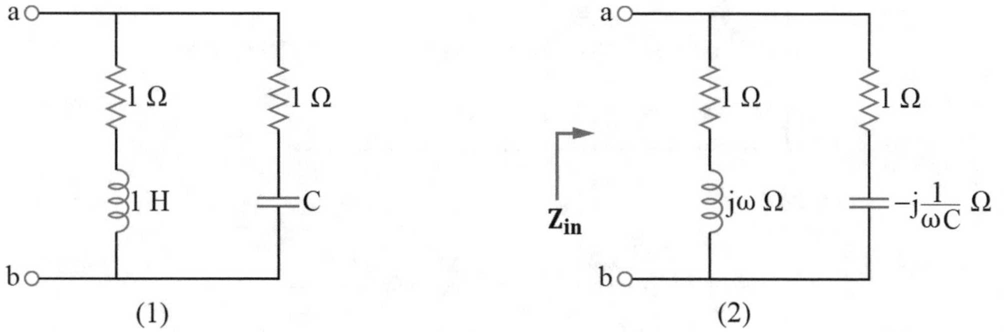

Figure 2.47 The circuit of solution of problem 2.49

2.50. The primary circuit is illustrated in frequency domain in Figure 2.48.2. The phasor of $\cos(2t)$, that is, $1\angle 0°$ is defined as the reference phasor, where "$\angle$ " is the symbol of phase angle. Therefore, the phasor of the voltage of the independent voltage source is $30\angle 0°\ V$ or $30\ V$. The impedances of the components are presented in the following:

$$\mathbf{Z_{0.1\ F}} = \frac{1}{j\omega C} = \frac{1}{j \times 2 \times 0.1} = -j5\ \Omega \quad (1)$$

$$\mathbf{Z_{5\ \Omega}} = 5\ \Omega \quad (2)$$

$$\mathbf{Z_{1\ \Omega}} = 1\ \Omega \quad (3)$$

$$\mathbf{Z_{1\,H}} = j\omega L = j \times 2 \times 1 = j2\ \Omega \tag{4}$$

This problem can be solved by using nodal analysis.

Defining $\mathbf{V_x}$ based on the nodal voltages:

$$\mathbf{V_x} = \mathbf{V_o} - 30 \tag{5}$$

Applying KCL in the indicated node:

$$\frac{\mathbf{V_o} - 30}{-j5} + \frac{\mathbf{V_o}}{5} + \frac{\mathbf{V_o} - 2\mathbf{V_x}}{1 + j2} = 0 \xrightarrow{\text{Using (5)}} \frac{\mathbf{V_o} - 30}{-j5} + \frac{\mathbf{V_o}}{5} + \frac{\mathbf{V_o} - 2(\mathbf{V_o} - 30)}{1 + j2} = 0$$

$$\Rightarrow \mathbf{V_o}\left(\frac{1}{-j5} + \frac{1}{5} - \frac{1}{1 + j2}\right) + \frac{30}{j5} + \frac{60}{1 + j2} = 0$$

$$\Rightarrow \mathbf{V_o}\left(\frac{-(1 + j2) + j(1 + j2) - j5}{j5(1 + j2)}\right) + \frac{30(1 + j2) + 60(\ j5)}{j5(1 + j2)} = 0$$

$$\Rightarrow \mathbf{V_o}\left(\frac{-3 - j6}{j5(1 + j2)}\right) + \frac{30 + j360}{j5(1 + j2)} = 0 \Rightarrow \mathbf{V_o}(3 + j6) = 30 + j360$$

$$\Rightarrow \mathbf{V_o} = \frac{30 + j360}{3 + j6} = (50 + j20)\ V$$

Choice (1) is the answer.

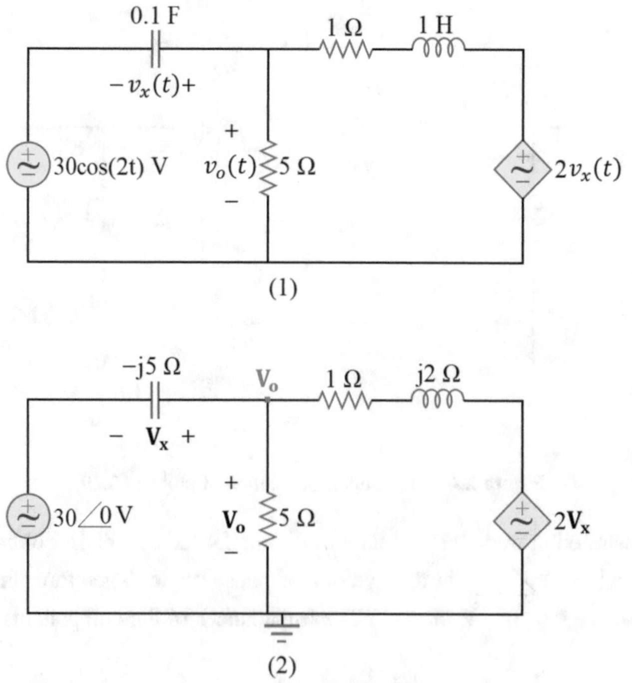

Figure 2.48 The circuit of solution of problem 2.50

2.51. The primary circuit is illustrated in frequency domain in Figure 2.49.2. The phasor of cos(100*t*), that is, $1\angle 0°$ is defined as the reference phasor, where "$\angle$" is the symbol of phase angle. Therefore, the phasor of the voltage of the independent voltage source is $10\angle 0°\ V$ or 10 *V*. The impedances of the components are presented in the following:

$$\mathbf{Z}_{\frac{2}{5}\,\mathbf{mF}} = \frac{1}{j\omega C} = \frac{1}{j \times 100 \times \frac{2}{5} \times 10^{-3}} = -j25\ \Omega \quad (1)$$

$$\mathbf{Z}_{50\ \Omega} = 50\ \Omega \quad (2)$$

$$\mathbf{Z}_{\frac{5}{6}\,\mathbf{mF}} = \frac{1}{j\omega C} = \frac{1}{j \times 100 \times \frac{5}{6} \times 10^{-3}} = -j12\ \Omega \quad (3)$$

$$\mathbf{Z}_{\frac{3}{5}\,\mathbf{H}} = j\omega L = j \times 100 \times \frac{3}{5} = j60\ \Omega \quad (4)$$

To simplify the problem, it is suggested to apply source transformation in the left-side part of the circuit as follows:

$$\mathbf{V} = (10\angle 0°) \times (-j25) = -j250\,V \quad (5)$$

In addition, the equivalent impedance of the right-side part of the circuit can be determined to simplify the problem:

$$\mathbf{Z} = (-j12)||(j60) = \frac{(-j12) \times (j60)}{(-j12) + (j60)} = \frac{720}{j48} = -j15\ \Omega \quad (6)$$

Now, by applying KVL in the only mesh of the circuit of Figure 2.49.3, we can write:

$$-(-j250) + (-j25)\mathbf{I} + 50\mathbf{I} - 0.4\mathbf{V_x} + (-j15)\mathbf{I} = 0 \Rightarrow j250 + (-j40 + 50)\mathbf{I} - 0.4\mathbf{V_x} = 0 \quad (7)$$

Applying Ohm's law for the resistor:

$$\mathbf{I} = \frac{\mathbf{V_x}}{50} \quad (8)$$

Solving (7) and (8):

$$j250 + (-j40 + 50)\frac{\mathbf{V_x}}{50} - 0.4\mathbf{V_x} = 0 \Rightarrow (-j0.8 + 0.6)\mathbf{V_x} = -j250$$

$$\Rightarrow \mathbf{V_x} = \frac{-j250}{0.6 - j0.8} = \left(250\angle\left(-90 + \left(tan^{-1}\left(\frac{4}{3}\right)\right)°\right)\right)V \quad (9)$$

As we know from trigonometry:

$$tan^{-1}(\alpha) + tan^{-1}\left(\frac{1}{\alpha}\right) = 90 \quad (10)$$

Solving (9) and (10) for $\alpha = \frac{4}{3}$:

$$\mathbf{V_x} = \left(250\angle - \left(tan^{-1}\left(\frac{3}{4}\right)\right)°\right)V \quad (11)$$

Transferring back to time domain:

$$v_x(t) = 250\cos\left(100t - \left(tan^{-1}\left(\frac{3}{4}\right)\right)°\right)$$

Choice (4) is the answer.

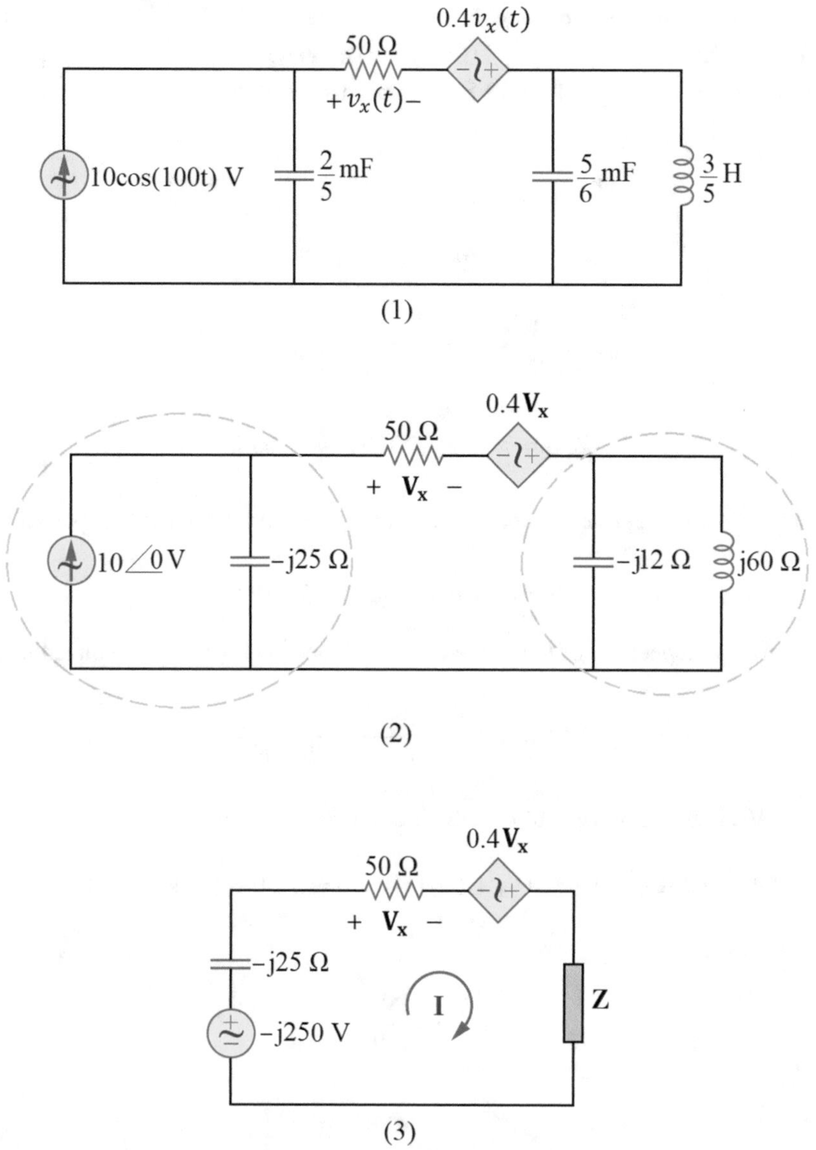

Figure 2.49 The circuit of solution of problem 2.51

2.52. The primary circuit is illustrated in frequency domain in Figure 2.50.2. The phasor of sin(t), that is, $1\angle 0^\circ$ is defined as the reference phasor, where "$\angle$" is the symbol of phase angle. Therefore, the phasor of the voltage of the independent voltage source is $100\angle 0^\circ$ V or 100 V. The impedances of the components are presented in the following:

$$\mathbf{Z}_{1\,\Omega} = 1\ \Omega \tag{1}$$

$$\mathbf{Z}_{1\,\mathrm{F}} = \frac{1}{j\omega C} = \frac{1}{j \times 1 \times 1} = -j\ \Omega \tag{2}$$

To simplify the circuit, it is suggested to apply source transformation theorem for the dependent current source and its parallel impedance, as is shown in Figure 2.50.3.

By assigning the current of **I** to the loop and applying KVL in that, we can write:

$$-100 + \mathbf{I} + \mathbf{I} - j\mathbf{I} - j\mathbf{V_1} = 0 \Rightarrow -100 + (2 - j)\mathbf{I} - j\mathbf{V_1} = 0 \quad (3)$$

Applying KVL in the left-side mesh, to define $\mathbf{V_1}$ based on the loop current:

$$-100 + \mathbf{I} + \mathbf{V_1} = 0 \Rightarrow \mathbf{V_1} = 100 - \mathbf{I} \quad (4)$$

Solving (3) and (4):

$$-100 + (2 - j)\mathbf{I} - j(100 - \mathbf{I}) = 0 \Rightarrow -100 + 2\mathbf{I} - j100 = 0$$

$$\Rightarrow \mathbf{I} = (50 + j50)\ A = (50\sqrt{2}\angle 45°)\ A \quad (5)$$

Now, we can calculate the average power of the independent voltage source by considering the associated reference direction for its voltage and current as follows:

$$P_{source} = \frac{1}{2} V_m I_m \cos(\theta_V - \theta_I) = \frac{1}{2} \times (100) \times \left(-50\sqrt{2}\right) \cos(0 - 45) = -2500\ W$$

Choice (4) is the answer.

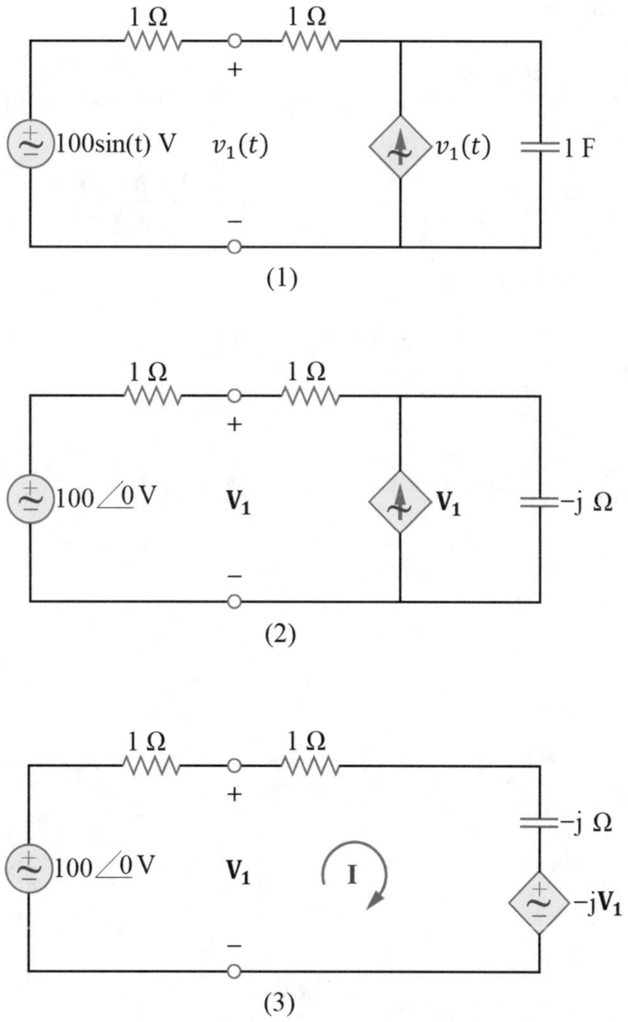

Figure 2.50 The circuit of solution of problem 2.52

2.53. The primary circuit is illustrated in frequency domain in Figure 2.51.2. The phasor of sin(t), that is, $1\angle 0^\circ$ is defined as the reference phasor, where "$\angle$" is the symbol of phase angle. Therefore, the phasor of the current of the current source is $1\angle 0^\circ\ A$ or $1\ A$. The impedances of the components are presented in the following:

$$\mathbf{Z}_{1\,\mathbf{H}} = j\omega L = j \times 1 \times 1 = j\,\Omega \tag{1}$$

$$\mathbf{Z}_{1\,\mathbf{F}} = \frac{1}{j\omega C} = \frac{1}{j \times 1 \times 1} = -j\,\Omega \tag{2}$$

$$\mathbf{Z}_{1\,\Omega} = 1\,\Omega \tag{3}$$

This problem can be solved by using nodal analysis as follows:

Defining $\mathbf{I_L}$ based on the nodal voltages:

$$\mathbf{I_L} = \frac{\mathbf{V_1}}{j} \tag{4}$$

Defining the voltage of the dependent voltage source based on the nodal voltages:

$$\alpha \mathbf{I_L} = \mathbf{V_1} - \mathbf{V_o} \xrightarrow{\text{Using (4)}} \alpha \frac{\mathbf{V_1}}{j} = \mathbf{V_1} - \mathbf{V_o} \Rightarrow \mathbf{V_1} = \frac{\mathbf{V_o}}{1+\alpha j} \tag{5}$$

Applying KCL in the supernode:

$$-1 + \frac{\mathbf{V_1}}{j} + \frac{\mathbf{V_o}}{1-j} = 0 \xrightarrow{\text{Using (5)}} -1 - j\frac{\mathbf{V_o}}{1+\alpha j} + \frac{\mathbf{V_o}}{1-j} = 0 \Rightarrow \mathbf{V_o}\left(\frac{-j}{1+\alpha j} + \frac{1}{1-j}\right) = 1$$

$$\Rightarrow \mathbf{V_o}\left(\frac{j(\alpha-1)}{1+\alpha+j(\alpha-1)}\right) = 1 \Rightarrow \mathbf{V_o} = \frac{1+\alpha+j(\alpha-1)}{j(\alpha-1)} \tag{6}$$

Based on the information given in the problem:

$$v_o(t) = \sin(t) \Rightarrow \mathbf{V_o} = 1\angle 0^\circ = 1 \text{ (7)} \tag{7}$$

Solving (6) and (7):

$$\frac{1+\alpha+j(\alpha-1)}{j(\alpha-1)} = 1 \Rightarrow 1+\alpha+j(\alpha-1) = j(\alpha-1) \Rightarrow \alpha = -1$$

Choice (3) is the answer.

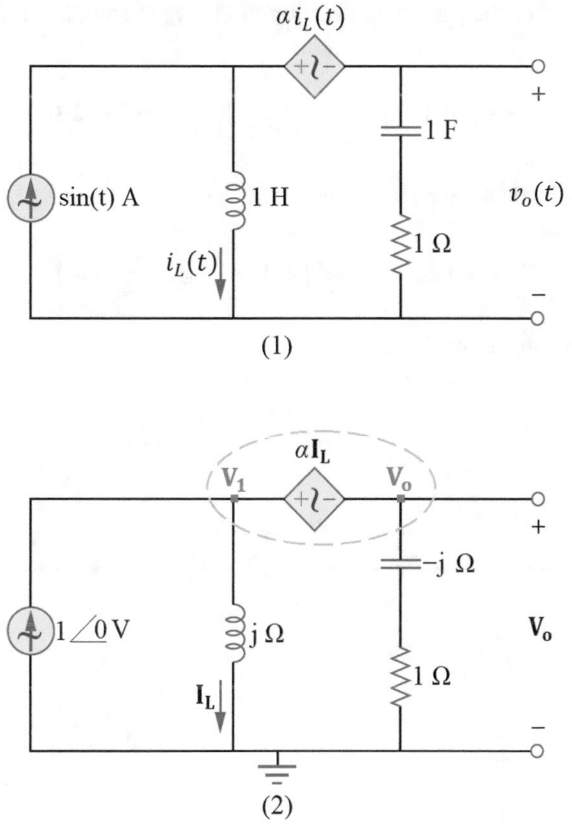

Figure 2.51 The circuit of solution of problem 2.53

2.54. The primary circuit is illustrated in frequency domain in Figure 2.52.2. The phasor of cos(t), that is, $1\angle 0°$ is defined as the reference phasor, where "$\angle$" is the symbol of phase angle. Therefore, the phasor of the current of the current source is $\left(3\angle \frac{\pi}{4}\, rad\right) A.$. The impedances of the components are presented in the following:

$$\mathbf{Z_{1\ \Omega}} = 1\,\Omega \tag{1}$$

$$\mathbf{Z_{1\ H}} = j\omega L = j \times 1 \times 1 = j\,\Omega \tag{2}$$

$$\mathbf{Z_{2\ F}} = \frac{1}{j\omega C} = \frac{1}{j \times 1 \times 2} = -j0.5\,\Omega \tag{3}$$

$$\mathbf{Z_{4\ H}} = j\omega L = j \times 1 \times 4 = j4\,\Omega \tag{4}$$

$$\mathbf{Z_{0.25\ F}} = \frac{1}{j\omega C} = \frac{1}{j \times 1 \times 0.25} = -j4\,\Omega \tag{5}$$

$$\mathbf{Z_{0.5\ H}} = j\omega L = j \times 1 \times 0.5 = j0.5\,\Omega \tag{6}$$

$$\mathbf{Z_{1\ F}} = \frac{1}{j\omega C} = \frac{1}{j \times 1 \times 1} = -j\,\Omega \tag{7}$$

This problem can be solved by simplifying the circuit. As can be seen in Figure 2.52.2, the total impedance of the indicated horizontal branch is zero ($j4 + (-j4) = 0\,\Omega$); therefore, it is replaced by a short circuit branch, as is shown in Figure 2.52.3.

Next, as can be seen in Figure 2.52.3, the total impedance of the parallel connection of $-j0.5\,\Omega$ and $j0.5\,\Omega$ is infinite (theoretically undefined). Thus, it is replaced by an open circuit, as is illustrated in Figure 2.52.4.

$$(-j0.5)||(j0.5) = \frac{(-j0.5) \times (j0.5)}{(-j0.5) + (j0.5)} = \frac{0.25}{0} = \infty \tag{8}$$

After that, the total impedance of the parallel connection of $(1 + j)\ \Omega$ and $(1 - j)\ \Omega$ is calculated which is $1\ \Omega$, as is shown in Figure 2.52.5.

$$(1+j)||(1-j) = \frac{(1+j)\times(1-j)}{(1+j)+(1-j)} = \frac{2}{2} = 1\ \Omega \tag{9}$$

Now, the output voltage can be determined by using Ohm's law:

$$\mathbf{V_o} = \left(3\angle\frac{\pi}{4}\ rad\right)\times 1 = \left(3\angle\frac{\pi}{4}\ rad\right)\ A \tag{10}$$

By transferring back to time domain, we have:

$$v_o(t) = 3\cos\left(t+\frac{\pi}{4}\right)V$$

Choice (4) is the answer.

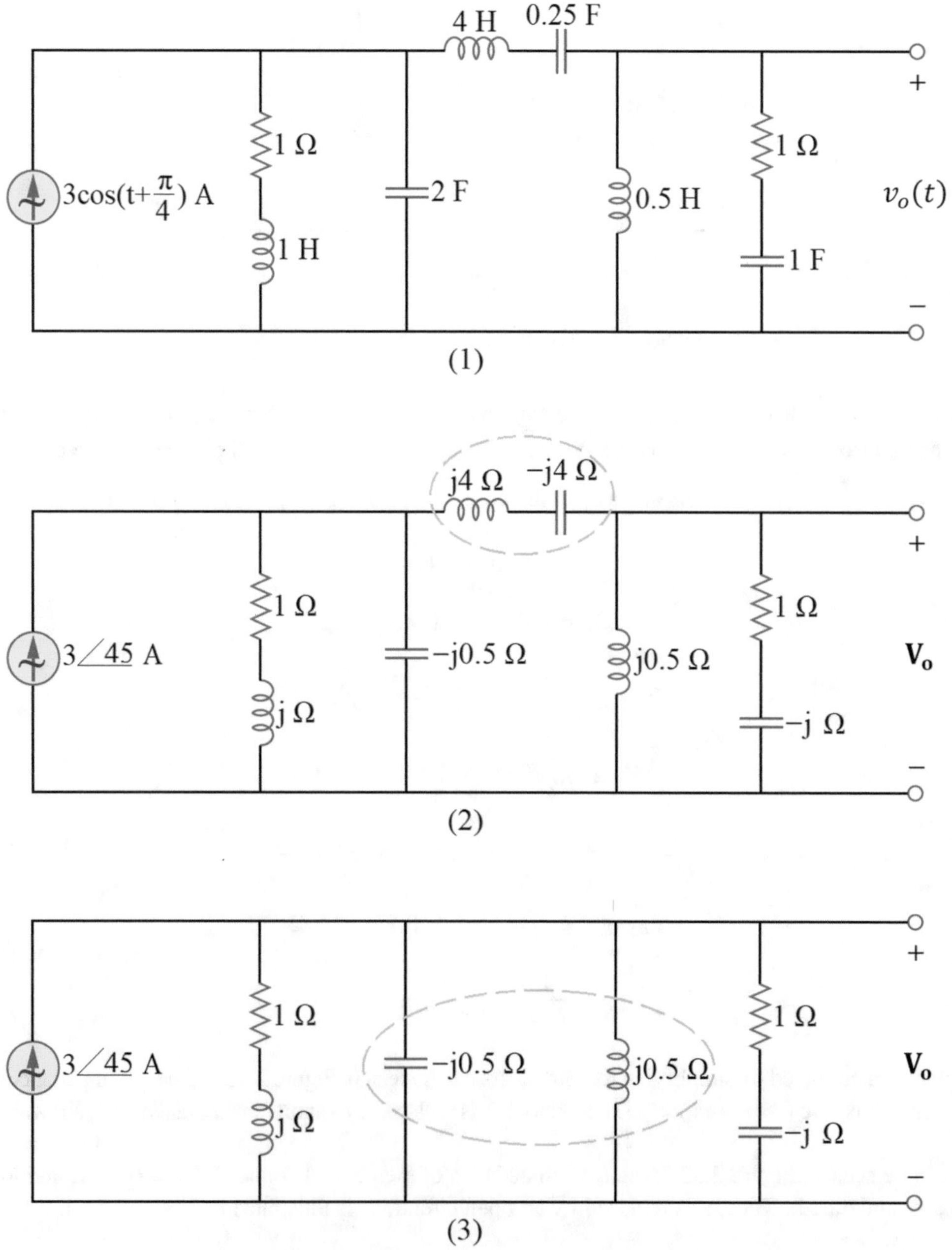

Figure 2.52 The circuit of solution of problem 2.54

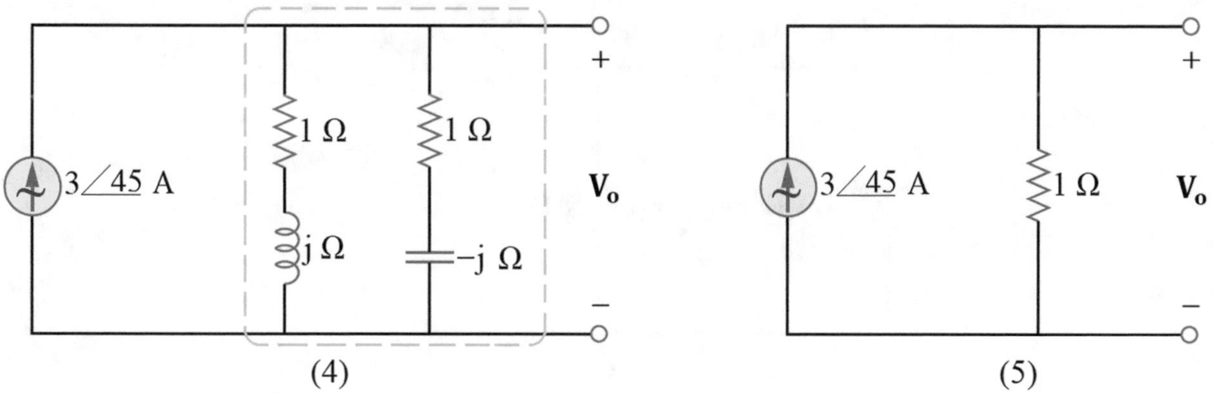

Figure 2.52 (continued)

2.55. The circuit is shown in frequency domain in Figure 2.53.2. The phasor of cos(10*t*), that is, $1\angle 0°$ is defined as the reference phasor, where "$\angle$" is the symbol of phase angle. Therefore, the phasor of the voltage of the independent voltage source is $(20\angle 0°)\ V$ or $20\ V$. Herein, "$\angle$" is the symbol of phase angle. The impedances of the components are as follows:

$$\mathbf{Z}_{10\ \Omega} = 10\ \Omega \tag{1}$$

$$\mathbf{Z}_{5\ \Omega} = 5\ \Omega \tag{2}$$

$$\mathbf{Z}_{\mathrm{L}} = j\omega L = j \times 10 \times 0.5 = j5\ \Omega \tag{3}$$

This problem can be solved by using mesh analysis as follows:

KVL in the left-side mesh:

$$-20 + 10\mathbf{I} + 10(\mathbf{I} - \mathbf{I_2}) = 0 \Rightarrow -20 + 20\mathbf{I} - 10\mathbf{I_2} = 0 \tag{4}$$

KVL in the right-side mesh:

$$10(\mathbf{I_2} - \mathbf{I}) - 10\mathbf{I} + (5 + j5)\mathbf{I_2} = 0 \Rightarrow -20\mathbf{I} + (15 + j5)\mathbf{I_2} = 0 \tag{5}$$

Solving (4) and (5):

$$-20 + 20\frac{(15 + j5)\mathbf{I_2}}{20} - 10\mathbf{I_2} = 0 \Rightarrow \mathbf{I_2} = \frac{4}{1 + j}\ A \tag{6}$$

Solving (4) and (6):

$$-20 + 20\mathbf{I} - 10\frac{4}{1 + j} = 0 \Rightarrow \mathbf{I} = 1 + \frac{2}{1 + j} = \frac{3 + j}{1 + j}\ A \tag{7}$$

Applying Ohm's law for the 10 Ω resistor in the vertical branch:

$$\mathbf{V} = 10\left(\mathbf{I} - \mathbf{I_2}\right) \xRightarrow{\text{Using (6), (7)}} \mathrm{V} = 10\left(\frac{3 + j}{1 + j} - \frac{4}{1 + j}\right) = 10\frac{-1 + j}{1 + j} = j10\ V = (10\angle 90°)\ V \tag{8}$$

Transferring to time domain:

$$v(t) = 10\cos\left(10t + 90^{\circ}\right)$$

Choice (4) is the answer.

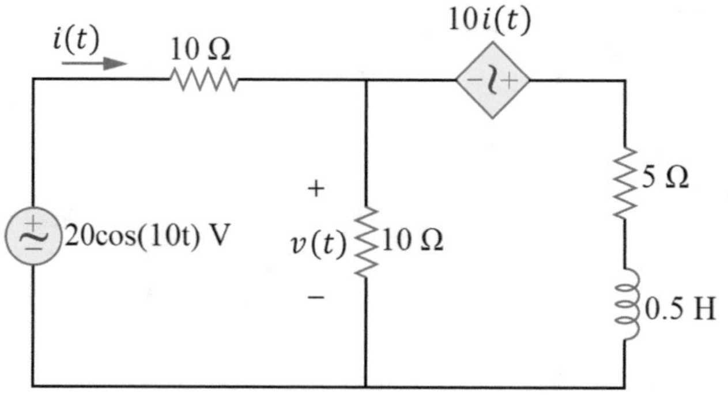

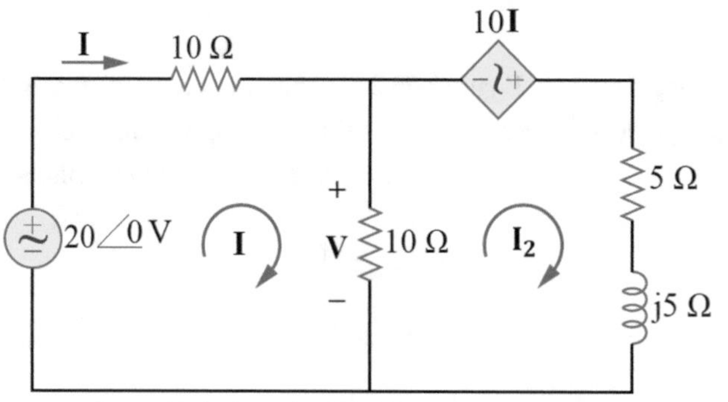

Figure 2.53 The circuit of solution of problem 2.55

2.56. Based on the information given in the problem, the power factor detected by the voltage source is one. Thus, the circuit is in the resonance state, and the imaginary part of the input impedance or admittance of the circuit is zero.

The primary circuit is illustrated in frequency domain in Figure 2.54.2. The impedances of the components are presented in the following:

$$\mathbf{Z}_{2\ \Omega} = 2\ \Omega \quad (1)$$

$$\mathbf{Z}_{R\ \Omega} = R\ \Omega \quad (2)$$

$$\mathbf{Z}_{0.5\ \mathrm{H}} = j\omega L = j \times 2 \times 0.5 = j\ \Omega \quad (3)$$

$$\mathbf{Z}_{1\ \mathrm{H}} = j\omega L = j \times 2 \times 1 = j2\ \Omega \quad (4)$$

$$\mathbf{Z}_{0.5\ \mathrm{F}} = \frac{1}{j\omega C} = \frac{1}{j \times 2 \times 0.5} = -j\ \Omega \quad (5)$$

The input admittance of the circuit is:

$$\mathbf{Z}_{\mathbf{in}} = 2 + (R + j)||(\ j2)||(-j) = 2 + \frac{1}{\frac{1}{R+j} + \frac{1}{j2} + \frac{1}{-j}} = 2 + \frac{1}{\frac{-R+j}{(R+j)(\ j2)}}$$

$$\Rightarrow \mathbf{Z}_{\mathbf{in}} = 2 + \frac{-2 + j2R}{-R + j} = \frac{-2 - 2R + j(2 + 2R)}{-R + j} \quad (6)$$

We do not have to find the imaginary part of $\mathbf{Z}_{\mathbf{in}}$ in (6). Instead, we can realize that, by putting $R = 1$ in (6), we will have a purely resistive impedance:

Solving (6) for $R = 1$:

$$\mathbf{Z_{in}} = \frac{-2 - 2 \times 1 + j(2 + 2 \times 1)}{-1 + j} = \frac{-4 + j4}{-1 + j} = 4\ \Omega \tag{7}$$

Therefore, for $R = 1$, the power factor seen by the voltage source is one.

Choice (1) is the answer.

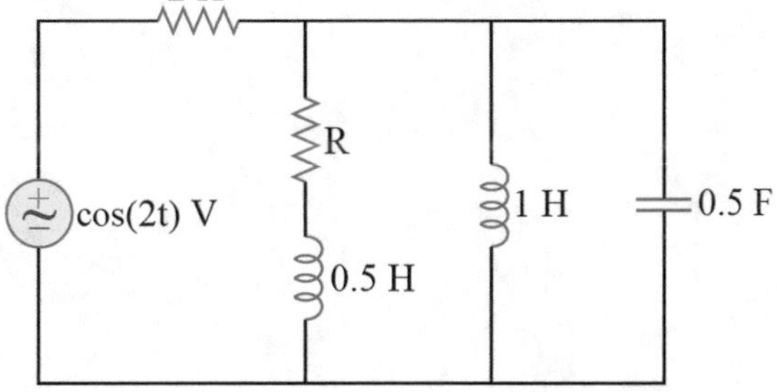

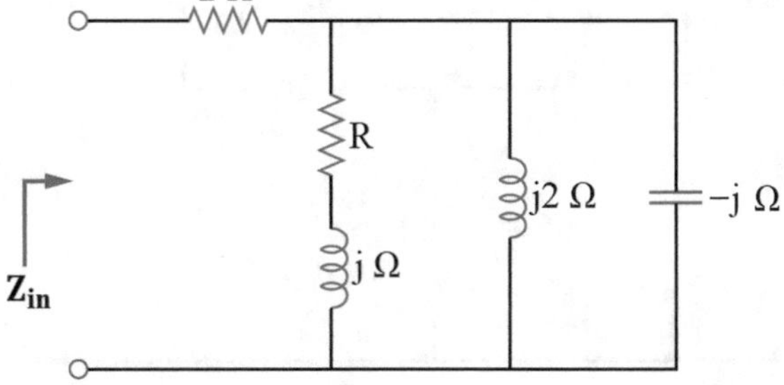

Figure 2.54 The circuit of solution of problem 2.56

2.57. Although the current sources have the same frequency, we should apply superposition theorem to simplify the problem solution.

Figure 2.55.2 shows the main circuit in frequency domain. The phasor of cos(2t), that is, $1\angle 0°$ is defined as the reference phasor, where "$\angle$" is the symbol of phase angle. Therefore, the phasors of the currents of the current sources are $2\angle 0°\ A$ and $1\angle 0°\ A$ or 2 A and 1 A. The impedances of the components are presented in the following:

$$\mathbf{Z_{1\ \Omega}} = 1\ \Omega \tag{1}$$

$$\mathbf{Z_{4\ H}} = j\omega L = j \times 2 \times 4 = j8\ \Omega \tag{2}$$

$$\mathbf{Z_{\frac{1}{2}\ F}} = \frac{1}{j\omega C} = \frac{1}{j \times 2 \times \frac{1}{2}} = -j\ \Omega \tag{3}$$

$$\mathbf{Z_{\frac{1}{4}\ F}} = \frac{1}{j\omega C} = \frac{1}{j \times 2 \times \frac{1}{4}} = -j2\ \Omega \tag{4}$$

Part 1: Applying current division formula in the circuit of Figure 2.55.3:

$$\mathbf{I_1} = \frac{1 + (-j)}{1 + (-j) + 1 + j8 + (-j2) + 1} \times 2 = \frac{2(1 - j)}{3 + j5}\ A \tag{5}$$

Part 2: Applying current division formula in the circuit of Figure 2.55.4:

$$\mathbf{I}_2 = \frac{(-j) + 1 + j8}{(-j) + 1 + j8 + (-j2) + 1 + 1} \times 1 = \frac{1 + j7}{3 + j5}\, A \tag{6}$$

Based on superposition theorem:

$$\mathbf{I} = \mathbf{I}_1 + \mathbf{I}_2 = \frac{2(1 - j)}{3 + j5} + \frac{1 + j7}{3 + j5} = \frac{3 + j5}{3 + j5} = 1\, A \tag{7}$$

Transferring to time domain:

$$i(t) = \cos(2t)\, A$$

Choice (4) is the answer.

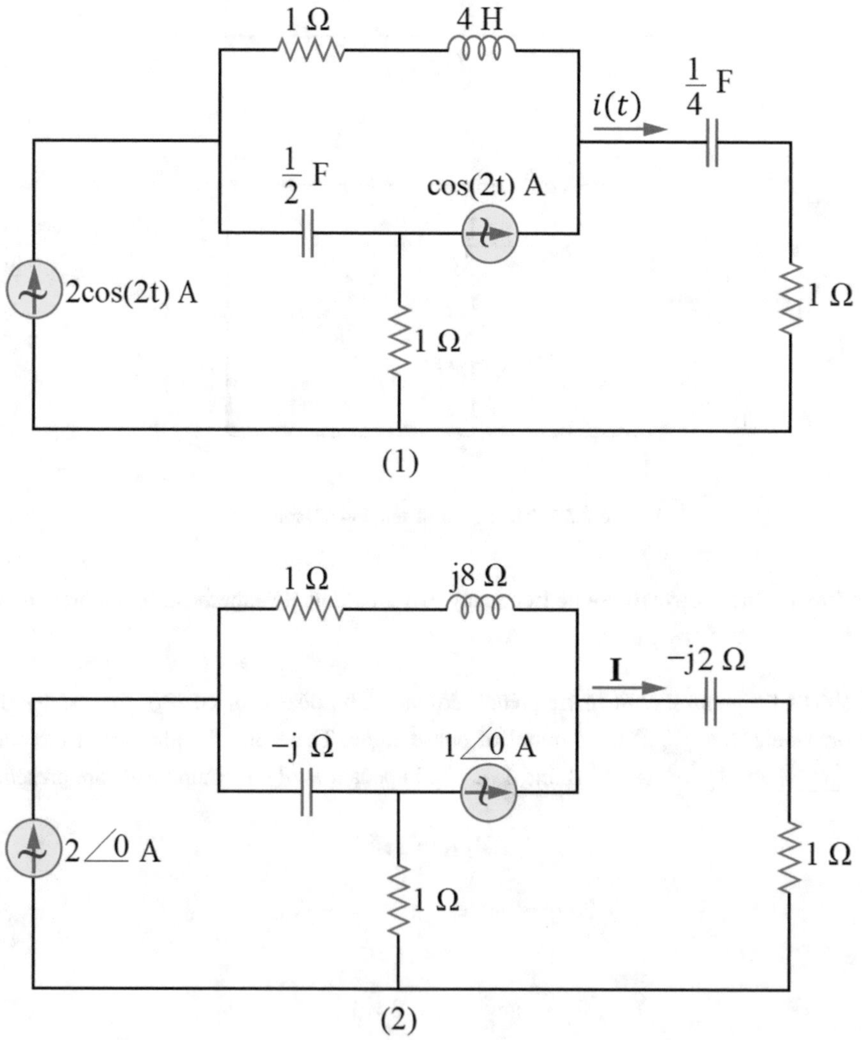

Figure 2.55 The circuit of solution of problem 2.57

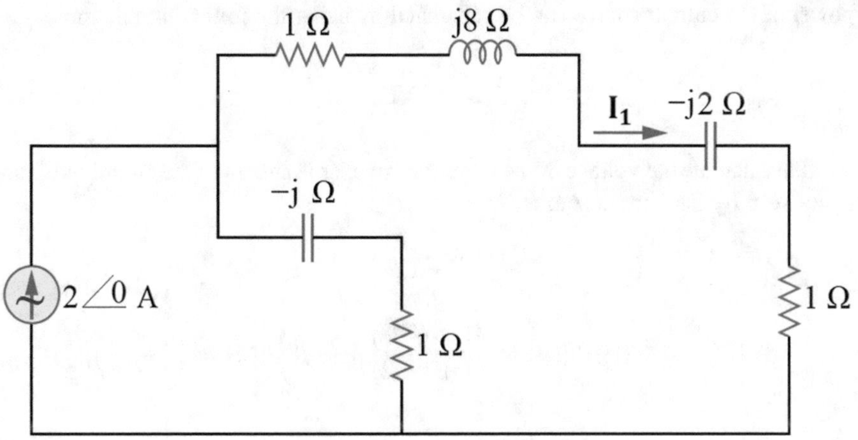

(3)

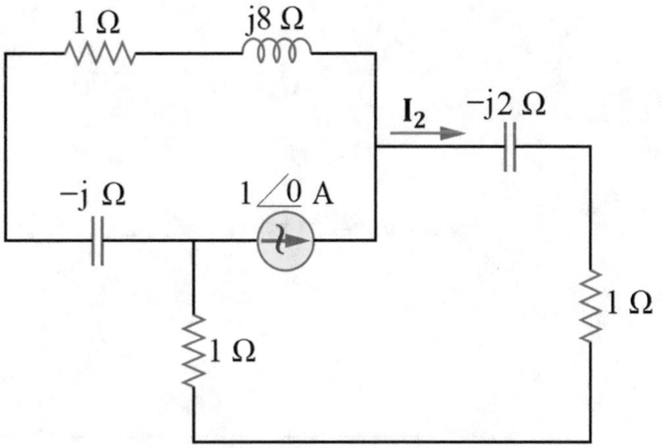

(4)

Figure 2.55 (continued)

2.58. Figure 2.56.2 shows the main circuit in frequency domain. The phasor of sin(t), that is, $1\angle 0^\circ$ is defined as the reference phasor, where "$\angle$" is the symbol of phase angle. Therefore, the phasor of the current of the current source is $1\angle 0^\circ\ A$ or $1\ A$. The impedances of the components are presented in the following:

$$\mathbf{Z}_{1\,\Omega} = 1\,\Omega \tag{1}$$

$$\mathbf{Z}_{0.5\ \mathrm{F}} = \frac{1}{j\omega C} = \frac{1}{j \times 1 \times 0.5} = -2j\,\Omega \tag{2}$$

$$\mathbf{Z}_{1\ \mathrm{F}} = \frac{1}{j\omega C} = \frac{1}{j \times 1 \times 1} = -j\,\Omega \tag{3}$$

$$\mathbf{Z}_{1\ \mathrm{H}} = j\omega L = j \times 1 \times 1 = j\,\Omega \tag{4}$$

The complex power of the current source can be calculated by using the following relation:

$$\mathbf{S} = \frac{1}{2}\mathbf{VI}^* \tag{5}$$

Therefore, we need to calculate the voltage of the current source, as is shown in the circuit of Figure 2.56.3. Herein, **Z** is the total impedance seen by the current source.

$$\mathbf{V} = \mathbf{IZ} \tag{6}$$

$$\mathbf{Z} = (1||(-j2) + (-j))||(j) = \left(\frac{1 \times (-j2)}{1 + (-j2)} + (-j)\right)||(j) = \left(\frac{4}{5} - j\frac{7}{5}\right)||(j)$$

$$\Rightarrow \mathbf{Z} = \frac{\left(\frac{4}{5} - j\frac{7}{5}\right) \times (j)}{\left(\frac{4}{5} - j\frac{7}{5}\right) + (j)} = \left(1 + j\frac{3}{2}\right)\Omega \tag{7}$$

Solving (6) and (7):

$$\mathbf{V} = 1 \times \left(1 + j\frac{3}{2}\right) = \left(1 + j\frac{3}{2}\right)\ V \tag{8}$$

Solving (5) and (8):

$$\mathrm{S} = \frac{1}{2}\mathbf{VI}^* = \frac{1}{2}\left(1 + j\frac{3}{2}\right) \times 1^* = \left(\frac{1}{2} + j\frac{3}{4}\right)\ VA$$

Choice (4) is the answer.

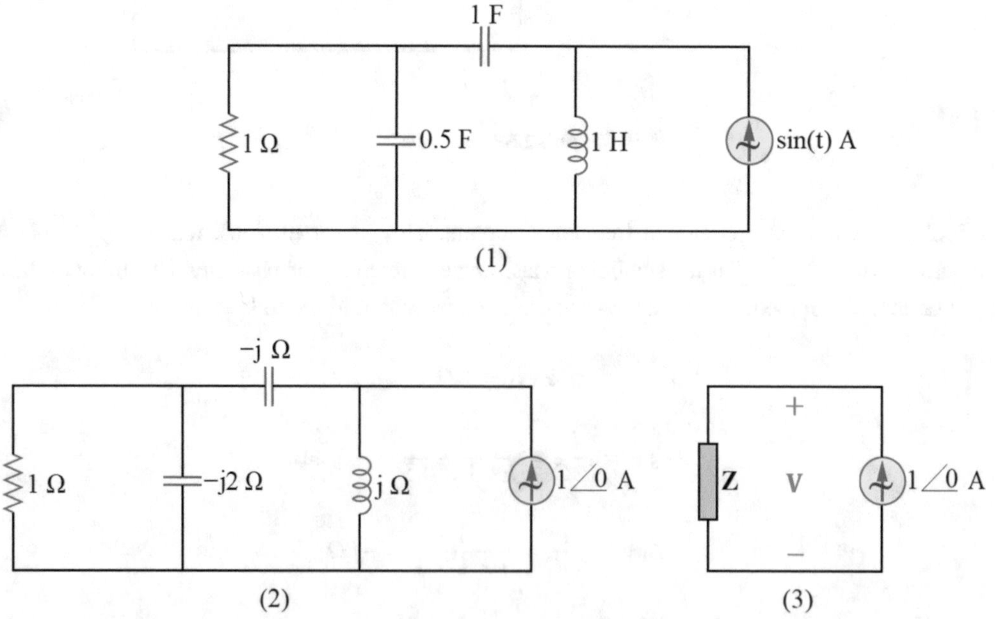

Figure 2.56 The circuit of solution of problem 2.58

2.59. The impedances of the components for the angular frequency of $\omega = 10^4$ *rad/sec* is presented in the following:

$$\mathbf{Z}_{10\ \Omega} = 10\ \Omega \tag{1}$$

$$\mathbf{Z}_{1\ \mathrm{mH}} = j\omega L = j \times 10^4 \times 10^{-3} = j10\ \Omega \tag{2}$$

$$\mathbf{Z}_{10\ \mu\mathrm{F}} = \frac{1}{j\omega C} = \frac{1}{j \times 10^4 \times 10 \times 10^{-6}} = -j10\ \Omega \tag{3}$$

Based on maximum average power transfer theorem, the load will absorb the maximum power if the complex conjugate of the Thevenin impedance, seen by the load, is equal to its impedance. In other words:

$$\mathbf{Z}_{\mathrm{L}} = \mathbf{Z}_{\mathrm{Th}}^{\ *} \tag{4}$$

To determine the Thevenin impedance, we need to turn off the independent voltage and current sources, as is shown in Figure 2.57.2.

$$\mathbf{Z}_{\mathrm{Th}} = (10 + j10)||(-j10) = \frac{(10 + j10) \times (-j10)}{(10 + j10) + (-j10)} = (10 - j10)\ \Omega \tag{5}$$

Solving (4) and (5):

$$\mathbf{Z}_{\mathrm{L}} = (10 - j10)^* = (10 + j10)\ \Omega \tag{6}$$

Figure 2.57.3 shows the main circuit in frequency domain, while the impedance of the load is known by using (6). The phasor of cos(10000t), that is, $1\angle 0°$ is defined as the reference phasor, where "$\angle$" is the symbol of phase angle. Therefore, the phasor of the voltage of the voltage source and the phasor of the current of the current source are $200\angle 0°\ V$ or $10\angle \frac{\pi}{2}\ A$, respectively. Although the circuit includes two power sources, the problem does not need to be solved by using superposition theorem because the circuit is working with a single frequency.
Applying KCL in the supernode of the circuit of Figure 2.57.3.

$$\frac{\mathbf{V_o} - 200}{10 + j10} - (10\angle 90°) + \frac{\mathbf{V_o}}{-j10} + \frac{\mathbf{V_o}}{10 + j10} = 0$$

$$\Rightarrow \mathbf{V_o}\left(\frac{1}{10 + j10} + \frac{1}{-j10} + \frac{1}{10 + j10}\right) - \frac{200}{10 + j10} - (10\angle 90°) = 0$$

$$\Rightarrow \mathbf{V_o}\left(\frac{1}{10}\right) - 10 = 0 \Rightarrow \mathbf{V_o} = 100\ V$$

Choice (4) is the answer.

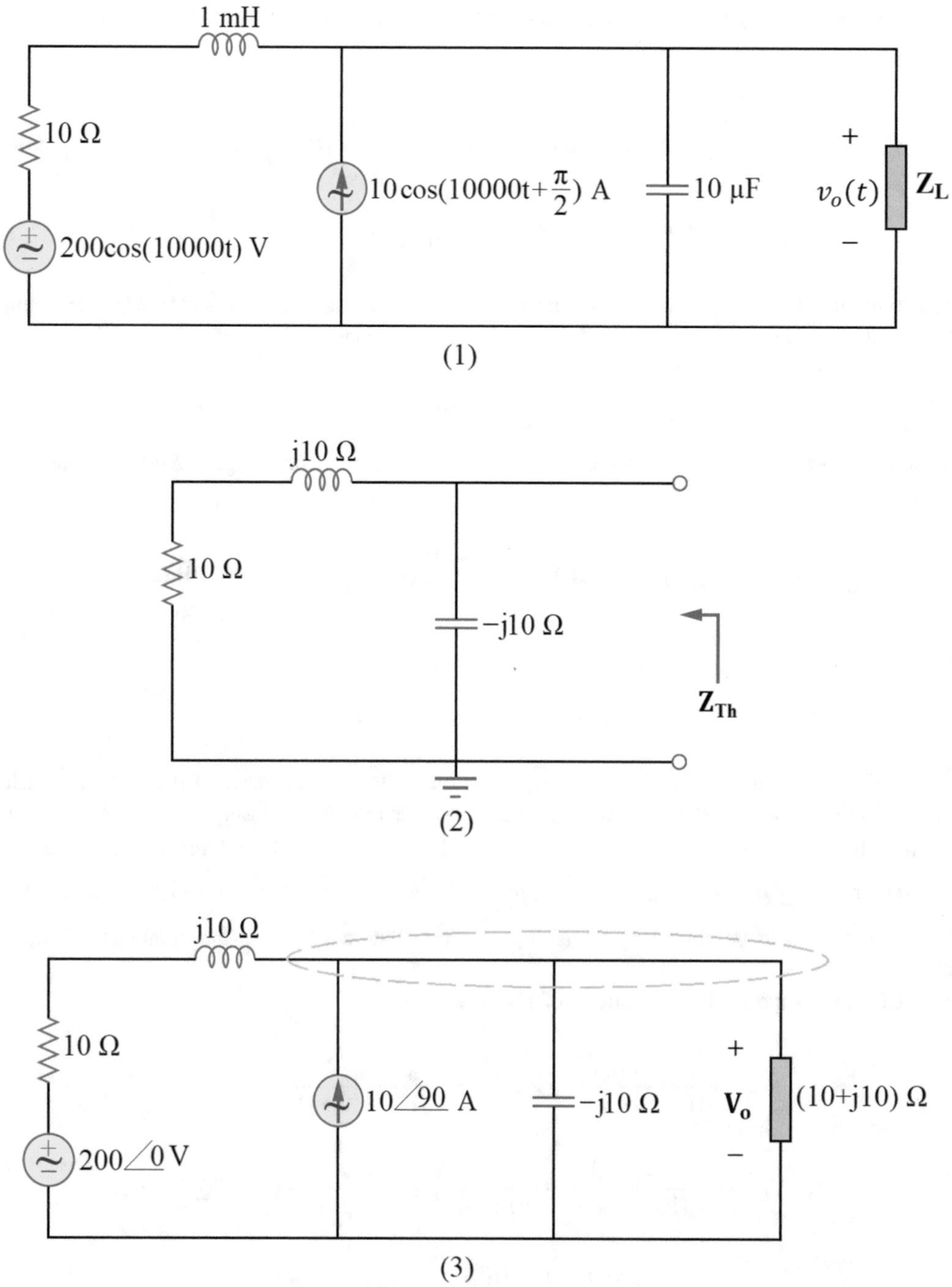

Figure 2.57 The circuit of solution of problem 2.59

2.60. The circuit is shown in frequency domain in Figure 2.58.2. The impedances of the components are as follows:

$$\mathbf{Z_{10\,H}} = j\omega L = j\omega \times 10 = j10\omega\ \Omega \tag{1}$$

$$\mathbf{Z_{1\,H}} = j\omega L = j\omega \times 1 = j\omega\ \Omega \tag{2}$$

$$\mathbf{Z_{0.1\,F}} = \frac{1}{j\omega C} = \frac{1}{j\omega \times 0.1} = -j\frac{10}{\omega}\ \Omega \tag{3}$$

$$\mathbf{Z_{6\,\Omega}} = 6\ \Omega \tag{4}$$

The phasor of the current of the voltage source can be calculated as follows:

$$\mathbf{I_s} = \frac{\mathbf{V_s}}{\mathbf{Z_{10\,H}} + (\mathbf{Z_{0.1\,F}})||(\mathbf{Z_{1\,H}} + \mathbf{Z_{6\,\Omega}})} = \frac{\mathbf{V_s}}{j10\omega + \left(-j\frac{10}{\omega}\right)||(j\omega + 6)}$$

$$\Rightarrow \mathbf{I_s} = \frac{\mathbf{V_s}}{j10\omega + \frac{\left(-j\frac{10}{\omega}\right)\times(j\omega+6)}{\left(-j\frac{10}{\omega}\right)+(j\omega+6)}} = \frac{\mathbf{V_s}}{j10\omega + \frac{10-j\frac{60}{\omega}}{6+j\left(\omega-\frac{10}{\omega}\right)}}$$

$$\Rightarrow \mathbf{I_s} = \frac{\mathbf{V_s}}{\frac{j60\omega+100-10\omega^2+10-j\frac{60}{\omega}}{6+j\left(\omega-\frac{10}{\omega}\right)}} = \frac{6+j\left(\omega-\frac{10}{\omega}\right)}{110-10\omega^2+j\left(60\omega-\frac{60}{\omega}\right)}\mathbf{V_s} \tag{5}$$

Using current division formula:

$$\mathbf{I} = \frac{-j\frac{10}{\omega}}{-j\frac{10}{\omega}+j\omega+6}\mathbf{I_s} = \frac{-j\frac{10}{\omega}}{6+j\left(\omega-\frac{10}{\omega}\right)}\mathbf{I_s} \tag{6}$$

Solving (5) and (6):

$$\mathbf{I} = \frac{-j\frac{10}{\omega}}{6+j\left(\omega-\frac{10}{\omega}\right)} \times \frac{6+j\left(\omega-\frac{10}{\omega}\right)}{110-10\omega^2+j\left(60\omega-\frac{60}{\omega}\right)}\mathbf{V_s} = \frac{-j\frac{10}{\omega}}{110-10\omega^2+j\left(60\omega-\frac{60}{\omega}\right)}\mathbf{V_s}$$

$$\Rightarrow \mathbf{I} = \frac{-j\frac{1}{\omega}}{11-\omega^2+j\left(6\omega-\frac{6}{\omega}\right)}\mathbf{V_s} \tag{7}$$

$$\Rightarrow |\mathbf{I}| = \left|\frac{-j\frac{1}{\omega}}{11-\omega^2+j\left(6\omega-\frac{6}{\omega}\right)}\mathbf{V_s}\right| = \frac{\frac{1}{\omega}}{\sqrt{\left(11-\omega^2\right)^2+\left(6\omega-\frac{6}{\omega}\right)^2}}|\mathbf{V_s}| \tag{8}$$

$$\Rightarrow \frac{|\mathbf{I}(\omega=2)|}{|\mathbf{I}(\omega=3)|} = \frac{\frac{\frac{1}{2}}{\sqrt{\left(11-2^2\right)^2+\left(6\times2-\frac{6}{2}\right)^2}}}{\frac{\frac{1}{3}}{\sqrt{\left(11-3^2\right)^2+\left(6\times3-\frac{6}{3}\right)^2}}} = \frac{\frac{\frac{1}{2}}{\sqrt{(7)^2+(9)^2}}}{\frac{\frac{1}{3}}{\sqrt{(2)^2+(16)^2}}} = \frac{\frac{\frac{1}{2}}{\sqrt{130}}}{\frac{\frac{1}{3}}{\sqrt{260}}} = 2.12$$

Therefore, decreasing ω from 3 rad/sec to 2 rad/sec will enlarge $|\mathbf{I}|$, 2.12 times as big. Choice (4) is the answer.

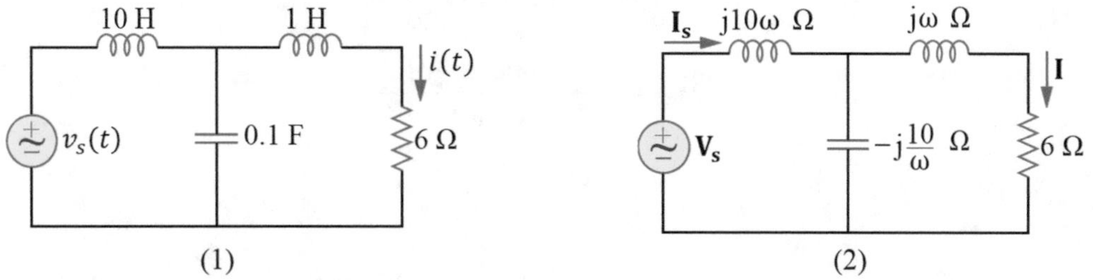

Figure 2.58 The circuit of solution of problem 2.60

2.61. Figure 2.59.2 illustrates the primary circuit in frequency domain. The impedances of the components are as follows:

$$\mathbf{Z_R} = R \tag{1}$$

$$\mathbf{Z_L} = j\omega L \tag{2}$$

$$\mathbf{Z_C} = \frac{1}{j\omega C} \tag{3}$$

The resonance frequency of a circuit can be determined by calculating the frequency that results in zero value for the imaginary part of the input admittance (or impedance) of the circuit. Therefore, we need to determine the admittance (or impedance) of the circuit and equate its imaginary part with zero to calculate the resonance frequency. In other words, the equation below must be solved:

$$Im\{\mathbf{Z_{in}}\} = 0 \tag{4}$$

Since the circuit includes at least one dependent source, we must apply a test source in the terminal to determine the value of $\frac{\mathbf{V_t}}{\mathbf{I_t}}$ to calculate the input impedance. The problem can be solved by using nodal analysis as follows:

Using Ohm's law to define **I** based on the desirable variable ($\mathbf{V_t}$):

$$\mathbf{I} = \frac{\mathbf{V_t}}{j\omega L} \tag{5}$$

Applying KCL in the supernode:

$$-\mathbf{I_t} + \frac{\mathbf{V_t}}{R} + \frac{\mathbf{V_t}}{\frac{1}{j\omega C}} + \frac{\mathbf{V_t}}{j\omega L} + \alpha\mathbf{I} = 0 \Rightarrow -\mathbf{I_t} + \mathbf{V_t}\left(\frac{1}{R} + j\omega C + \frac{1}{j\omega L}\right) + \alpha\mathbf{I} = 0 \tag{6}$$

Solving (5) and (6):

$$-\mathbf{I_t} + \mathbf{V_t}\left(\frac{1}{R} + j\omega C + \frac{1}{j\omega L}\right) + \alpha\frac{\mathbf{V_t}}{j\omega L} = 0 \Rightarrow -\mathbf{I_t} + \mathbf{V_t}\left(\frac{1}{R} + j\omega C + \frac{1}{j\omega L} + \frac{\alpha}{j\omega L}\right) = 0$$

$$\Rightarrow \frac{\mathbf{V_t}}{\mathbf{I_t}} = \frac{1}{\frac{1}{R} + j\left(\omega C - \frac{\alpha+1}{\omega L}\right)} = \frac{\frac{1}{R} - j\left(\omega C - \frac{\alpha+1}{\omega L}\right)}{\left(\frac{1}{R}\right)^2 + \left(\omega C - \frac{\alpha+1}{\omega L}\right)^2} \tag{7}$$

$$\Rightarrow \mathbf{Z_{in}} = \frac{\frac{1}{R}}{\left(\frac{1}{R}\right)^2 + \left(\omega C - \frac{\alpha+1}{\omega L}\right)^2} - j\frac{\left(\omega C - \frac{\alpha+1}{\omega L}\right)}{\left(\frac{1}{R}\right)^2 + \left(\omega C - \frac{\alpha+1}{\omega L}\right)^2} \tag{8}$$

$$Im\{\mathbf{Z_{in}}\} = -\frac{\left(\omega C - \frac{\alpha+1}{\omega L}\right)}{\left(\frac{1}{R}\right)^2 + \left(\omega C - \frac{\alpha+1}{\omega L}\right)^2} \tag{9}$$

Solving (4) and (9):

$$\frac{\left(\omega C - \frac{\alpha+1}{\omega L}\right)}{\left(\frac{1}{R}\right)^2 + \left(\omega C - \frac{\alpha+1}{\omega L}\right)^2} = 0 \Rightarrow \left(\omega C - \frac{\alpha+1}{\omega L}\right) = 0 \Rightarrow \omega^2 = \frac{\alpha+1}{LC} \Rightarrow \omega = \sqrt{\frac{\alpha+1}{LC}}$$

Choice (4) is the answer.

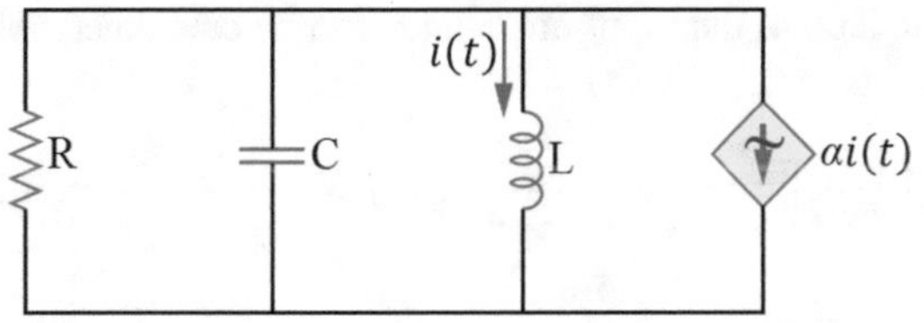

(1)

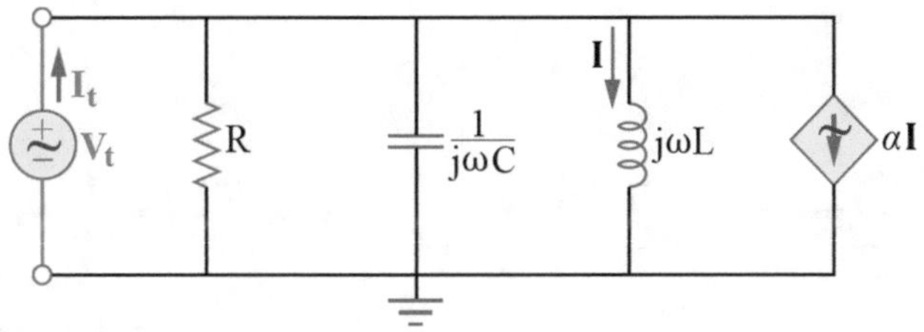

(2)

Figure 2.59 The circuit of solution of problem 2.61

2.62. To determine the resonance frequency of a circuit, we need to find the input impedance (or the input admittance) of the circuit, equate its imaginary part with zero, and solve it. In other words, we need to solve (1) or (2).

$$Im\{\mathbf{Z_{in}}(j\omega)\} = 0 \Rightarrow \omega = \omega_0 \tag{1}$$

$$Im\{\mathbf{Y_{in}}(j\omega)\} = 0 \Rightarrow \omega = \omega_0 \tag{2}$$

Figures 2.60.3 and 2.60.4 show the primary circuits in frequency domain. The impedances of the components are presented in the following:

$$\mathbf{Z_R} = R \tag{3}$$

$$\mathbf{Z_L} = j\omega L \tag{4}$$

$$\mathbf{Z_C} = \frac{1}{j\omega C} \tag{5}$$

Part 1: In the circuit of Figure 2.60.3, the input impedance can be calculated as follows:

$$\mathbf{Z_{in,1}} = \frac{1}{j\omega C}||(R + j\omega L) = \frac{\frac{1}{j\omega C} \times (R + j\omega L)}{\frac{1}{j\omega C} + (R + j\omega L)} = \frac{\frac{L}{C} - j\frac{R}{\omega C}}{R + j\left(\omega L - \frac{1}{\omega C}\right)} \tag{6}$$

$$\Rightarrow \mathbf{Z_{in,1}} = \frac{\frac{LR}{C} - \frac{R}{\omega C}\left(\omega L - \frac{1}{\omega C}\right) + j\left(-\frac{L}{C}\left(\omega L - \frac{1}{\omega C}\right) - \frac{R^2}{\omega C}\right)}{R^2 + \left(\omega L - \frac{1}{\omega C}\right)^2} \tag{7}$$

$$\Rightarrow \mathbf{Z_{in,1}} = \frac{\frac{LR}{C} - \frac{R}{\omega C}\left(\omega L - \frac{1}{\omega C}\right)}{R^2 + \left(\omega L - \frac{1}{\omega C}\right)^2} + j\frac{\left(-\frac{L}{C}\left(\omega L - \frac{1}{\omega C}\right) - \frac{R^2}{\omega C}\right)}{R^2 + \left(\omega L - \frac{1}{\omega C}\right)^2} \tag{8}$$

Solving (1) and (8):

$$\frac{-\frac{L}{C}\left(\omega L - \frac{1}{\omega C}\right) - \frac{R^2}{\omega C}}{R^2 + \left(\omega L - \frac{1}{\omega C}\right)^2} = 0 \Rightarrow -\frac{L}{C}\left(\omega L - \frac{1}{\omega C}\right) - \frac{R^2}{\omega C} = 0 \Rightarrow -\frac{L^2}{\omega C}\left(\omega^2 - \frac{1}{LC} + \frac{R^2}{L^2}\right) = 0$$

$$\Rightarrow \omega^2 - \frac{1}{LC} + \frac{R^2}{L^2} = 0 \Rightarrow \omega^2 = \frac{1}{LC} - \frac{R^2}{L^2} \Rightarrow \omega_{01} = \sqrt{\frac{1}{LC} - \left(\frac{R}{L}\right)^2} \tag{9}$$

Part 2: In the circuit of Figure 2.60.4, the input impedance can be calculated as follows:

$$\mathbf{Z}_{\mathbf{in},2} = j\omega L + \frac{1}{j\omega C}||R = j\omega L + \frac{\frac{1}{j\omega C} \times R}{\frac{1}{j\omega C} + R} = j\omega L + \frac{-j\frac{R}{\omega C}}{R - j\frac{1}{\omega C}} = j\omega L + \frac{\frac{R}{\omega^2 C^2} - j\frac{R^2}{\omega C}}{R^2 + \frac{1}{\omega^2 C^2}}$$

$$\Rightarrow \mathbf{Z}_{\mathbf{in},2} = \frac{\frac{R}{\omega^2 C^2} + j\left(-\frac{R^2}{\omega C} + R^2 L\omega + \frac{L}{\omega C^2}\right)}{R^2 + \frac{1}{\omega^2 C^2}} = \frac{\frac{R}{\omega^2 C^2}}{R^2 + \frac{1}{\omega^2 C^2}} + j\frac{-\frac{R^2}{\omega C} + R^2 L\omega + \frac{L}{\omega C^2}}{R^2 + \frac{1}{\omega^2 C^2}} \quad (10)$$

Solving (1) and (10):

$$\frac{-\frac{R^2}{\omega C} + R^2 L\omega + \frac{L}{\omega C^2}}{R^2 + \frac{1}{\omega^2 C^2}} = 0 \Rightarrow -\frac{R^2}{\omega C} + R^2 L\omega + \frac{L}{\omega C^2} = 0 \Rightarrow \frac{R^2 L}{\omega}\left(-\frac{1}{LC} + \omega^2 + \frac{1}{R^2 C^2}\right) = 0$$

$$\Rightarrow -\frac{1}{LC} + \omega^2 + \frac{1}{R^2 C^2} = 0 \Rightarrow \omega^2 = \frac{1}{LC} - \frac{1}{R^2 C^2} \Rightarrow \omega_{02} = \sqrt{\frac{1}{LC} - \frac{1}{(RC)^2}} \quad (11)$$

Based on the information given in the problem:

$$\omega_{01} = \omega_{02} \quad (12)$$

Solving (9), (11), and (12):

$$\sqrt{\frac{1}{LC} - \left(\frac{R}{L}\right)^2} = \sqrt{\frac{1}{LC} - \frac{1}{(RC)^2}} \Rightarrow \frac{1}{LC} - \left(\frac{R}{L}\right)^2 = \frac{1}{LC} - \frac{1}{(RC)^2} \Rightarrow \left(\frac{R}{L}\right)^2 = \frac{1}{(RC)^2}$$

$$\Rightarrow \frac{R}{L} = \frac{1}{RC} \Rightarrow R = \sqrt{\frac{L}{C}}$$

Choice (3) is the answer.

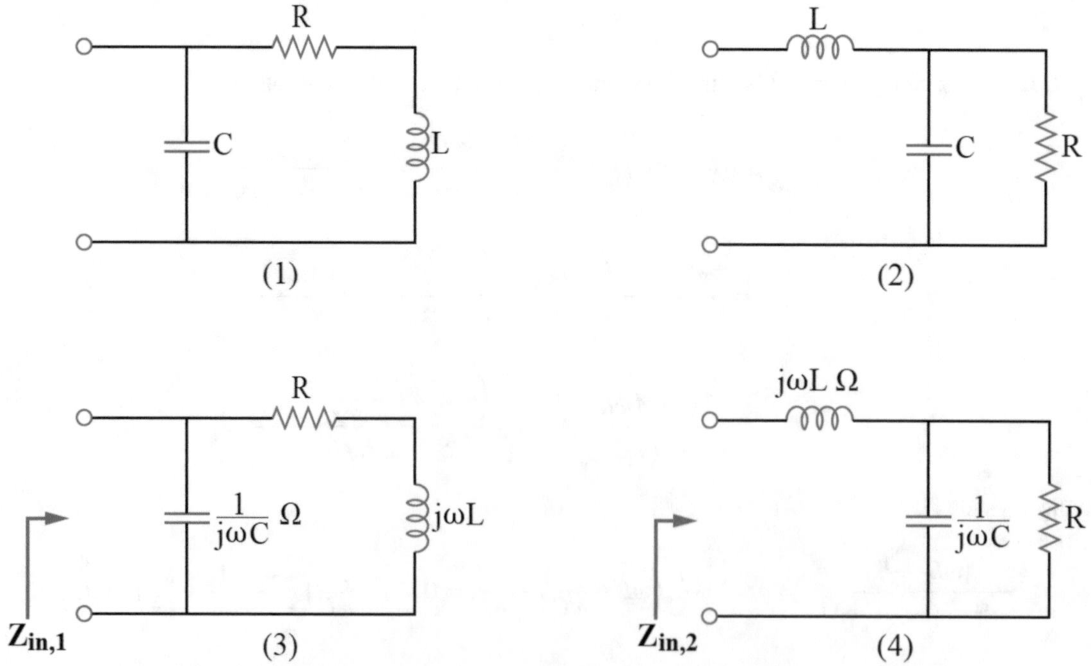

Figure 2.60 The circuit of solution of problem 2.62

2.63. This problem needs to be solved by using superposition theorem, since the circuit is operated with more than one frequency.

Part 1: As can be seen in the circuit of Figure 2.61.2, the right-side voltage source is turned off (short circuit), but the left-side voltage source is kept in the circuit. In addition, the circuit has been transferred to frequency domain based on the angular frequency of 1 rad/sec. The phasor of cos(t), that is, $1\angle 0^\circ$ is defined as the reference phasor, where "$\angle$" is the symbol of phase angle. Therefore, the phasor of the voltage of the voltage source is $2\angle 30^\circ\ V$. The impedances of the components are presented in the following:

$$\mathbf{Z_{1\,H}} = j\omega L = j \times 1 \times 1 = j\,\Omega \tag{1}$$

$$\mathbf{Z_{2\,H}} = j\omega L = j \times 1 \times 2 = j2\,\Omega \tag{2}$$

$$\mathbf{Z_{0.5\,F}} = \frac{1}{j\omega C} = \frac{1}{j \times 1 \times 0.5} = -j2\,\Omega \tag{3}$$

$$\mathbf{Z_{0.02\,H}} = j\omega L = j \times 1 \times 0.02 = j0.02\,\Omega \tag{4}$$

$$\mathbf{Z_{5\,\Omega}} = 5\,\Omega \tag{5}$$

As can be seen in Figure 2.61.2, the indicated part of the circuit has zero impedance ($j2 + (-j2) = 0\,\Omega$); therefore, it can be replaced by a short circuit branch, as is shown in Figure 2.61.3. By applying KVL in the left-side mesh of the circuit of Figure 2.61.3, we can write:

$$-2\angle 30^\circ + j\mathbf{I_L} = 0 \Rightarrow \mathbf{I_L} = \frac{2\angle 30^\circ}{j} = (2\angle -60^\circ)A \tag{6}$$

Now by applying KVL in right-side mesh, we have:

$$-5\mathbf{I_{X1}} - 5\mathbf{I_L} = 0 \Rightarrow \mathbf{I_{X1}} = -\mathbf{I_L} \xrightarrow{Using\ (6)} \mathbf{I_{x1}} = (-2\angle -60^\circ)\,A \tag{7}$$

By transferring back to time domain, we have:

$$i_{x1}(t) = -2\cos\left(t - 60^\circ\right)A \tag{8}$$

Part 2: As is shown in Figure 2.61.4, the left-side voltage source is turned off (short circuit), but the right-side voltage source is kept. In addition, the circuit has been transferred to frequency domain based on the angular frequency of 10 rad/sec. The phasor of sin($10t$), that is, $1\angle 0^\circ$ is defined as the reference phasor, where "$\angle$" is the symbol of phase angle. Therefore, the phasor of the voltage of the voltage source is $1\angle 0^\circ\ V$ or $1\ V$. The impedances of the components are presented in the following:

$$\mathbf{Z_{1\,H}} = j\omega L = j \times 10 \times 1 = j10\,\Omega \tag{9}$$

$$\mathbf{Z_{2\,H}} = j\omega L = j \times 10 \times 2 = j20\,\Omega \tag{10}$$

$$\mathbf{Z_{0.5\,F}} = \frac{1}{j\omega C} = \frac{1}{j \times 10 \times 0.5} = -j0.2\,\Omega \tag{11}$$

$$\mathbf{Z_{0.02\,H}} = j\omega L = j \times 10 \times 0.02 = j0.2\,\Omega \tag{12}$$

$$\mathbf{Z_{5\,\Omega}} = 5\,\Omega \tag{13}$$

The indicated part of the circuit, shown in Figure 2.61.4, has infinite impedance $(-j0.2)\|(j0.2) = \infty$; thus it can be replaced by an open circuit branch, as is shown in Figure 2.61.5. Therefore:

$$\mathbf{I_L} = 0\,A \tag{14}$$

Consequently, the dependent voltage source will be shut down (short-circuited), as is shown in Figure 2.61.5.

By applying KVL in the right-side mesh, we can write:

$$-5\mathbf{I}_{x2} + 1 = 0 \Rightarrow \mathbf{I}_{x2} = \frac{1}{5}\ A \tag{15}$$

By transferring back to time domain, we can write:

$$i_{x2}(t) = 0.2 \sin(10t) A \tag{16}$$

Now, based on superposition theorem and using (8) and (16), we have:

$$i_x(t) = i_{x1}(t) + i_{x2}(t) = -2\cos(t - 60^\circ) + 0.2\sin(10t)\ A$$

Choice (1) is the answer.

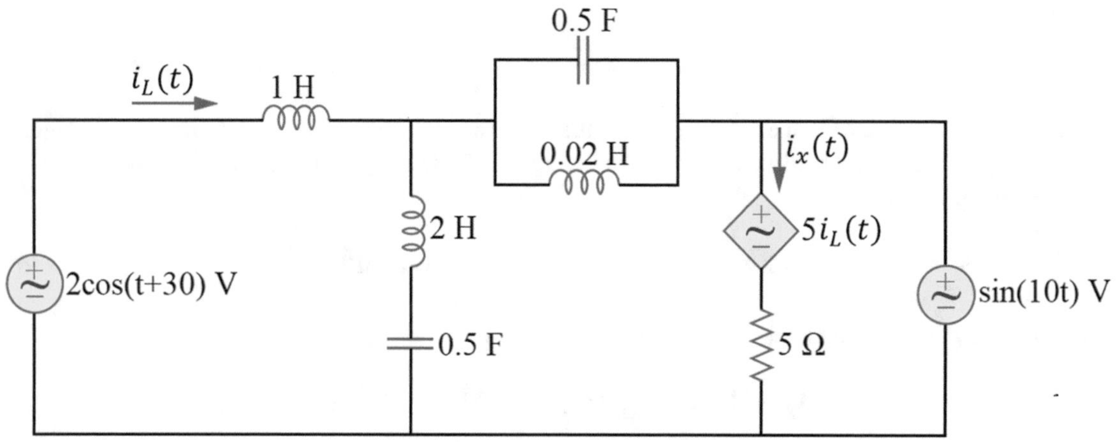

(1)

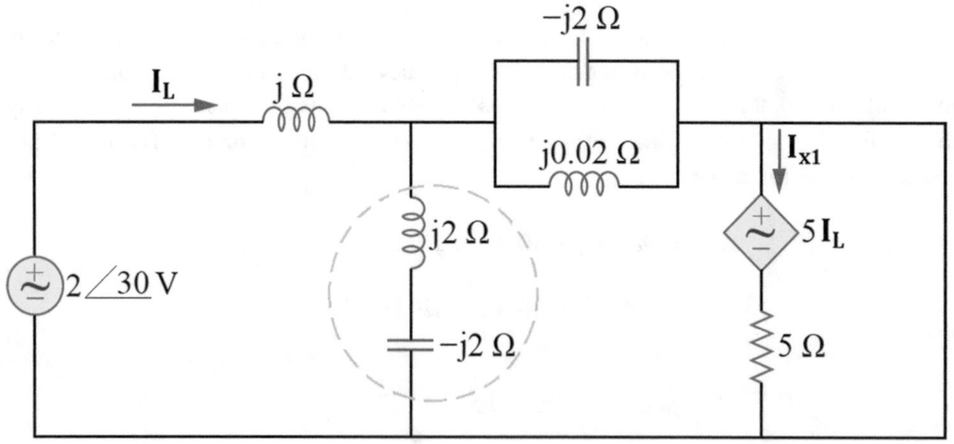

(2)

Figure 2.61 The circuit of solution of problem 2.63

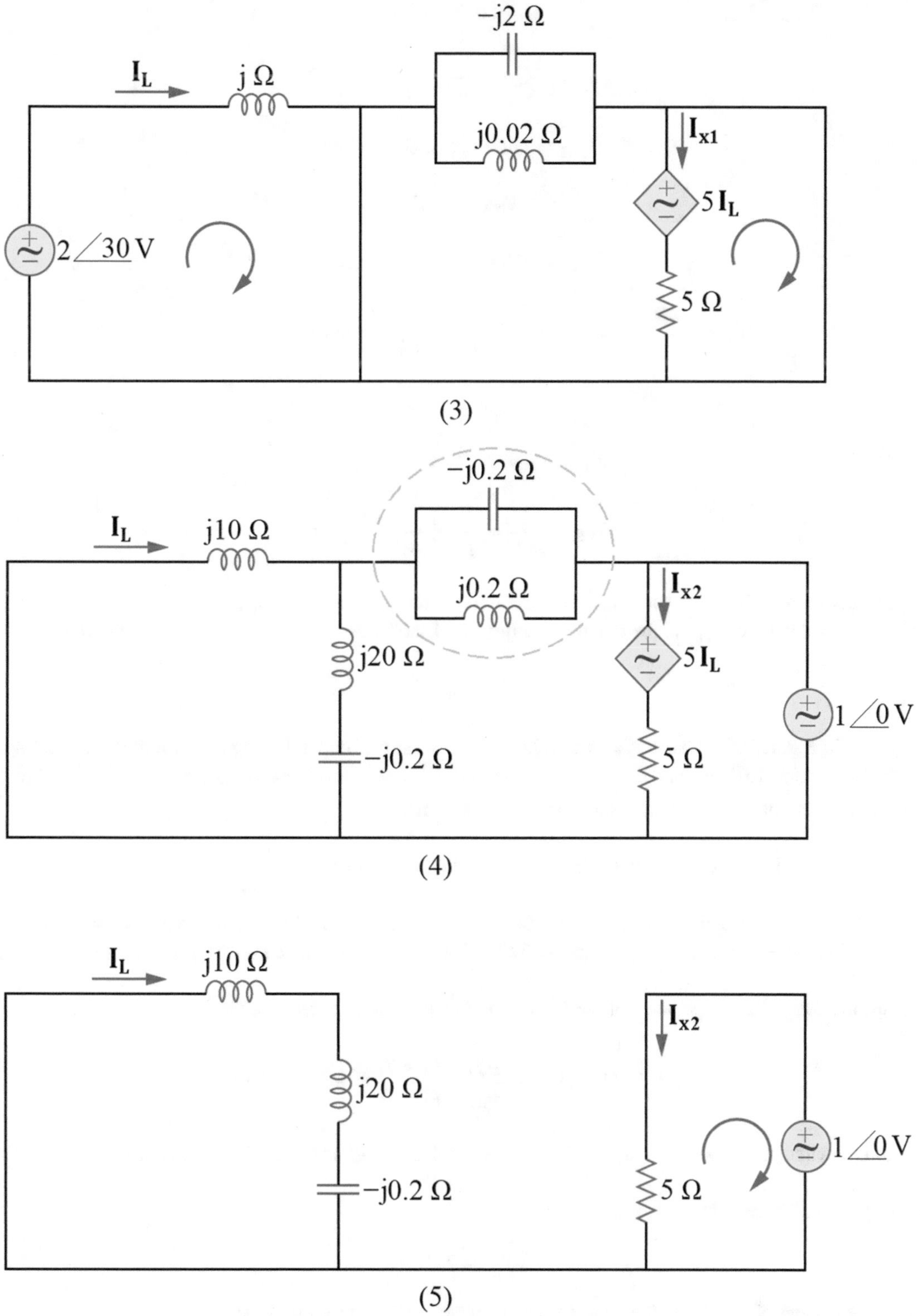

Figure 2.61 (continued)

2.64. The primary circuit is illustrated in frequency domain in Figure 2.62.2. Although several independent power sources exit in the circuit, their frequencies are the same. Hence, we do not need to apply superposition theorem. The impedances of the components are presented in the following:

$$\mathbf{Z}_{\frac{1}{3}\,\mathbf{F}} = \frac{1}{j\omega C} = \frac{1}{j \times 1 \times \frac{1}{3}} = -j3\ \Omega \tag{1}$$

$$\mathbf{Z}_{2\ \Omega} = 2\ \Omega \tag{2}$$

$$\mathbf{Z}_{2\ \mathbf{F}} = \frac{1}{j\omega C} = \frac{1}{j \times 1 \times 2} = -j0.5\ \Omega \tag{3}$$

$$\mathbf{Z}_{1\ \mathbf{H}} = j\omega L = j \times 1 \times 1 = j\ \Omega \tag{4}$$

$$\mathbf{Z}_{1\ \Omega} = 1\ \Omega \tag{5}$$

$$\mathbf{Z}_{2\ \mathbf{H}} = j\omega L = j \times 1 \times 2 = j2\ \Omega \tag{6}$$

$$\mathbf{Z}_{\frac{1}{2}\ \mathbf{F}} = \frac{1}{j\omega C} = \frac{1}{j \times 1 \times \frac{1}{2}} = -j2\ \Omega \tag{7}$$

$$\mathbf{Z}_{3\ \Omega} = 3\ \Omega \tag{8}$$

$$\mathbf{Z}_{4\ \mathbf{H}} = j\omega L = j \times 1 \times 4 = j4\ \Omega \tag{9}$$

$$\mathbf{Z}_{1\ \mathbf{F}} = \frac{1}{j\omega C} = \frac{1}{j \times 1 \times 1} = -j\ \Omega \tag{10}$$

Based on maximum average power transfer theorem, the relation below must be held to transfer the maximum average power to the load, where $\mathbf{Z}_{\mathbf{Th}}^*$ is the complex conjugate of the Thevenin impedance seen by the load.

$$\mathbf{Z}_{\mathbf{L}} = \mathbf{Z}_{\mathbf{Th}}^* \tag{11}$$

As is shown in Figure 2.62.2, since the circuit includes at least one dependent source, we must connect a test voltage source to the terminal to determine the value of $\frac{\mathbf{V}_t}{\mathbf{I}_t}$, which will give us the Thevenin impedance of the circuit. Herein, we need to turn off all the independent voltage and current sources.

Before solving the problem, the circuit should be simplified as follows:

The equivalent impedance of the series connection of $j2\ \Omega$ and $-j2\ \Omega$ is zero. Thus, it can be replaced by a short circuit branch. Therefore, the branch including the impedances of $3\ \Omega$ and $j4\ \Omega$ will be eliminated since it is parallel to a short circuit branch.

Moreover, the equivalent impedance of the bottom part of the circuit is as follows:

$$(1 + j)||(1 - j) = \frac{(1 + j) \times (1 - j)}{(1 + j) + (1 - j)} = \frac{1 + 1}{2} = 1\ \Omega \tag{12}$$

The simplified circuit is shown in Figure 2.62.3. In this problem, mesh analysis is the best approach to solve the problem.

Defining $\mathbf{V}$ based on the mesh current:

$$\mathbf{V} = (j + 1)\mathbf{I}_2 \tag{13}$$

Defining the current of the dependent current source based on the mesh currents:

$$\mathbf{V} = \mathbf{I}_2 - \mathbf{I}_t \xrightarrow{Using\ (13)} (j + 1)\mathbf{I}_2 = \mathbf{I}_2 - \mathbf{I}_t \Rightarrow \mathbf{I}_2 = j\mathbf{I}_t \tag{14}$$

KVL in the supermesh:

$$-\mathbf{V}_t + (-j3)\mathbf{I}_t + (j + 1)\mathbf{I}_2 + \mathbf{I}_2 = 0 \xrightarrow{Using\ (14)} -\mathbf{V}_t + (-3j)\mathbf{I}_t + (j + 2)(j\mathbf{I}_t) = 0 \tag{15}$$

$$\Rightarrow -\mathbf{V}_t + (-1 - j)\mathbf{I}_t = 0 \Rightarrow \frac{\mathbf{V}_t}{\mathbf{I}_t} = -1 - j \Rightarrow \mathbf{Z}_{\mathbf{Th}} = (-1 - j)\ \Omega \tag{16}$$

Solving (11) and (16):

$$Z_L = (-1 - j)^* = (-1 + j)\ \Omega$$

Choice (2) is the answer.

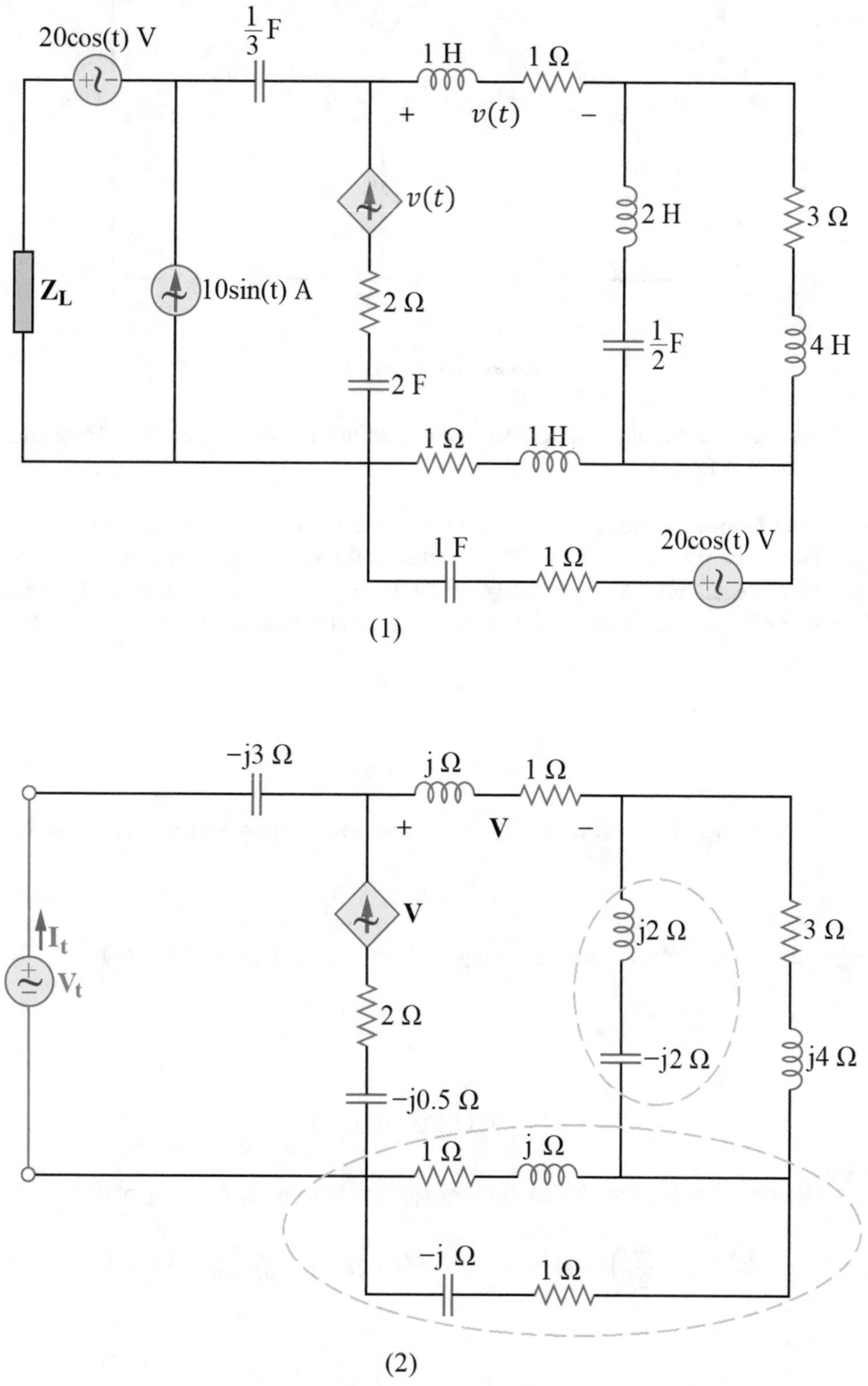

Figure 2.62 The circuit of solution of problem 2.64

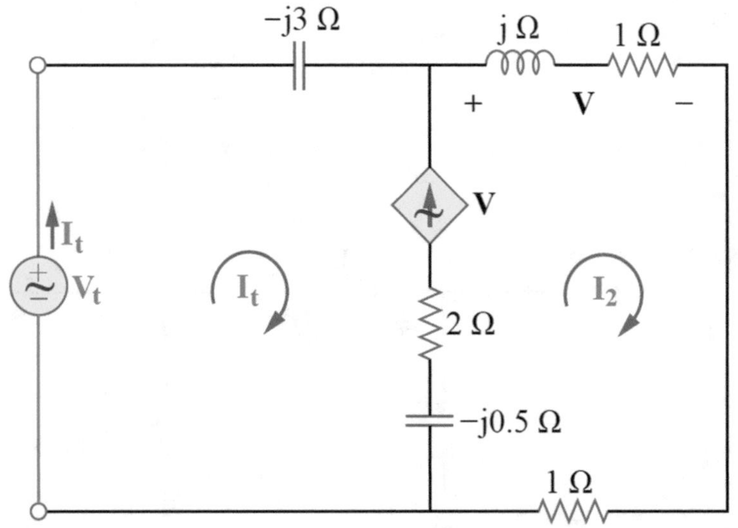

(3)

Figure 2.62 (continued)

2.65. The circuit is working with two different angular frequencies, that is, $\omega = 1,\ 2\ rad/sec$. Therefore, we must apply superposition theorem as follows.

Part 1: Figure 2.63.2 shows the circuit in frequency domain for $\omega = 1\ rad/sec$. Herein, only the bottom voltage source is turned off. The phasor of cos(t), that is, $1\angle 0°$ is defined as the reference phasor, where "$\angle$" is the symbol of phase angle. Therefore, the phasors of the voltage of the voltage source and the current of the current source are $(2\angle 0°)V$ and $(2\angle 0°)A$, respectively. The impedances of the components are presented in the following:

$$\mathbf{Z}_{1\,\Omega} = 1\,\Omega \quad (1)$$

$$\mathbf{Z}_{1\,\mathrm{F}} = \frac{1}{j\omega C} = \frac{1}{j \times 1 \times 1} = -j\,\Omega \quad (2)$$

The power of a resistor can be calculated by using the relation below. Hence, we need to calculate its current.

$$P_R = RI_{R,rms}^2 \quad (3)$$

By applying source transformation on the indicated part of the circuit of Figure 2.63.2, we have:

$$\mathbf{Z} = 1||(-j) = \frac{1 \times (-j)}{1 + (-j)} = \frac{1-j}{2}\,\Omega \quad (4)$$

$$\mathbf{V} = \left(\frac{1-j}{2}\right)(2\angle 0) = (1-j)\ V \quad (5)$$

The updated circuit is shown in Figure 2.63.3. Now, by applying KVL in the only loop of the circuit, we can write:

$$-2 + \mathbf{I} + \left(\frac{1-j}{2}\right)\mathbf{I} - (1-j) = 0 \Rightarrow \left(\frac{3-j}{2}\right)\mathbf{I} - (3-j) = 0 \Rightarrow \mathbf{I} = 2\,A \quad (6)$$

Solving (3) and (6):

$$P_1 = 1 \times \left(\frac{|2|}{\sqrt{2}}\right)^2 = 1 \times \left(\frac{2}{\sqrt{2}}\right)^2 = 2\,W \quad (7)$$

Part 2: Figure 2.63.4 illustrates the circuit in frequency domain for $\omega = 2\ rad/sec$. Herein, the top voltage source and the current source are turned off. The impedances of the components are presented in the following:

$$\mathbf{Z}_{1\,\Omega} = 1\,\Omega \tag{8}$$

$$\mathbf{Z}_{1\,\mathbf{F}} = \frac{1}{j\omega C} = \frac{1}{j \times 2 \times 1} = -j0.5\,\Omega \tag{9}$$

The current of the horizontal resistor can be calculated as follows:

$$\mathbf{I} = \frac{4}{1 + 1||(-j0.5)} = \frac{4}{1 + \frac{1\times(-j0.5)}{1+(-j0.5)}} = \frac{4}{1.2 - j0.4} = (3 + j)\ A \tag{10}$$

Solving (3) and (10):

$$P_2 = 1 \times \left(\frac{|\,3 + j\,|}{\sqrt{2}}\right)^2 = 1 \times \left(\frac{\sqrt{10}}{\sqrt{2}}\right)^2 = 5\ W \tag{11}$$

Based on superposition theorem and using (7) and (11):

$$P = P_1 + P_2 = 2 + 5 = 7\ W$$

Choice (4) is the answer.

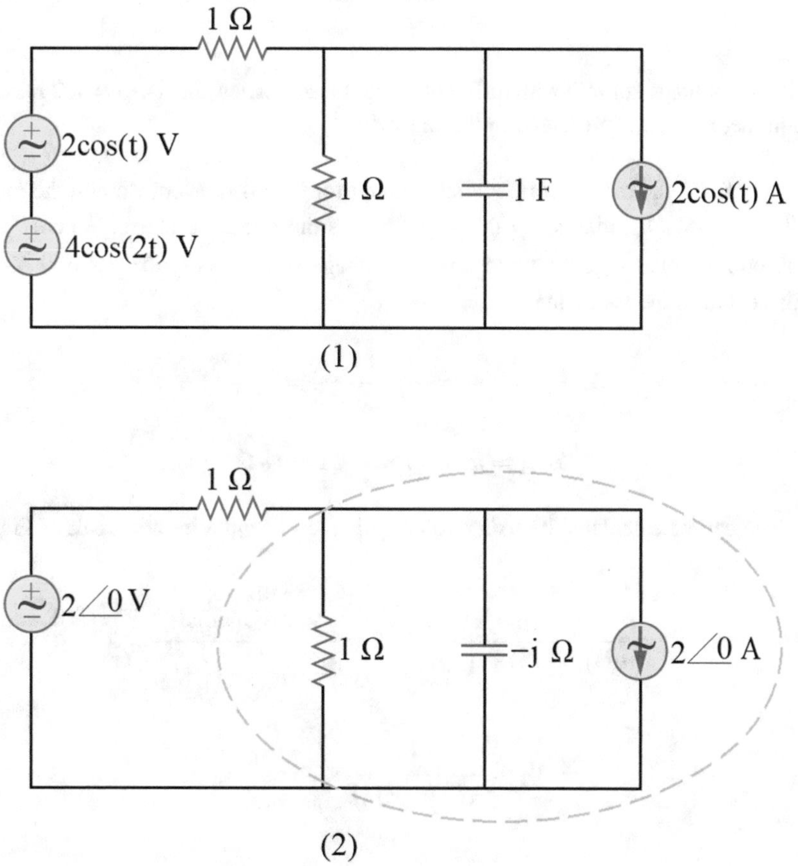

Figure 2.63 The circuit of solution of problem 2.65

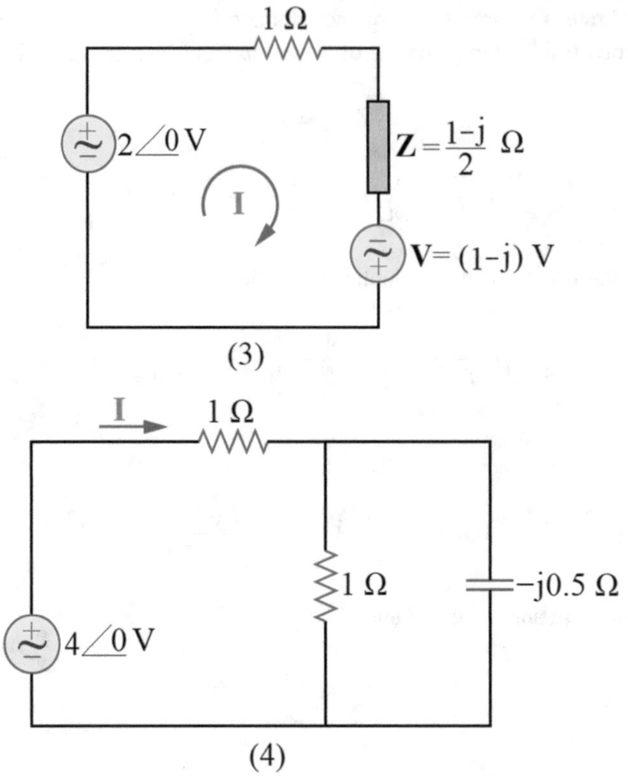

Figure 2.63 (continued)

2.66. The circuit includes two voltage sources with different angular frequencies, that is, $\omega = 4, 2$ *rad/sec*. Therefore, we must apply superposition theorem, as is presented in the following.

Part 1: Figure 2.64.2 shows the circuit in frequency domain for $\omega = 4$ *rad/sec*. Herein, the left-side voltage source is turned off. The phasor of cos(4*t*), that is, $1\angle 0^\circ$ is defined as the reference phasor, where "$\angle$" is the symbol of phase angle. Therefore, the phasor of the voltage of the voltage source is $(2\angle 0^\circ)V$ or 2 *V*. The impedances of the components are presented in the following:

$$\mathbf{Z}_{0.5\ \mathbf{F}} = \frac{1}{j\omega C} = \frac{1}{j \times 4 \times 0.5} = -j\frac{1}{2}\ \Omega \tag{1}$$

$$\mathbf{Z}_{1\ \mathbf{H}} = j\omega L = j \times 4 \times 1 = j4\ \Omega \tag{2}$$

The requested voltage can be calculated by using voltage division formula in the circuit of Figure 2.64.2 as follows:

$$\mathbf{V}_1 = \frac{(j4)||\left(-j\frac{1}{2}\right)}{(j4)||\left(-j\frac{1}{2}\right) + \left(-j\frac{1}{2}\right)} \times 2 = \frac{\frac{(j4)\times\left(-j\frac{1}{2}\right)}{(j4)+\left(-j\frac{1}{2}\right)}}{\frac{(j4)\times\left(-j\frac{1}{2}\right)}{(j4)+\left(-j\frac{1}{2}\right)} + \left(-j\frac{1}{2}\right)} \times 2$$

$$\Rightarrow \mathbf{V}_1 = \frac{-j\frac{4}{7}}{-j\frac{4}{7} + \left(-j\frac{1}{2}\right)} \times 2 = \frac{16}{15}\ V \tag{3}$$

Transferring to time domain:

$$v_1(t) = \frac{16}{15} \cos(4t) \tag{4}$$

Part 2: Figure 2.64.3 illustrates the circuit in frequency domain for $\omega = 2$ *rad/sec*. Therefore, the right-side voltage source is turned off. The phasor of sin(2*t*), that is, $1\angle 0^\circ$ is defined as the reference phasor, where "$\angle$" is the symbol of phase angle. Therefore, the phasor of the voltage of the voltage source is $1\angle 0^\circ$ *V* or 1 *V*. The impedances of the components are presented in the following:

$$\mathbf{Z}_{0.5\ \mathbf{F}} = \frac{1}{j\omega C} = \frac{1}{j \times 2 \times 0.5} = -j\,\Omega \tag{5}$$

$$\mathbf{Z}_{1\ \mathbf{H}} = j\omega L = j \times 2 \times 1 = j2\,\Omega \tag{6}$$

The requested voltage can be calculated by using voltage division relation in the circuit of Figure 2.64.3, as is presented in the following:

$$\mathbf{V}_2 = \frac{(j2)||(-j)}{(-j) + (j2)||(-j)} \times 1 = \frac{\frac{(j2)\times(-j)}{(j2)+(-j)}}{(-j) + \frac{(j2)\times(-j)}{(j2)+(-j)}} \times 1$$

$$\Rightarrow \mathbf{V}_2 = \frac{-j2}{(-j) - j2} \times 1 = \frac{2}{3}\ V \tag{7}$$

Transferring to time domain:

$$v_2(t) = \frac{2}{3} \sin(2t) \tag{8}$$

Based on superposition theorem and using (4) and (8):

$$v(t) = v_1(t) + v_2(t) = \frac{16}{15} \cos(4t) + \frac{2}{3} \sin(2t)\ V$$

Choice (2) is the answer.

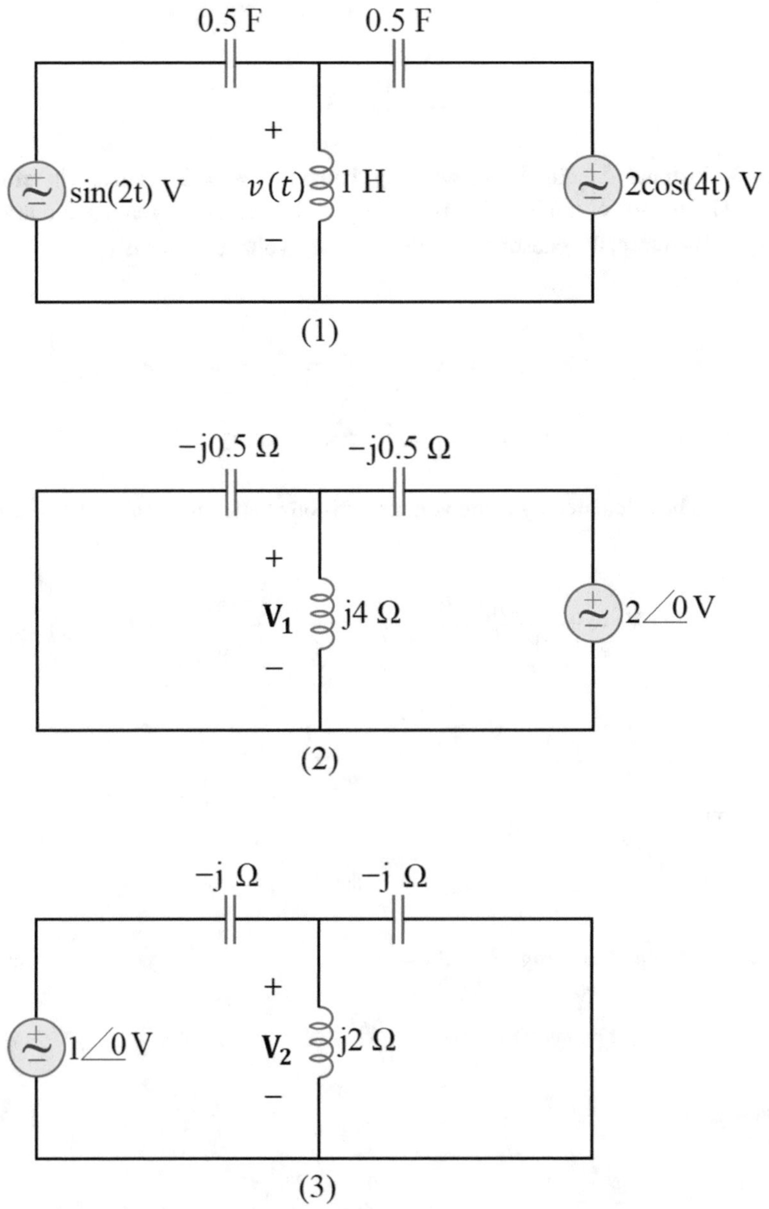

Figure 2.64 The circuit of solution of problem 2.66

2.67. Based on the information given in the problem, we have:

$$|\mathbf{Z_L}| = 26\ \Omega\ \textit{inductive} \tag{1}$$

$$P_L = 13\ kW\ \textit{consumed} \tag{2}$$

$$P_{Source} = 13.5\ kW\ \textit{generated} \tag{3}$$

From (1), we can conclude:

$$\sqrt{{R_L}^2 + {X_L}^2} = 26 \tag{4}$$

In addition, based on energy conservation theorem, we can conclude that the difference between the average powers, generated and consumed by the voltage source and the load, is consumed in the 0.8 Ω resistor. Now, by using this fact and average power formula for the 0.8 Ω resistor, we can write:

$$P_{Source} - P_{\mathrm{L}} = \frac{1}{2} R_{0.8\ \Omega} |\mathbf{I}|^2 \xrightarrow{Using\ (2)(3)} 13.5\ kW - 13\ kW = \frac{1}{2} \times 0.8 \times |\mathbf{I}|^2 \Rightarrow |\mathbf{I}| = 25\sqrt{2}\,A \quad (5)$$

Solving (2) and (5):

$$P_{\mathrm{L}} = \frac{1}{2} R_L |\mathbf{I}|^2 \Rightarrow 13000 = \frac{1}{2} \times R_L \times \left(25\sqrt{2}\right)^2 \Rightarrow R_L = 20.8\ \Omega \quad (6)$$

Solving (4) and (6):

$$\sqrt{20.8^2 + {X_L}^2} = 26 \Rightarrow X_L = 15.6\ \Omega$$

Choice (1) is the answer.

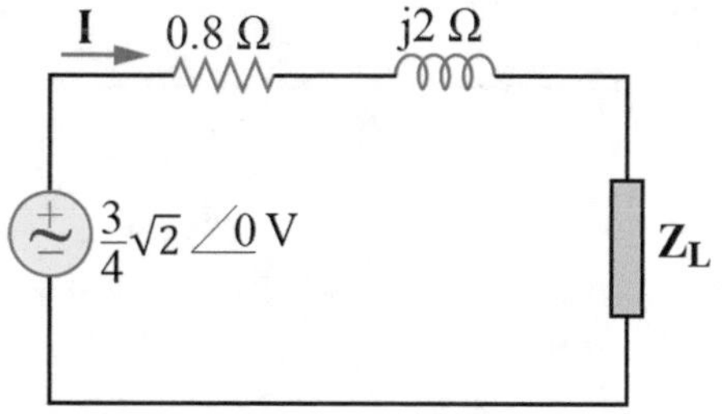

Figure 2.65 The circuit of solution of problem 2.67

2.68. The primary circuit is illustrated in frequency domain in Figure 2.66.2. The phasor of sin(t), that is, $1\angle 0°$ is defined as the reference phasor, where "$\angle$" is the symbol of phase angle. Therefore, the phasor of the voltage of the voltage source is $1\angle\varphi°\ V$. The impedances of the components are presented in the following:

$$\mathbf{Z}_{1\ \Omega} = 1\ \Omega \quad (1)$$

$$\mathbf{Z}_{2\ \mathrm{H}} = j\omega L = j \times 1 \times 2 = j2\ \Omega \quad (2)$$

$$\mathbf{Z}_{\frac{3}{2}\ \mathrm{F}} = \frac{1}{j\omega C} = \frac{1}{j \times 1 \times \frac{3}{2}} = -j\frac{2}{3}\ \Omega \quad (3)$$

We need to calculate the output current of the circuit. First, we should simplify the circuit by determining the equivalent impedance of the right-side part of the circuit as follows:

$$\mathbf{Z} = j2 || \left(-j\frac{2}{3}\right) = \frac{j2 \times \left(-j\frac{2}{3}\right)}{j2 + \left(-j\frac{2}{3}\right)} = -j\ \Omega \quad (4)$$

Now, by applying Ohm's law for the $\mathbf{Z}$, we can write:

$$\mathbf{I_o} = \frac{1\angle\varphi°}{1 - j} = \frac{1\angle\varphi°}{\sqrt{2}\angle{-45°}} = \frac{1}{\sqrt{2}}\angle(\varphi + 45)° \quad (5)$$

Based on the information given in the problem:

$$i_o = I_m \cos(t) \Rightarrow \mathbf{I_o} = I_m \angle 90^\circ \quad (6)$$

By solving (5) and (6), we can conclude that:

$$\varphi = 45^\circ = \frac{\pi}{4}\ rad$$

$$I_m = \frac{1}{\sqrt{2}}\ A$$

Choice (3) is the answer.

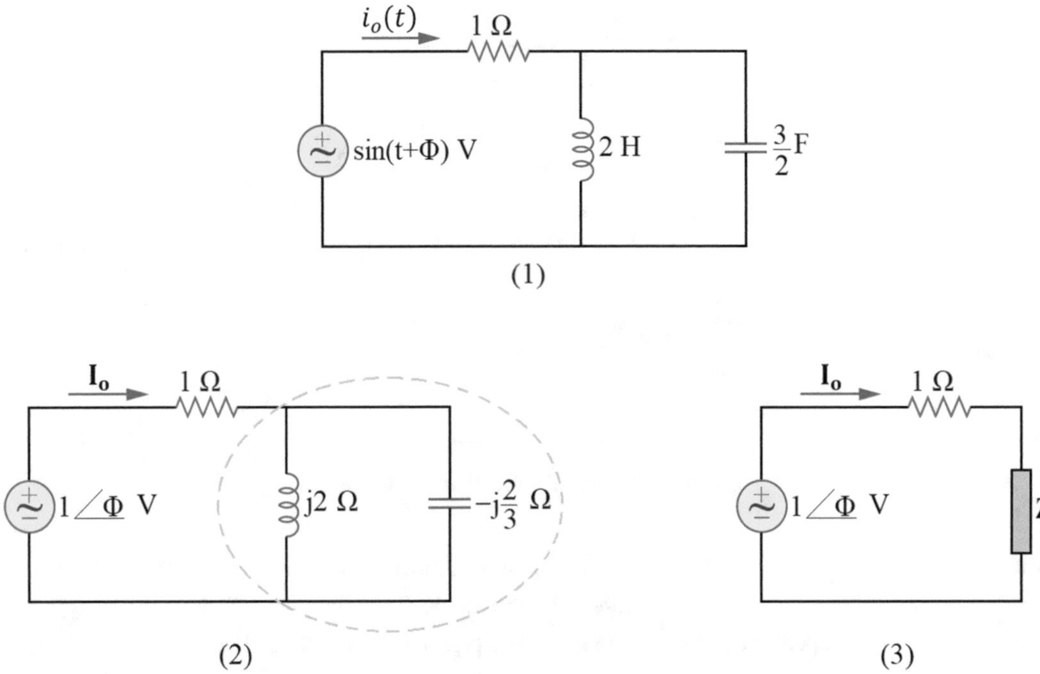

Figure 2.66 The circuit of solution of problem 2.68

2.69. The network can be modeled by its Thevenin equivalent circuit including Thevenin impedance and Thevenin voltage source. However, since we are interested in the determination of the input impedance, the Thevenin voltage source needs to be turned off (short-circuited). Thus, the network can be replaced by a single impedance ($\mathbf{Z_N} = R_N + jX_N$), as is shown in Figs. 2.67.3–4.

In the first test, illustrated in Figure 2.67.3, we can write:

$$\mathbf{Z_a} = R_N + jX_N \quad (1)$$

In the second test, shown in Figure 2.67.4, we have:

$$\mathbf{Z_b} = \frac{1}{j\omega C} + R_N + jX_N = \frac{1}{j \times 100 \times 100 \times 10^{-6}} + R_N + jX_N = R_N + j(X_N - 100) \quad (2)$$

Based on the given information:

$$|\mathbf{Z_a}| = 100\ \Omega \tag{3}$$

$$|\mathbf{Z_b}| = 50\ \Omega \tag{4}$$

By solving (1) and (3), we have:

$$\sqrt{R_N{}^2 + X_N{}^2} = 100 \Rightarrow R_N{}^2 + X_N{}^2 = 10000 \tag{5}$$

Solving (2) and (4) gives:

$$\sqrt{R_N{}^2 + (X_N - 100)^2} = 50 \Rightarrow R_N{}^2 + (X_N - 100)^2 = 2500 \tag{6}$$

Solving (5) and (6):

$$R_N = 48.4\ \Omega, X_N = 87.5\ \Omega \tag{7}$$

Solving (1) and (7):

$$\mathbf{Z_a} = 48.4 + j87.5 = 100\ \angle 61° \Rightarrow \angle \mathbf{Z_a} = 61°$$

Choice (3) is the answer.

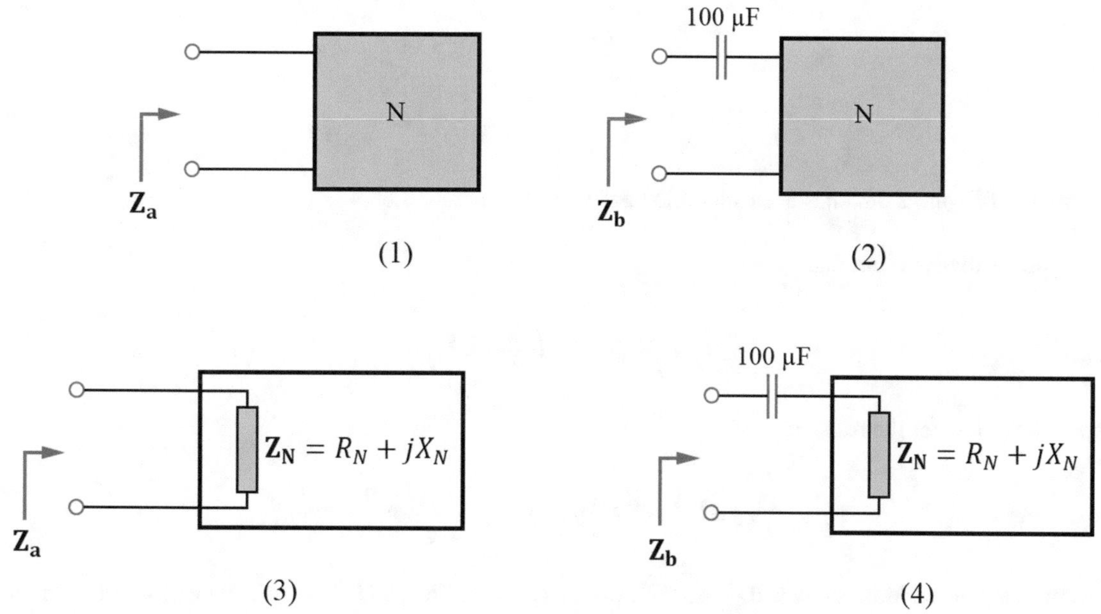

Figure 2.67 The circuit of solution of problem 2.69

2.70. This problem must be solved by using superposition theorem, since the circuit is operated with more than one frequency.

Part 1: As can be seen in the circuit of Figure 2.68.2, the current source has been turned off (open circuit), but the voltage source is kept in the circuit. In addition, the circuit has been transferred to frequency domain based on the angular frequency of 2 rad/sec. The phasor of cos(2*t*), that is, $1\angle 0°$ is defined as the reference phasor, where "$\angle$" is the symbol of phase angle. Therefore, the phasor of the voltage of the voltage source is $\sqrt{5}\angle 0°\ V$ or $\sqrt{5}\ V$. The impedances of the components are as follows:

$$\mathbf{Z}_{1\ \Omega} = 1\ \Omega \tag{1}$$

$$\mathbf{Z}_{1\ \mathrm{H}} = j\omega L = j \times 2 \times 1 = j2\ \Omega \tag{2}$$

$$\mathbf{Z}_{1\ \mathrm{F}} = \frac{1}{j\omega C} = \frac{1}{j \times 2 \times 1} = -j0.5\ \Omega \tag{3}$$

Using voltage division formula in the circuit of Figure 2.68.2:

$$\mathbf{V_1} = \frac{1}{1+2j} \times \sqrt{5} = (1\angle 63°)V \Rightarrow V_{1rms} = \frac{1}{\sqrt{2}} \times 1 = \frac{1}{\sqrt{2}}\ V \tag{4}$$

Part 2: Figure 2.68.3 shows the primary circuit that includes the current source but excludes the voltage source, since it has been shut down (short-circuited). As can be seen, the circuit has been transferred to frequency domain by using the angular frequency of 1 rad/sec. The phasor of cos(*t*), that is, $1\angle 0°$ is defined as the reference phasor, where "$\angle$" is the symbol of phase angle. Therefore, the phasor of the current of the current source is $1\angle 0°\ A$ or 1 *A*. The impedances of the components are as follows:

$$\mathbf{Z}_{1\ \Omega} = 1\ \Omega \tag{5}$$

$$\mathbf{Z}_{1\ \mathrm{H}} = j\omega L = j \times 1 \times 1 = j\ \Omega \tag{6}$$

$$\mathbf{Z}_{1\ \mathrm{F}} = \frac{1}{j\omega C} = \frac{1}{j \times 1 \times 1} = -j\ \Omega \tag{7}$$

The circuit of Figure 2.68.3 has been simplified and shown in Figure 2.68.4.

Using current division formula:

$$\mathbf{I_2} = \frac{j}{j+1} \times 1 = \left(\frac{1}{\sqrt{2}}\angle 45°\right) A \tag{8}$$

Using Ohm's law for the resistor:

$$\mathbf{V_2} = \mathbf{I_2} \times 1 = \left(\frac{1}{\sqrt{2}}\angle 45°\right) V \Rightarrow V_{2rms} = \frac{1}{\sqrt{2}} \times \frac{1}{\sqrt{2}} = \frac{1}{\sqrt{2}} V \tag{9}$$

Now, by using the relation below and (4) and (9), we can calculate the total root mean square (rms) value of the voltage of the resistor:

$$V_{rms} = \sqrt{V_{1rms}^2 + V_{2rms}^2} = \sqrt{\left(\frac{1}{\sqrt{2}}\right)^2 + \left(\frac{1}{2}\right)^2} = \sqrt{\frac{1}{2} + \frac{1}{4}} = \sqrt{\frac{3}{4}} = \frac{\sqrt{3}}{2}\ V$$

Choice (2) is the answer.

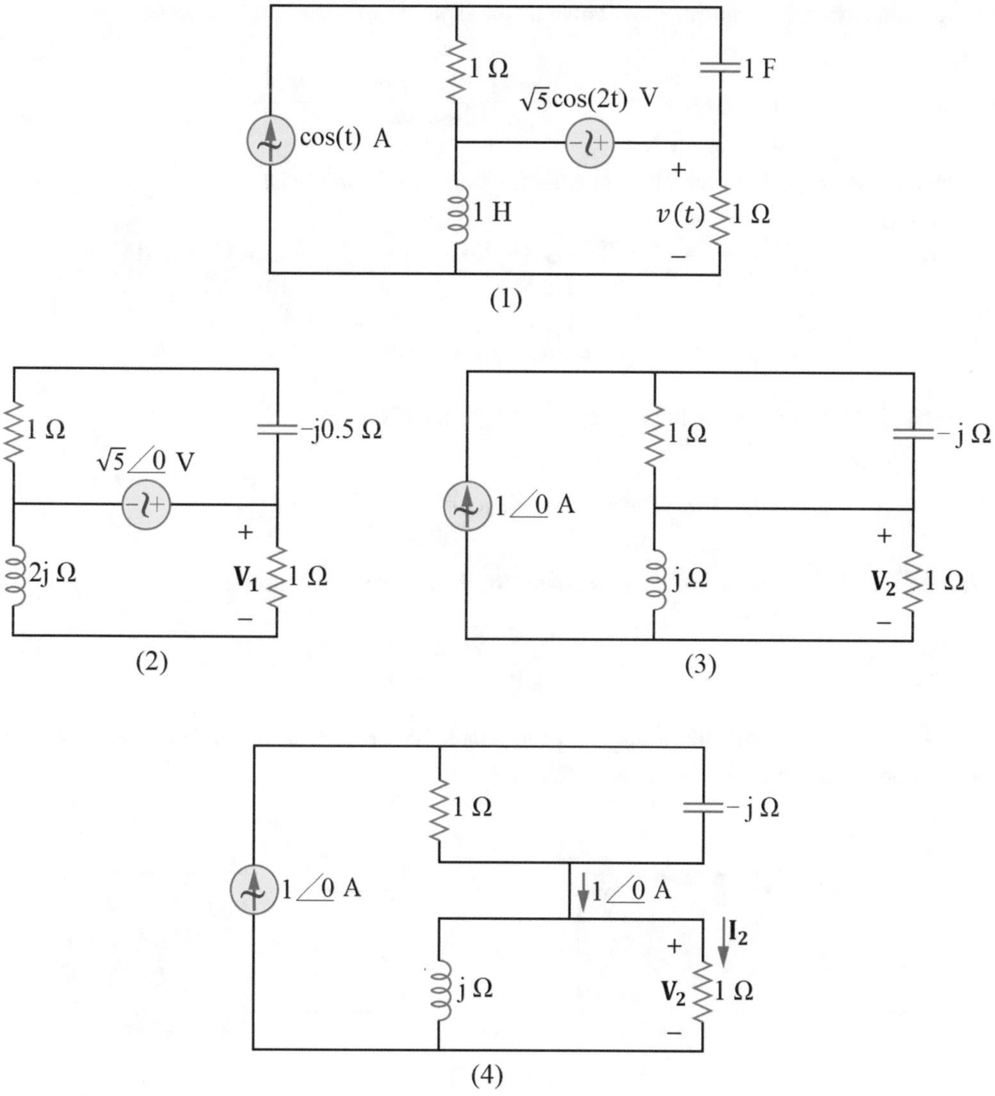

Figure 2.68 The circuit of solution of problem 2.70

2.71. Based on the problem, the relation below between the voltage and current of the linear time-invariant (LTI) circuit exists:

$$\frac{d^4}{dt^4}i(t) + 10\frac{d^3}{dt^3}i(t) + 40\frac{d^2}{dt^2}i(t) + 60\frac{d}{dt}i(t) + 784i(t) = 10\frac{d}{dt}v_s(t) + 40v_s(t) \tag{1}$$

As we know, the first derivative operator in time domain, that is, $\frac{d}{dt}$ is equivalent to $j\omega$ in frequency domain. Therefore, (1) can be transferred to frequency domain as follows:

$$(j\omega)^4\mathbf{I} + 10(j\omega)^3\mathbf{I} + 40(j\omega)^2\mathbf{I} + 60(j\omega)\mathbf{I} + 784\mathbf{I} = 10(j\omega)\mathbf{V_s} + 40\mathbf{V_s}$$

$$\Rightarrow \omega^4\mathbf{I} - j10\omega^3\mathbf{I} - 40\omega^2\mathbf{I} + j60\omega\mathbf{I} + 784\mathbf{I} = j10\omega\mathbf{V_s} + 40\mathbf{V_s}$$

$$\Rightarrow \left(\omega^4 - 40\omega^2 + 784 + j\left(60\omega - 10\omega^3\right)\right)\mathbf{I} = (40 + j10\omega)\mathbf{V_s} \tag{2}$$

Based on the main circuit, the value of $\frac{\mathbf{V_s}}{\mathbf{I}}$ will give us the input impedance of the network.

$$\mathbf{Z_{in}} = \frac{\mathbf{V_s}}{\mathbf{I}} = \frac{\omega^4 - 40\omega^2 + 784 + j(60\omega - 10\omega^3)}{40 + j10\omega} \tag{3}$$

The impedance of the network for the given angular frequency of 4 *rad/sec* is:

$$\mathbf{Z_{in}}(\omega = 4) = \frac{4^4 - 40 \times 4^2 + 784 + j(60 \times 4 - 10 \times 4^3)}{40 + j10 \times 4} = \frac{400 - j400}{40 + j40} = \frac{10(1 - j)}{1 + j}$$

$$\Rightarrow \mathbf{Z_{in}}(\omega = 4) = -j10\ \Omega \tag{4}$$

On the other hand, we know that the impedance of a single capacitor is:

$$\mathbf{Z_C} = \frac{1}{j\omega C} \xrightarrow{for\ \omega = 4} \mathbf{Z_C}(\omega = 4) = \frac{1}{j4C} = -j\frac{1}{4C} \tag{5}$$

By comparing (4) and (5), we can find the capacitance of the capacitor as follows:

$$-j10 = -j\frac{1}{4C} \Rightarrow C = \frac{1}{40}\ F$$

Therefore, the network is behaving like a single capacitor with the capacitance of $\frac{1}{40}$ *F* in the sinusoidal steady state with the angular frequency of 4 *rad/sec*. Choice (2) is the answer.

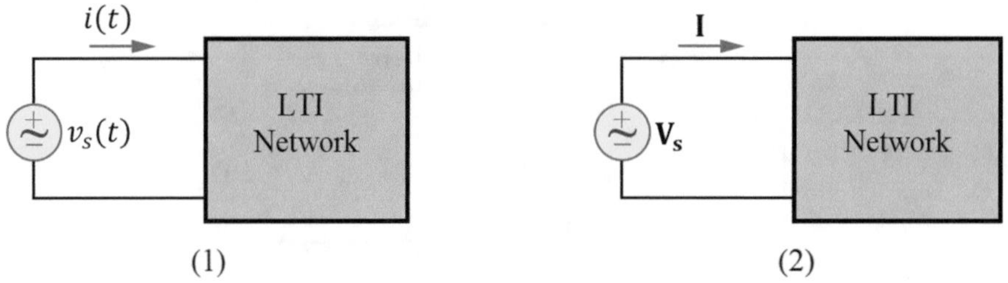

Figure 2.69 The circuit of solution of problem 2.71

2.72. Figures 2.70.2–3 show the left-side and right-side parts of the circuit in frequency domain. The impedances of the components are as follows:

$$\mathbf{Z_{50\ \Omega}} = 50\ \Omega \tag{1}$$

$$\mathbf{Z_L} = j\omega L\ \Omega \tag{2}$$

$$\mathbf{Z_C} = \frac{1}{j\omega C}\ \Omega \tag{3}$$

$$\mathbf{Z_{100\ \Omega}} = 100\ \Omega \tag{4}$$

Based on maximum average power transfer theorem, to transfer maximum average power to the right-side of terminal a–b, the relation below must be held between the impedances seen from the left side and the right side of the terminal:

$$\mathbf{Z_{in-right}} = \mathbf{Z_{in-left}}^* \tag{5}$$

As can be seen in the circuit of Figure 2.70.2:

$$\mathbf{Z}_{\text{in-left}} = 50\ \Omega \tag{6}$$

$$\mathbf{Z}_{\text{in-right}} = j\omega L + \frac{1}{j\omega C}||100 = j\omega L + \frac{\frac{1}{j\omega C}\times 100}{\frac{1}{j\omega C}+100} = j\omega L + \frac{\frac{100}{\omega C}}{\frac{1}{\omega C}+j100}$$

$$\Rightarrow \mathbf{Z}_{\text{in-right}} = j\omega L + \frac{1}{0.01 + j\omega C} = j\omega L + \frac{0.01 - j\omega C}{10^{-4}+\omega^2C^2}$$

$$\Rightarrow \mathbf{Z}_{\text{in-right}} = \frac{0.01}{10^{-4}+\omega^2C^2} + j\left(\omega L - \frac{\omega C}{10^{-4}+\omega^2C^2}\right)\ \Omega \tag{7}$$

Solving (5)–(7):

$$\frac{0.01}{10^{-4}+\omega^2C^2} + j\left(\omega L - \frac{\omega C}{10^{-4}+\omega^2C^2}\right) = 50^* \Rightarrow \begin{cases} \dfrac{0.01}{10^{-4}+\omega^2C^2} = 50 & (8) \\ \omega L - \dfrac{\omega C}{10^{-4}+\omega^2C^2} = 0 & (9) \end{cases}$$

From (8), we can write:

$$2\times 10^{-4} = 10^{-4} + \omega^2C^2 \Rightarrow \omega^2C^2 = 10^{-4} \Rightarrow \omega C = 0.01 \tag{10}$$

Solving (9) and (10):

$$\omega L = \frac{10^{-2}}{10^{-4}+10^{-4}} \Rightarrow \omega L = 50$$

Choice (1) is the answer.

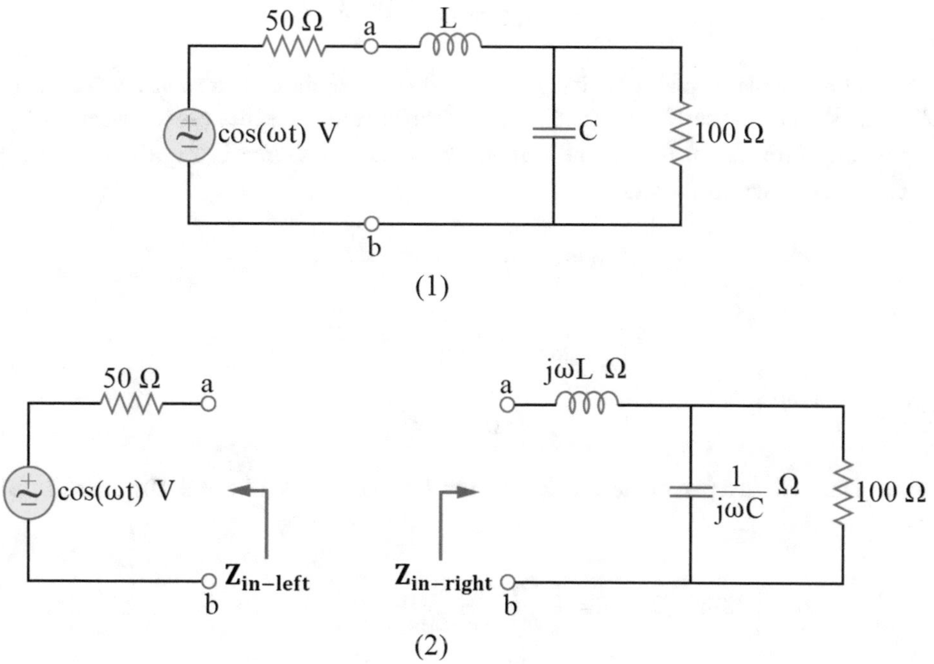

Figure 2.70 The circuit of solution of problem 2.72

2.73. The circuit includes two power sources with different angular frequencies. Therefore, superposition theorem must be applied in this problem.

Part 1: Figure 2.71.2 shows the main circuit in frequency domain for the voltage source with the angular frequency of 1 rad/sec. In this regard, the phasor of cos(t), that is, $1\angle 0°$ is defined as the reference phasor, where "$\angle$" is the symbol of phase angle. Thus, the phasor of the voltage of the voltage source is $8\angle 0°$ V or 8 V. The impedances of the components are presented in the following:

$$\mathbf{Z_{1\,H}} = j\omega L = j \times 1 \times 1 = j\,\Omega \tag{1}$$

$$\mathbf{Z_{\frac{1}{4}\,F}} = \frac{1}{j\omega C} = \frac{1}{j \times 1 \times \frac{1}{4}} = -j4\,\Omega \tag{2}$$

$$\mathbf{Z_{2\,\Omega}} = 2\,\Omega \tag{3}$$

By applying KCL in the indicated node, we can write:

$$\frac{\mathbf{V_x} - 8}{j} + \frac{\mathbf{V_x}}{-j4} + \frac{\mathbf{V_x}}{2+2} = 0 \Rightarrow \mathbf{V_x}\left(\frac{1}{4} - j\frac{3}{4}\right) = -8j$$

$$\Rightarrow \mathbf{V_x} = \frac{-8j}{\frac{1}{4} - j\frac{3}{4}} = \left(\frac{16\sqrt{10}}{5}\angle 18.43°\right)V \tag{4}$$

Applying voltage division formula:

$$\mathbf{V_{o1}} = \frac{2}{2+2} \times \left(\frac{16\sqrt{10}}{5}\angle 18.43°\right) = \left(\frac{8\sqrt{10}}{5}\angle 18.43°\right)V \tag{5}$$

The output voltage in time domain is:

$$v_{o1}(t) = \frac{8\sqrt{10}}{5}\cos\left(t - 18.43°\right)V \tag{6}$$

Part 2: Figure 2.71.3 exhibits the main circuit in frequency domain for the current source with the angular frequency of 2 rad/sec. Herein, the phasor of cos($2t$), that is, $1\angle 0°$ is defined as the reference phasor, where "$\angle$" is the symbol of phase angle. Therefore, the phasor of the current of the current source is $2\angle 30°$ A. The impedances of the components can be calculated as follows:

$$\mathbf{Z_{1\,H}} = j\omega L = j \times 2 \times 1 = j2\,\Omega \tag{7}$$

$$\mathbf{Z_{\frac{1}{4}\,F}} = \frac{1}{j\omega C} = \frac{1}{j \times 2 \times \frac{1}{4}} = -j2\,\Omega \tag{8}$$

$$\mathbf{Z_{2\,\Omega}} = 2\,\Omega \tag{9}$$

The impedance of the indicated part of the circuit of Figure 2.71.3 is infinite (theoretically undefined), as can be seen in the following:

$$\mathbf{Z} = (j2)||(-j2) + 2 = \frac{(j2)\times(-j2)}{(j2)+(-j2)} + 2 = \frac{4}{0} + 2 = \infty + 2 = \infty \tag{10}$$

Therefore, that part of the circuit is an open circuit branch. Thus, the whole current of the current source will flow through the 2 Ω resistor in the vertical branch.

$$\mathbf{V_{o2}} = (2\angle 30^\circ) \times 2 = (4\angle 30^\circ) V \tag{11}$$

The output voltage in time domain is:

$$v_{o2}(t) = 4\cos\left(2t + 30^\circ\right) V \tag{12}$$

Based on superposition theorem and using (6) and (12):

$$v_o(t) = v_{o1}(t) + v_{o2}(t) = \frac{8\sqrt{10}}{5}\cos\left(t - 18.43^\circ\right) + 4\cos\left(2t + 30^\circ\right) V$$

Choice (4) is the answer.

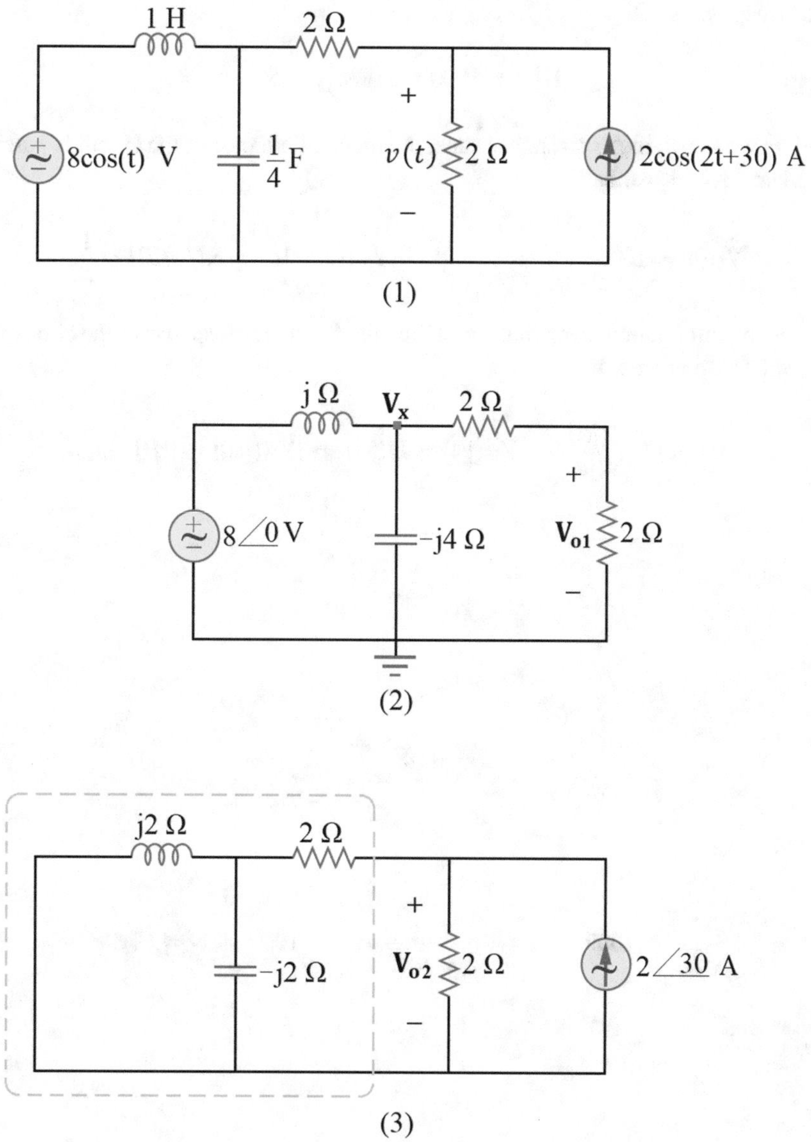

Figure 2.71 The circuit of solution of problem 2.73

2.74. In a series RLC circuit, the $-3\ dB$ bandwidth and the resonance frequency are determined by using the following relations:

$$BW_{Series} = \frac{R}{L}\ rad/sec \quad (1)$$

$$\omega_0 = \frac{1}{\sqrt{LC}}\ rad/sec \quad (2)$$

The $-3\ dB$ bandwidth and the resonance frequency of the circuit can be extracted from the graph shown in Figure 2.72.2.

$$BW_{Series} = 1100 - 900 = 200\ rad/sec \quad (3)$$

$$\omega_0 = 1000\ rad/sec \quad (4)$$

Moreover, as can be noticed from the graph of Figure 2.72.2, the amplitude of the frequency response of the circuit in the resonance frequency is:

$$|\mathbf{H}(\omega = 1000\ rad/sec)| = 50 \quad (5)$$

On the other hand, as we know, in the resonance state, the input admittance of the circuit is purely resistive. Therefore, for the circuit of Figure 2.72.1, we have:

$$\mathbf{Y_{in}}(\omega = 1000\ rad/sec) = \frac{1}{R} \Rightarrow |\mathbf{Y_{in}}(\omega = 1000\ rad/sec)| = \frac{1}{R} \quad (6)$$

In addition, based on the information given in the problem, the frequency response of the circuit is the same as the input admittance of the circuit. In other words:

$$\mathbf{H}(j\omega) = \frac{\mathbf{I}(j\omega)}{\mathbf{V}(j\omega)} \Rightarrow \mathbf{Y_{in}}(j\omega) = \mathbf{H}(j\omega) \Rightarrow |\mathbf{Y_{in}}(j\omega)| = |\mathbf{H}(j\omega)| \quad (7)$$

Solving (5)–(7):

$$\frac{1}{R} = 50 \Rightarrow R = 0.02\ \Omega \quad (8)$$

Solving (1) and (3):

$$\frac{R}{L} = 200 \xrightarrow{Using\ (8)} \frac{0.02}{L} = 200 \Rightarrow L = 0.1\ mH \quad (9)$$

Solving (2) and (4):

$$\frac{1}{\sqrt{LC}} = 1000 \xrightarrow{Using\ (9)} \frac{1}{\sqrt{10^{-4} \times C}} = 1000 \Rightarrow C = 10\ mF$$

Choice (4) is the answer.

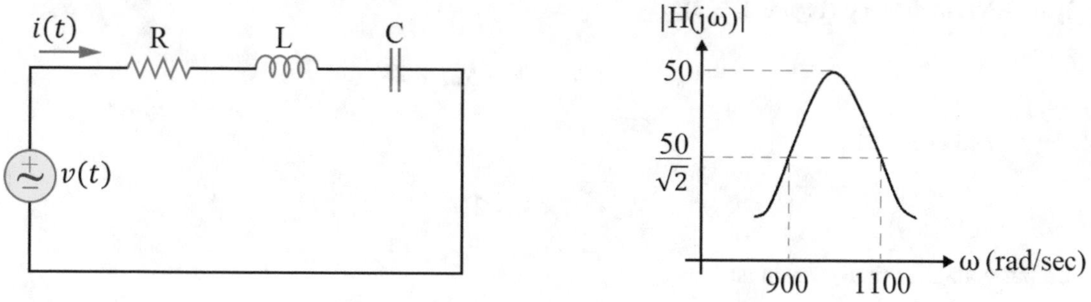

(1) (2)

Figure 2.72 The circuit of solution of problem 2.74

2.75. This problem needs to be solved by using superposition theorem, since the circuit is working with more than one frequency.

Part 1: The AC voltage source is left intact, but the DC current source is turned off (changed to an open circuit branch), as is shown in the circuit of Figure 2.73.2. The circuit is transferred to frequency domain based on the 1 rad/sec angular frequency of the voltage source. The phasor of the voltage of the voltage source, that is, $1\angle 0^\circ$ is defined as the reference phasor, where "$\angle$" is the symbol of phase angle. The impedances of the components are as follows:

$$\mathbf{Z}_{1\,\Omega} = 1\,\Omega \tag{1}$$

$$\mathbf{Z}_{1\,\mathrm{F}} = \frac{1}{j\omega C} = \frac{1}{j \times 1 \times 1} = -j\,\Omega \tag{2}$$

Applying voltage division formula for the capacitor (Figure 2.73.2):

$$\mathbf{V} = \frac{-j}{-j+1} \times (1\angle 0^\circ) = \frac{1}{1+j}\,V \tag{3}$$

Applying KVL in the mesh (Figure 2.73.2):

$$-\mathbf{V}_{o1} + \mathbf{V} + \mathbf{V} = 0 \Rightarrow \mathbf{V}_{o1} = 2\mathbf{V} \tag{4}$$

Solving (3) and (4):

$$\mathbf{V}_{o1} = (2 \times \frac{1}{1+j} = \frac{2}{1+j} = \sqrt{2}\angle 45^\circ\,V \tag{5}$$

Thus, the output voltage in time domain is:

$$v_{o1}(t) = \sqrt{2}\cos\left(t - 45^\circ\right)V \tag{6}$$

Part 2: The DC current source is kept in the circuit; however, the AC voltage source is shut down (short-circuited), as can be seen in the circuit of Figure 2.73.3. Moreover, as we know, a capacitor behaves like an open circuit in the steady-state condition if the circuit includes only DC power source.

The 1 A current of the current source passes through the 1 Ω resistor. Thus, by applying Ohm's law for the resistor (Figure 2.73.3), we can write:

$$\mathbf{V} = 1 \times 1 = 1\,\Omega \tag{7}$$

Applying KVL in the loop (Figure 2.73.3):

$$-\mathbf{V}_{o2} + \mathbf{V} + \mathbf{V} = 0 \Rightarrow \mathbf{V}_{o2} = 2\mathbf{V} \quad (8)$$

Solving (7) and (8):

$$\mathbf{V}_{o2} = 2 \times 1 = 2 \text{ V} \quad (9)$$

The output voltage in time domain is:

$$v_{o2}(t) = 2 \text{ V} \quad (10)$$

Based on superposition theorem and using (6) and (10), we can determine the total output voltage as follows:

$$v_o(t) = v_{o1}(t) + v_{o2}(t) = \sqrt{2}\cos\left(t - 45^\circ\right) + 2\ V$$

Choice (3) is the answer.

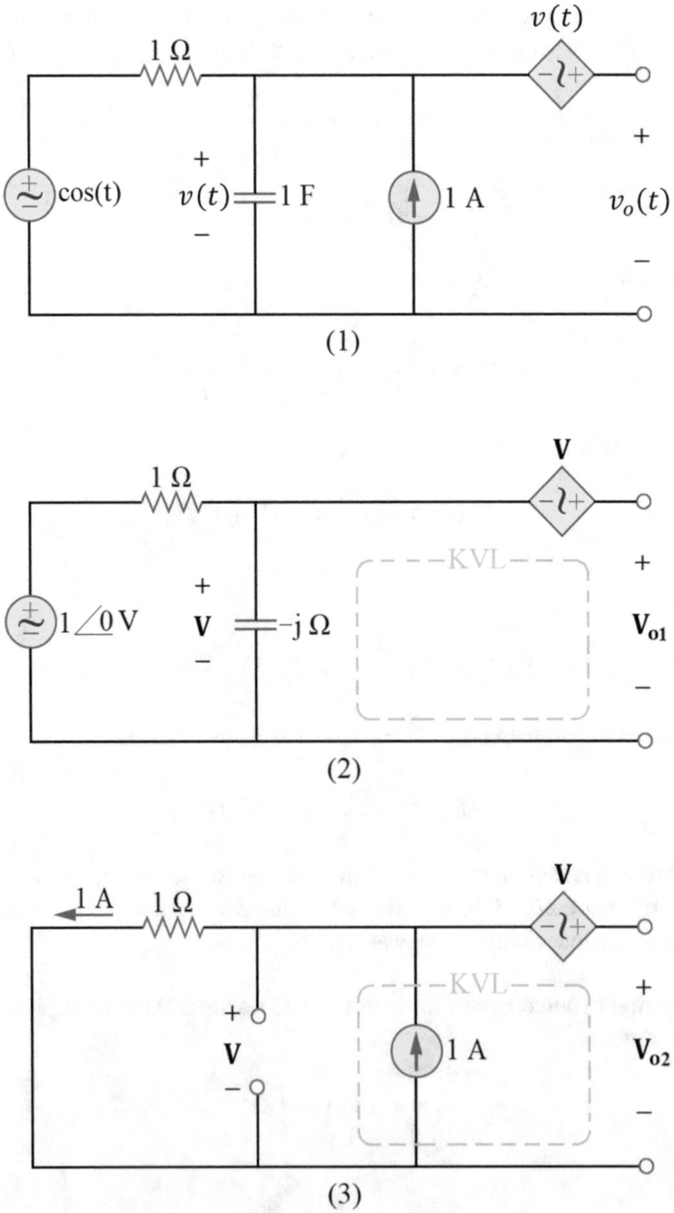

Figure 2.73 The circuit of solution of problem 2.75

2.76. Based on the information given in the problem, we know that:

$$P_1 = 25\ kW \tag{1}$$

$$Q_1 = 25\ kVAr \tag{2}$$

$$S_2 = 15\ kVA \tag{3}$$

$$PF_2 = \cos(\theta_2) = 0.8 \text{ Leading} \tag{4}$$

$$P_3 = 11\ kW \tag{5}$$

$$PF_3 = \cos(\theta_3) = 1 \tag{6}$$

By using power triangle relation and (3) and (4), we can calculate the active and reactive powers of the second load.

$$P_2 = S_2 \cos(\theta_2) = 15\ kW \times 0.8 = 12\ kW \tag{7}$$

$$Q_2 = -S_2 \sin(\theta_2) = -15\ kVAr \times \sqrt{1 - 0.8^2} = -15\ kVAr \times 0.6 = -9\ kVAr \tag{8}$$

In (8), minus sign was applied for the reactive power of the second load, since it has a leading power factor.

Based on (6), we can conclude that:

$$Q_3 = 0 \tag{9}$$

Now, we can calculate the total active and reactive powers of the circuit as follows:

$$\xrightarrow{Using\ (1),(5),(7)} P_{Total} = P_1 + P_2 + P_3 = 25\ kW + 12\ kW + 11\ kW = 48\ kW \tag{10}$$

$$\xrightarrow{Using\ (2),(8),(9)} Q_{Total} = Q_1 + Q_2 + Q_3 = 25\ kVAr + (-9\ kVAr) + 0 = 16\ kVAr \tag{11}$$

The total apparent power of the circuit can be calculated by using the following relation:

$$S_{Total} = \sqrt{(P_{Total})^2 + (Q_{Total})^2} = \sqrt{(48\ kW)^2 + (16\ kVAr)^2} = 16\sqrt{10}\ kVA \tag{12}$$

Finally, the total power factor of the circuit is:

$$PF_{Total} = \cos(\theta_{Total}) = \frac{P_{Total}}{S_{Total}} = \frac{48\ kW}{16\sqrt{10}\ kVA} = \frac{3}{\sqrt{10}} \text{ Lagging}$$

The total power factor of the circuit is lagging because $Q_{Total} > 0$. Choice (4) is the answer.

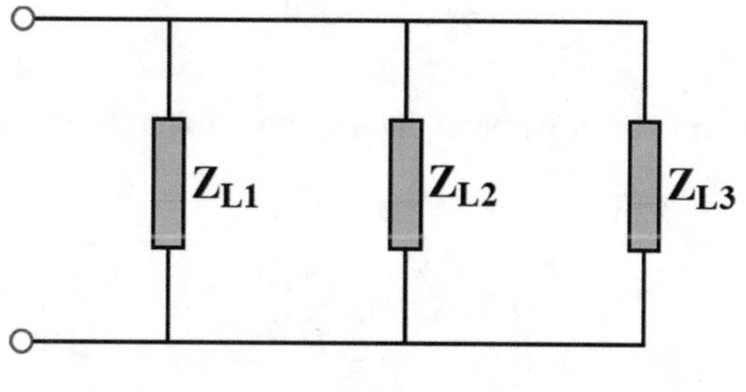

Figure 2.74 The circuit of solution of problem 2.76

2.77. Figure 2.75.2 shows the primary circuit in frequency domain. The phasor of cos(t), that is, $1\angle 0^\circ$ is defined as the reference phasor, where "$\angle$" is the symbol of phase angle. Thus, the phasor of the voltage of the voltage source is $A\angle 0^\circ$ *Volt*. The impedances of the components are as follows:

$$\mathbf{Z}_{1\,\Omega} = 1\,\Omega \tag{1}$$

$$\mathbf{Z}_{1\,\mathbf{H}} = j\omega L = j \times 1 \times 1 = j\,\Omega \tag{2}$$

$$\mathbf{Z}_{\mathbf{R}} = R \tag{3}$$

$$\mathbf{Z}_{1\,\mathbf{F}} = \frac{1}{j\omega C} = \frac{1}{j \times 1 \times 1} = -j\,\Omega \tag{4}$$

Based on maximum average power transfer theorem, to transfer the maximum average power to the network, the impedance of the network must be equal to the complex conjugate of the Thevenin impedance seen by the network.

$$\mathbf{Z}_{\mathbf{N}} = \mathbf{Z}_{\mathbf{Th}}^{\;*} \tag{5}$$

As can be noticed from the circuit of Figure 2.75.2, 1 Ω is achieved as the Thevenin impedance seen by the network.

Herein, the voltage source needs to be turned off (short-circuited). Therefore:

$$\mathbf{Z}_{\mathbf{Th}} = 1\,\Omega \tag{6}$$

Solving (5) and (6):

$$\mathbf{Z}_{\mathbf{N}} = 1^* = 1\,\Omega \tag{7}$$

The input impedance of the network can be directly calculated by using Figure 1.75.2.

$$\mathbf{Z}_{\mathbf{N}} = j + (R+1)||(-j2) = j + \frac{(R+1) \times (-j2)}{(R+1) + (-j2)} = j + \frac{-j2(R+1)}{(R+1) - j2}$$

$$\Rightarrow \mathbf{Z}_{\mathbf{N}} = j + \frac{4(R+1) - j2(R+1)^2}{(R+1)^2 + 4} = \frac{4(R+1)}{(R+1)^2 + 4} + j\left(1 - \frac{2(R+1)^2}{(R+1)^2 + 4}\right) \tag{8}$$

Solving (7) and (8):

$$\frac{4(R+1)}{(R+1)^2 + 4} = 1 \Rightarrow 4(R+1) = R^2 + 2R + 1 + 4 \Rightarrow R^2 - 2R + 1 = 0 \Rightarrow (R-1)^2 = 0$$

$$\Rightarrow R = 1\,\Omega \tag{9}$$

Based on the information given in the problem:

$$P_{N,max} = \frac{1}{8}\,W \tag{10}$$

Based on maximum average power transfer theorem, the maximum average power of the network can be calculated by using following relation:

$$P_{N,max} = \frac{1}{8}\frac{|\mathbf{V}_{\mathbf{Th}}|^2}{R_N} \tag{11}$$

where $\mathbf{V_{Th}}$ and R_N are the phasor (peak value) of the Thevenin voltage seen by the network and the resistance of the impedance of the network.

By using the circuit of Figure 2.75.2, the open circuit voltage or the Thevenin voltage seen by the network is:

$$\mathbf{V_{Th}} = \mathbf{V_{oc}} = A\angle 0° = A\, Volt \tag{12}$$

Solving (7), (9)–(12):

$$\frac{1}{8} = \frac{1}{8}\frac{A^2}{1} \Rightarrow A = 1$$

Choice (2) is the answer.

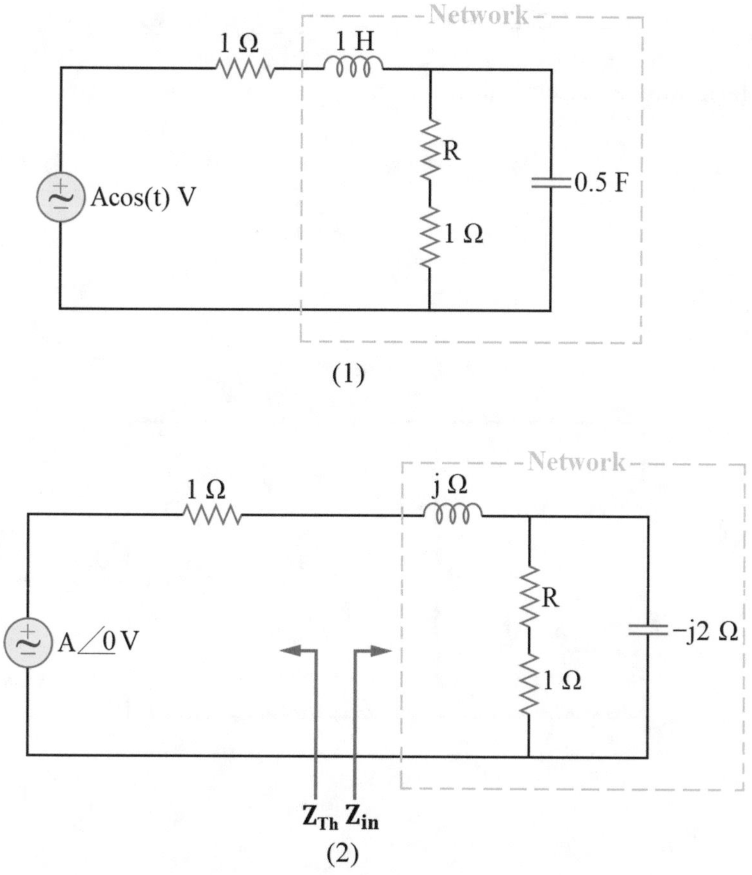

Figure 2.75 The circuit of solution of problem 2.77

2.78. The circuit of Figure 2.76.1 is in frequency domain. To determine the Thevenin equivalent circuit seen from terminal a–b, we can connect a test source (e.g. voltage source with the voltage and current of $\mathbf{V_1}$ and $\mathbf{I_1}$) to the terminal and find the relation below, where $\boldsymbol{\alpha} = \mathbf{Z_{Th}}$ and $\boldsymbol{\beta} = \mathbf{V_{Th}}$. In this method, we do not need to turn off the independent power sources.

$$\mathbf{V_1} = \boldsymbol{\alpha}\mathbf{I_1} + \boldsymbol{\beta} \tag{1}$$

Applying KCL in the indicated upper-left node:

$$-\mathbf{I} + \mathbf{I_2} - \mathbf{I_1} = 0 \Rightarrow \mathbf{I} = \mathbf{I_2} - \mathbf{I_1} \tag{2}$$

Applying KVL in the left-side mesh:

$$-1\angle 90^\circ + \mathbf{I} + \mathbf{I_2} \times (-j) = 0 \xrightarrow{Using\ (2)} -1\angle 90^\circ + \mathbf{I_2} - \mathbf{I_1} - j\mathbf{I_2} = 0$$

$$\Rightarrow -1\angle 90^\circ + (1-j)\mathbf{I_2} - \mathbf{I_1} = 0 \Rightarrow \mathbf{I_2} = \frac{\mathbf{I_1} + 1\angle 90^\circ}{1-j} \tag{3}$$

Applying KVL in the right-side mesh:

$$-\mathbf{V_1} + \mathbf{I_2} + \mathbf{I_2} \times (-j) = 0 \Rightarrow -\mathbf{V_1} + (1-j)\mathbf{I_2} = 0 \tag{4}$$

Solving (3) and (4):

$$-\mathbf{V_1} + (1-j)\frac{\mathbf{I_1} + 1\angle 90^\circ}{1-j} = 0 \Rightarrow -\mathbf{V_1} + \mathbf{I_1} + 1\angle 90^\circ = 0 \Rightarrow \mathbf{V_1} + \mathbf{I_1} + 1\angle 90^\circ \tag{5}$$

By comparing (1) and (5), we can conclude that:

$$\mathbf{Z_{Th}} = 1\ \Omega$$

$$\mathbf{V_{Th}} = (1\angle 90^\circ)\ V = j\ V$$

Choice (1) is the answer.

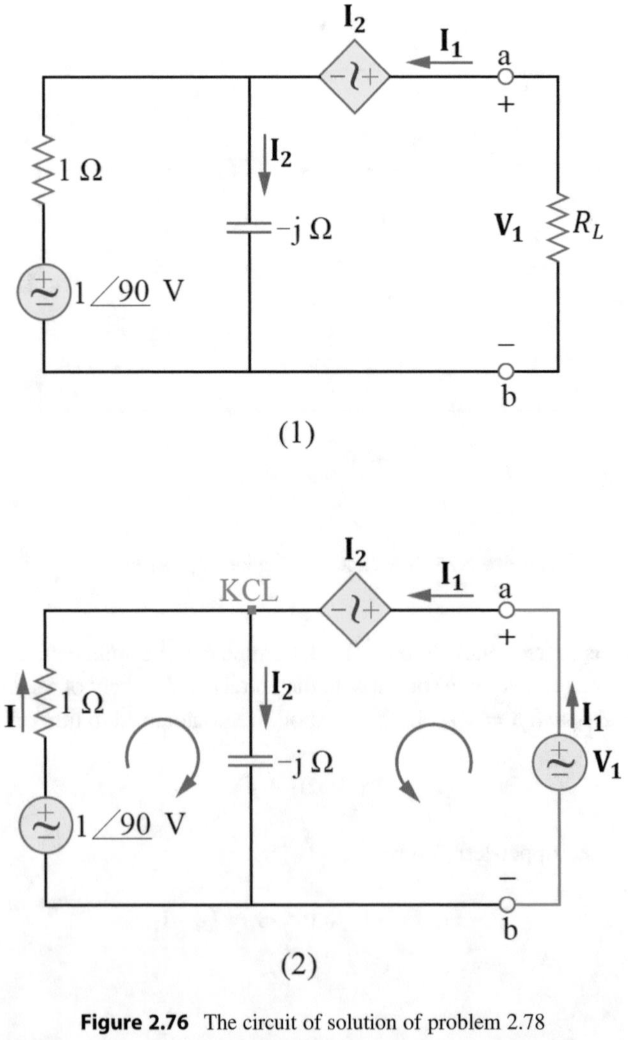

Figure 2.76 The circuit of solution of problem 2.78

2.79. Based on the information given in the problem, the power factor of the indicated part of the circuit is one; hence, the circuit is in the resonance state. Therefore, the imaginary part of the input impedance or the input admittance of that section of the circuit is zero.

$$Im\{\mathbf{Y_{in}}\} = 0 \quad (1)$$

The circuit is illustrated in frequency domain in Figure 2.77.2. In this problem, the phase angle of the current of the current source is defined as the reference phase angle. Therefore, its phasor is $(2\sqrt{2}\angle 0^\circ)\,A$ or $2\sqrt{2}\,A$. Herein, "$\angle$" is the symbol of phase angle. The impedances of the components are presented in the following:

$$\mathbf{Z_{250\,\Omega}} = 250\,\Omega \quad (2)$$

$$\mathbf{Z_{5\,H}} = j\omega L = j5\omega\,\Omega \quad (3)$$

$$\mathbf{Z_{5\,\Omega}} = 5\,\Omega \quad (4)$$

$$\mathbf{Z_{4\,mF}} = \frac{1}{j\omega C} = \frac{1}{j\omega \times 4 \times 10^{-3}} = -j\frac{10^3}{4\omega}\,\Omega \quad (5)$$

The input admittance of the circuit is:

$$\mathbf{Y_{in}} = \frac{1}{250} + \frac{1}{5+j5\omega} + \frac{1}{-j\frac{10^3}{4\omega}} = \frac{1}{250} + \frac{5-j5\omega}{25+25\omega^2} + j4\omega \times 10^{-3}$$

$$\Rightarrow \mathbf{Y_{in}} = \frac{1}{250} + \frac{5}{25+25\omega^2} + j\omega\left(-\frac{5}{25+25\omega^2} + 4 \times 10^{-3}\right) \quad (6)$$

Solving (1) and (6) results in the resonance frequency.

$$-\frac{5}{25+25\omega^2} + 4 \times 10^{-3} = 0 \Rightarrow 0.1 + 0.1\omega^2 = 5 \Rightarrow \omega^2 = 49 \Rightarrow \omega_0 = 7\ rad/sec \quad (7)$$

By solving (6) and (7), we can calculate the input resistance of the circuit as follows:

$$\mathbf{Y_{in}} = \frac{1}{250} + \frac{5}{25 + 25 \times 7^2} = 0.008\,\Omega^{-1} \quad (8)$$

$$\Rightarrow \mathbf{Z_{in}} = \frac{1}{\mathbf{Y_{in}}} = 125\,\Omega \Rightarrow R_{in} = 125\,\Omega \quad (9)$$

The average power of the indicated part of the circuit can be calculated as follows:

$$P = R_{in} I_{rms}^2 = 125 \times \left(\frac{2\sqrt{2}}{\sqrt{2}}\right)^2 = 500\ W$$

Choice (4) is the answer.

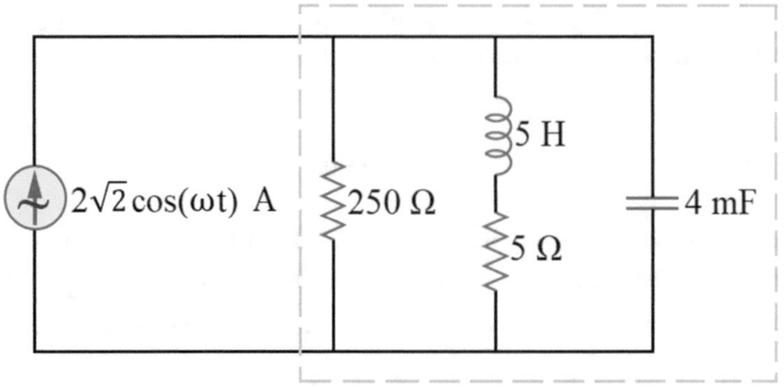

(1)

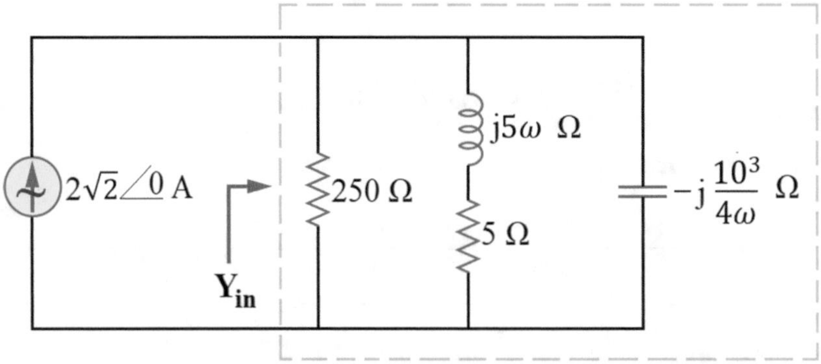

(2)

Figure 2.77 The circuit of solution of problem 2.79

2.80. The primary circuit is illustrated in frequency domain in Figure 2.78.2. In this problem, the phase angle of the voltage of the voltage source is defined as the reference phase angle. Therefore, its phasor is $(2\angle 0^\circ)V$ or 2 V. Herein, "$\angle$" is the symbol of phase angle. The impedances of the components are presented in the following:

$$\mathbf{Z_L} = j\omega L = j2L\,\Omega \tag{1}$$

$$\mathbf{Z_C} = \frac{1}{j\omega C} = \frac{1}{j2C} \tag{2}$$

$$\mathbf{Z_{1\,\Omega}} = 1\,\Omega \tag{3}$$

$$\mathbf{Z_{\frac{1}{2}\,H}} = j\omega L = j \times 2 \times \frac{1}{2} = j\,\Omega \tag{4}$$

$$\mathbf{Z_{2\,\Omega}} = 2\,\Omega \tag{5}$$

$$\mathbf{Z_{\frac{1}{4}\,F}} = \frac{1}{j\omega C} = \frac{1}{j \times 2 \times \frac{1}{4}} = -j2\,\Omega \tag{6}$$

The average power of the 1 Ω resistor can be calculated as follows (See Figure 2.78.2):

$$P_{1\Omega} = \frac{1}{2} R_{1\Omega} |\mathbf{I_1}|^2 = \frac{1}{2} |\mathbf{I_1}|^2 \tag{7}$$

Likewise, the average power of the voltage source, which is equal to sum of the average powers of the resistors, can be calculated as follows:

$$P_S = \frac{1}{2}R_{1\Omega}|\mathbf{I}_1|^2 + \frac{1}{2}R_{2\Omega}|\mathbf{I}_2|^2 = \frac{1}{2}|\mathbf{I}_1|^2 + |\mathbf{I}_2|^2 \tag{8}$$

By applying Ohm's law for the left-side vertical branch of the circuit, we can write (See Figure 2.78.2):

$$|\mathbf{I}_1| = \left|\frac{\mathbf{V}}{1+j}\right| = \frac{|\mathbf{V}|}{\sqrt{2}} \tag{9}$$

By applying Ohm's law for the right-side vertical branch of the circuit, we can write (See Figure 2.78.2):

$$|\mathbf{I}_2| = \left|\frac{\mathbf{V}}{2-j2}\right| = \frac{|\mathbf{V}|}{2\sqrt{2}} \tag{10}$$

Solving (9) and (10):

$$|\mathbf{I}_1| = 2|\mathbf{I}_2| \tag{11}$$

Therefore, the requested ratio can be calculated by solving (7), (8), and (11) as follows:

$$\frac{P_{1\Omega}}{P_S} = \frac{\frac{1}{2}|\mathbf{I}_1|^2}{\frac{1}{2}|\mathbf{I}_1|^2 + |\mathbf{I}_2|^2} = \frac{\frac{1}{2}\times 4|\mathbf{I}_2|^2}{\frac{1}{2}\times 4|\mathbf{I}_2|^2 + |\mathbf{I}_2|^2} = \frac{2}{2+1} = \frac{2}{3}$$

Choice (2) is the answer.

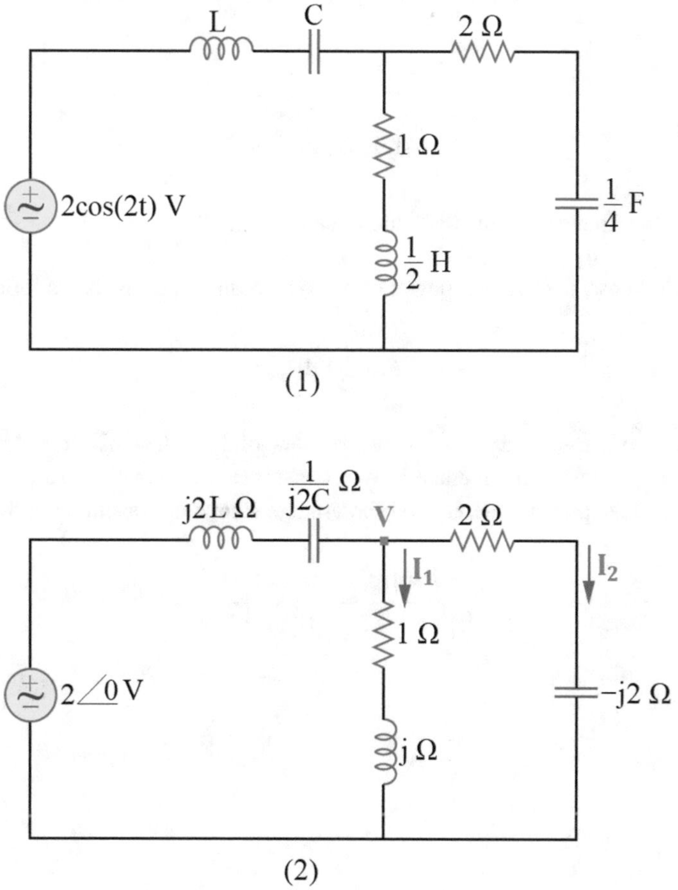

Figure 2.78 The circuit of solution of problem 2.80

2.81. Based on maximum average power transfer theorem, to transfer the maximum average power to R, the relation below must be held, where $\mathbf{Z_{Th}}$ is the Thevenin impedance seen by R:

$$R = |\mathbf{Z_{Th}}| \tag{1}$$

The circuit is shown in frequency domain in Figure 2.79.2, while it is intended to determine the Thevenin equivalent circuit seen by the R. Herein, the phasor of the voltage of the voltage source has been defined as the reference phasor. The impedances of the components are presented in the following:

$$\mathbf{Z}_{2\sqrt{2}\ \mathbf{H}} = j\omega L = j \times 1 \times 2\sqrt{2} = j2\sqrt{2}\ \Omega \tag{2}$$

$$\mathbf{Z}_{1\ \Omega} = 1\ \Omega \tag{3}$$

$$\mathbf{Z}_{1\ \mathbf{H}} = j\omega L = j \times 1 \times 1 = j\ \Omega \tag{4}$$

$$\mathbf{Z}_{1\ \mathbf{F}} = \frac{1}{j\omega C} = \frac{1}{j \times 1 \times 1} = -j\ \Omega \tag{5}$$

The Thevenin impedance, seen by the R, is calculated as follows:

$$\mathbf{Z_{Th}} = \mathbf{Z_{in}} = j2\sqrt{2} + (1+j)||(1-j) = j2\sqrt{2} + \frac{(1+j)\times(1-j)}{(1+j)+(1-j)} = \left(1 + j2\sqrt{2}\right)\ \Omega \tag{6}$$

In (6), we assumed that we have turned off the independent voltage source of Figure 2.79.2. In addition, the Thevenin voltage can be calculated by applying KVL in the left-side mesh of the circuit of Figure 2.79.2 as follows:

$$-1\angle 90^\circ + \mathbf{V_{oc}} + 0 = 0 \Rightarrow \mathbf{V_{Th}} = \mathbf{V_{oc}} = (1\angle 90^\circ)V \tag{7}$$

Solving (1) and (6):

$$R = \left|1 + j2\sqrt{2}\right| = 3\ \Omega \tag{8}$$

Figure 2.79.3 illustrates the Thevenin equivalent circuit seen by the R.

On the other hand, as we know, the average power of a resistor can be calculated as follows:

$$P = \frac{1}{2}R|\mathbf{I}|^2 \tag{9}$$

Now, let us define $P_{Z_{Th}}$ and P_{Source} as the average powers of the Thevenin impedance and the voltage source, respectively. Herein, $P_{Z_{Th}}$ includes the average power consumed by the two 1 Ω resistors of the main circuit. In addition, P_{Source} is the average power generated by the voltage source or consumed in the whole circuit. Therefore:

$$\frac{P_{Z_{Th}}}{P_{Source}} = \frac{\frac{1}{2}\times 1 \times |\mathbf{I}|^2}{\frac{1}{2}\times 1 \times |\mathbf{I}|^2 + \frac{1}{2}\times 3 \times |\mathbf{I}|^2} = \frac{1}{1+3} = 0.25 = 25\ \%$$

Choice (1) is the answer.

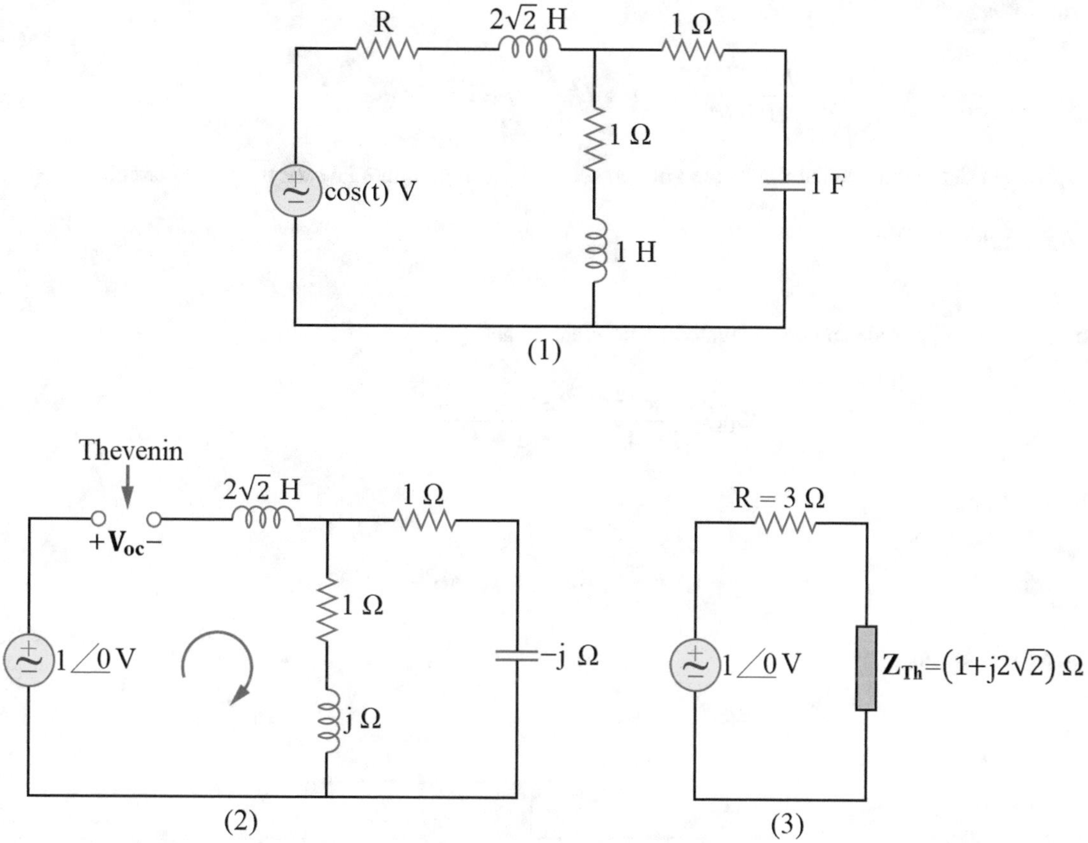

Figure 2.79 The circuit of solution of problem 2.81

2.82. Figure 2.80.3 shows the circuit in frequency domain. Herein, the phasor of sin(2*t*), that is, $1\angle 0°$ is defined as the reference phasor, where "$\angle$" is the symbol of phase angle. Therefore, the phasor of the voltage of the voltage source is $40\angle 0°$ *V* or 40 *V*. The impedances of the components are as follows:

$$\mathbf{Z}_{0.5\ \mathbf{F}} = \frac{1}{j\omega C} = \frac{1}{j \times 2 \times 0.5} = -j\ \Omega \tag{1}$$

$$\mathbf{Z}_{1\ \mathbf{H}} = j\omega L = j2L\ \Omega \tag{2}$$

$$\mathbf{Z}_{2\ \Omega} = 2\ \Omega \tag{3}$$

In order to have $v_s(t)$ and $i_s(t)$ in phase in sinusoidal steady state, the circuit must be in resonance state, and the Thevenin impedance seen by the voltage source must be purely resistive. In other words:

$$Im\{\mathbf{Z}_{\mathbf{Th}}\} = 0 \tag{4}$$

Therefore, we need to determine the Thevenin impedance seen by the voltage source (see Figure 2.80.2).

$$\mathbf{Z}_{\mathbf{Th}} = -j + (2)||(j2L) = -j + \frac{(2) \times (j2L)}{(2) + (j2L)} = -j + \frac{j2L}{1 + jL} = -j + \frac{2L^2 + j2L}{1 + L^2}$$

$$\Rightarrow \mathbf{Z}_{\mathbf{Th}} = \frac{-j - jL^2 + 2L^2 + j2L}{1 + L^2} = \frac{2L^2}{1 + L^2} + j\frac{2L - L^2 - 1}{1 + L^2} \tag{5}$$

Solving (4) and (5):

$$\frac{2L - L^2 - 1}{1 + L^2} = 0 \Rightarrow L^2 - 2L + 1 = 0 \Rightarrow L = 1\ H \quad (6)$$

Now, we can determine the current of the voltage source by using Ohm's law for the whole circuit.

$$\mathbf{I_s} = \frac{40}{\mathbf{Z_{Th}}} \quad (7)$$

The Thevenin impedance can be calculated by solving (5) and (6):

$$\mathbf{Z_{Th}} = \frac{2 \times 1^2}{1 + 1^2} + j\frac{2 \times 1 - 1^2 - 1}{1 + 1^2} = 1\ \Omega \quad (8)$$

Solving (7) and (8):

$$\mathbf{I_s} = \frac{40}{1} = 40\ A \Rightarrow |\mathbf{I_s}| = 40\ A$$

Choice (3) is the answer.

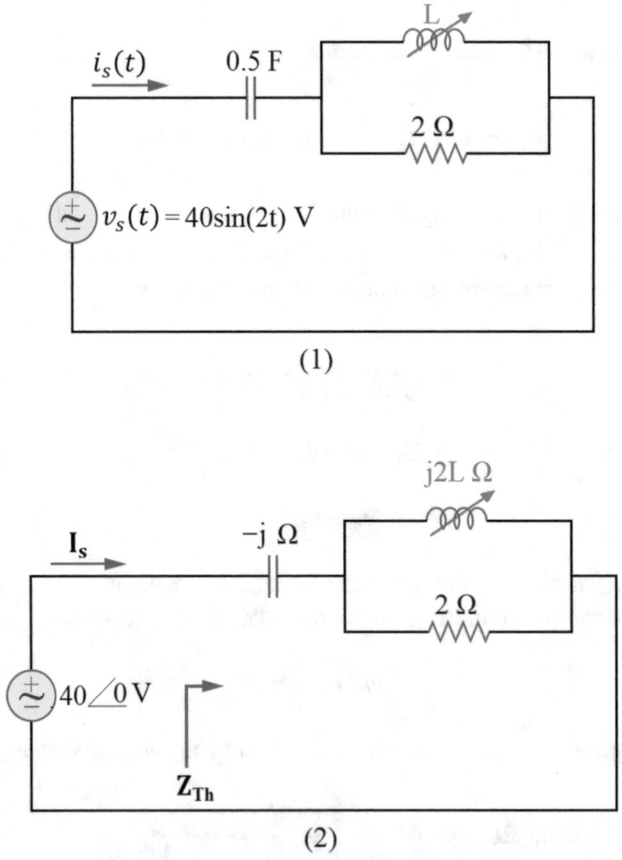

Figure 2.80 The circuit of solution of problem 2.82

2.83. The primary circuit is illustrated in frequency domain in Figure 2.81.2. The impedances of the components are as follows:

$$\mathbf{Z}_{2\sqrt{2}\ \mathbf{H}} = j\omega L = j2\sqrt{2}\omega\ \Omega \tag{1}$$

$$\mathbf{Z}_{1\ \Omega} = 1\ \Omega \tag{2}$$

$$\mathbf{Z}_{2\ \mathbf{H}} = j\omega L = j2\omega\ \Omega \tag{3}$$

$$\mathbf{Z}_{0.5\ \mathbf{F}} = \frac{1}{j\omega C} = \frac{1}{j0.5\omega}\ \Omega \tag{4}$$

$$\mathbf{Z}_{\mathbf{R}} = R \tag{5}$$

Based on the information given in the problem, the network is in resonance state. Hence, its input impedance is purely resistive. In other words:

$$Im\{\mathbf{Z}_{\mathbf{in}}\} = 0 \tag{6}$$

The input impedance of the network can be calculated from the circuit of Figure 2.81.2 as follows:

$$\mathbf{Z}_{\mathbf{in}} = j2\omega + \frac{1}{j0.5\omega} + R = R + j\left(2\omega - \frac{2}{\omega}\right) \tag{7}$$

By solving (6) and (7), we can calculate the resonance frequency.

$$2\omega - \frac{2}{\omega} = 0 \Rightarrow \omega_0 = 1\ rad/\ sec \tag{8}$$

Based on the information given in the problem, the maximum average power of the network is 3 W.

$$P_N = 3\ W \tag{9}$$

To transfer the maximum average power to the network, the input impedance of the network must match with the impedance seen by itself. Herein, due to the resonance state, the input impedance of the network is purely resistive and equal to R as can be noticed from the solution of (6) and (7). Therefore, to absorb the maximum average power, the relation below must be held. See Figure 2.81.3 that shows the circuit in the resonance frequency, that is, $\omega = \omega_0 = 1\ rad/sec$.

$$R = \left|\mathbf{Z}_{\mathbf{Seen\ by\ N}}\right| \tag{10}$$

The impedance seen by the network can be calculated by using the circuit of Figure 2.81.3. Herein, the independent current source needs to be turned off (open circuit).

$$\left|\mathbf{Z}_{\mathbf{Seen\ by\ N}}\right| = \left|1 + j2\sqrt{2}\right| = \sqrt{1^2 + \left(2\sqrt{2}\right)^2} = \sqrt{9} = 3\ \Omega \tag{11}$$

Solving (10) and (11)

$$R = 3\,\Omega \tag{12}$$

The updated circuit is illustrated in Figure 2.81.4. On the other hand, we know that the average power of the resistor can be calculated by using the following relation:

$$P_N = P_R = RI_{R,rms}{}^2 \tag{13}$$

Solving (9), (12), and (13):

$$3 = 3I_{R,rms}{}^2 \Rightarrow I_{R,rms} = 1\,A \tag{14}$$

Applying Ohm's law for the series connection of the 3 Ω and 1 Ω resistors in the circuit of Figure 2.81.4 will give us **V** indicated in the circuit:

$$\mathbf{V} = (1+3) \times I_{R,rms} \xrightarrow{\text{Using (14)}} \mathbf{V} = 4 \times 1 = 4\,V \tag{15}$$

Applying current division relation in the circuit of Figure 2.81.4:

$$I_{R,rms} = \frac{j2\sqrt{2}}{j2\sqrt{2}+4} I_{S,rms} \xrightarrow{\text{Using (14)}} I_{S,rms} = \frac{j2\sqrt{2}+4}{j2\sqrt{2}} \times 1 = \left(1 - j\sqrt{2}\right) A \tag{16}$$

Applying complex power relation for the current source:

$$\mathbf{S_S} = \mathbf{V_{S,rms}}\mathbf{I_{S,rms}}^* = -\mathbf{V}\mathbf{I_{S,rms}}^* \xrightarrow{\text{Using (15), (16)}} \mathbf{S_S} = -4 \times \left(1 - j\sqrt{2}\right)^* = \left(-4 - j4\sqrt{2}\right) VA \tag{17}$$

The minus sign has been used in (17) to apply the associate reference direction for the current source.

From (17), we can conclude that:

$$Q_S = -4\sqrt{2}\ VAr$$

Choice (2) is the answer.

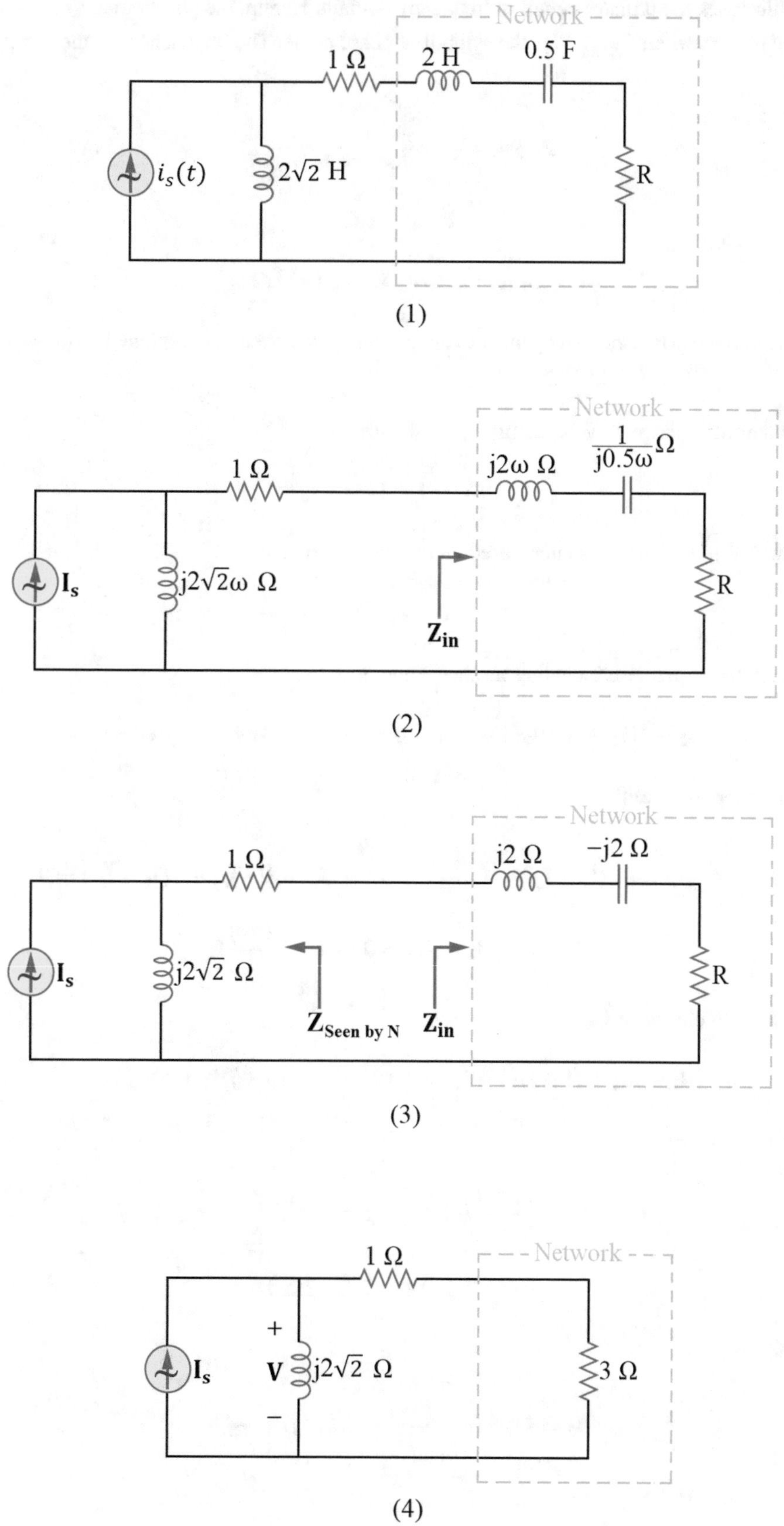

Figure 2.81 The circuit of solution of problem 2.83

2.84. Figure 2.82.2 illustrates the primary circuit in frequency domain. Herein, the phasor of sin(t), that is, $1\angle 0°$ is defined as the reference phasor, where "$\angle$" is the symbol of phase angle. The impedances of the components are presented as follows:

$$\mathbf{Z}_{1\ \mathbf{F}} = \frac{1}{j\omega C} = \frac{1}{j \times 1 \times 1} = -j\,\Omega \tag{1}$$

$$\mathbf{Z}_{1\ \Omega} = 1\,\Omega \tag{2}$$

$$\mathbf{Z}_{1\ \mathbf{H}} = j\omega L = j \times 1 \times 1 = j\,\Omega \tag{3}$$

To calculate the average power of the dependent voltage source, we need to determine its voltage and current. To solve this problem, mesh analysis is suggested.

Defining the unknown voltage of $\mathbf{V}$ based on the mesh current:

$$\mathbf{V} = -1 \times \mathbf{I}_2 = -\mathbf{I}_2 \tag{4}$$

Defining the current of the current source based on the mesh currents:

$$1 = \mathbf{I}_3 - \mathbf{I}_2 \tag{5}$$

Applying KVL in the supermesh including meshes 2 and 3:

$$\mathbf{I}_2 + 1(\mathbf{I}_2 - \mathbf{I}_1) + (1+j)\mathbf{I}_3 = 0 \Rightarrow -\mathbf{I}_1 + 2\mathbf{I}_2 + (1+j)\mathbf{I}_3 = 0 \tag{6}$$

Applying KVL in the top mesh:

$$-j\mathbf{I}_1 + 3\mathbf{V} + 1(\mathbf{I}_1 - \mathbf{I}_2) = 0 \xrightarrow{Using\ (4)} -j\mathbf{I}_1 + 3(-\mathbf{I}_2) + 1(\mathbf{I}_1 - \mathbf{I}_2) = 0$$

$$\Rightarrow (1-j)\mathbf{I}_1 - 4\mathbf{I}_2 = 0 \Rightarrow \mathbf{I}_2 = \frac{1-j}{4}\mathbf{I}_1 \tag{7}$$

Solving (5) and (6) to eliminate $\mathbf{I}_3$:

$$-\mathbf{I}_1 + 2\mathbf{I}_2 + (1+j)(1+\mathbf{I}_2) = 0 \Rightarrow -\mathbf{I}_1 + (3+j)\mathbf{I}_2 + 1 + j = 0 \tag{8}$$

Solving (7) and (8):

$$-\mathbf{I}_1 + (3+j)\left(\frac{1-j}{4}\mathbf{I}_1\right) + 1 + j = 0 \Rightarrow \mathbf{I}_1\left(\frac{-j}{2}\right) = -1 - j$$

$$\Rightarrow \mathbf{I}_1 = 2(1\text{-}j)A = (2\sqrt{2}\angle{-45°})A \tag{9}$$

Solving (7) and (9):

$$\mathbf{I}_2 = \frac{1-j}{4}\mathbf{I}_1 = \frac{1-j}{4} \times 2(1-j) = -j\,A \tag{10}$$

Solving (4) and (10):

$$\mathbf{V} = -\mathbf{I}_2 = j\,V = (1\angle 90°)\,V \tag{11}$$

Now, we can calculate the average power of the dependent voltage source as follows:

$$P = \frac{1}{2}|3\mathbf{V}||\mathbf{I}_1|\cos(\theta_V - \theta_I) = \frac{1}{2}|3\angle 90^\circ||2\sqrt{2}\angle -45^\circ|\cos(90-(-45))$$

$$\Rightarrow P = \frac{1}{2}\times 3\times 2\sqrt{2}\cos(135) \Rightarrow P = -3\ W$$

Since negative value was achieved for the average power, the dependent voltage source is generating that power. Choice (1) is the answer.

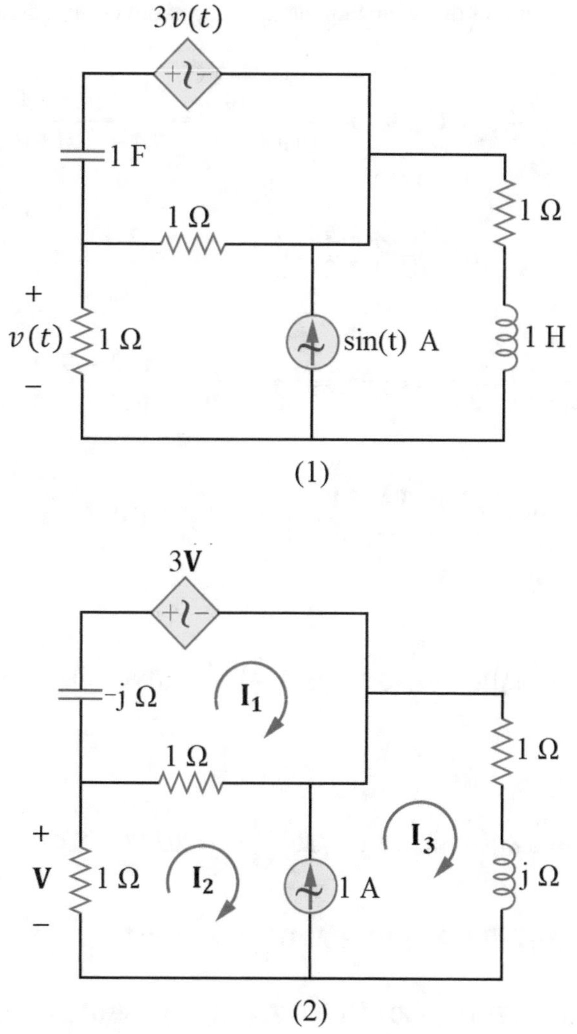

Figure 2.82 The circuit of solution of problem 2.84

2.85. The complex power of a component can be calculated by using the following relations, where **V** and **I** are the phasors (peak value) of the voltage and the current of the component, respectively.

$$\mathbf{S} = \frac{1}{2}\mathbf{V}\mathbf{I}^* = \frac{1}{2}\frac{|\mathbf{V}|^2}{\mathbf{Z}^*} = \frac{1}{2}\mathbf{Z}|\mathbf{I}|^2 \quad (1)$$

Moreover, the magnitude of the complex power of a component is defined as its apparent power.

$$S = |\mathbf{S}| \tag{2}$$

In addition, the real part and the imaginary part of the complex power of a component are defined as the average power (active power) and the reactive power of the component, respectively.

$$P = Re\{\mathbf{S}\} \tag{3}$$

$$Q = Im\{\mathbf{S}\} \tag{4}$$

Now, we need to calculate the current of each component. By applying current division formula, we can write:

$$\mathbf{I}_1 = \frac{\mathbf{Z}_2 + \mathbf{Z}_3}{\mathbf{Z}_1 + \mathbf{Z}_2 + \mathbf{Z}_3} \times (1\angle 0^\circ) = \frac{(0.4 - j0.2) + (0.2 + j0.4)}{(0.3 + j0.1) + (0.4 - j0.2) + (0.2 + j0.4)}$$

$$\mathbf{I}_1 = \frac{0.6 + j0.2}{0.9 + j0.3} = \frac{2}{3}\ A \Rightarrow |\mathbf{I}_1| = \frac{2}{3}\ A \tag{5}$$

$$\mathbf{I}_2 = \mathbf{I}_3 = \frac{\mathbf{Z}_1}{\mathbf{Z}_1 + \mathbf{Z}_2 + \mathbf{Z}_3} \times (1\angle 0^\circ) = \frac{(0.3 + j0.1)}{(0.3 + j0.1) + (0.4 - j0.2) + (0.2 + j0.4)}$$

$$\mathbf{I}_2 = \mathbf{I}_3 = \frac{0.3 + j0.1}{0.9 + j0.3} = \frac{1}{3}\ A \Rightarrow |\mathbf{I}_2| = |\mathbf{I}_3| = \frac{1}{3}\ A \tag{6}$$

Solving (1) and (5):

$$\mathbf{S}_1 = \frac{1}{2}\mathbf{Z}_1|\mathbf{I}_1|^2 = \frac{1}{2}(0.3 + j0.1)\left(\frac{2}{3}\right)^2 = (0.066 + j0.022)\ VA \tag{7}$$

Solving (1) and (6):

$$\mathbf{S}_2 = \frac{1}{2}\mathbf{Z}_2|\mathbf{I}_2|^2 = \frac{1}{2}(0.4 - j0.2)\left(\frac{1}{3}\right)^2 = (0.022 - j0.011)\ VA \tag{8}$$

$$\mathbf{S}_3 = \frac{1}{2}\mathbf{Z}_3|\mathbf{I}_3|^2 = \frac{1}{2}(0.2 + j0.4)\left(\frac{1}{3}\right)^2 = (0.011 + j0.022)\ VA \tag{9}$$

Choice (1) is correct, since $|\mathbf{S}_2| = |0.022 - j0.011| = 0.024\ VA$, $|\mathbf{S}_3| = |0.011 + j0.022| = 0.024\ VA$, and $|\mathbf{S}_2| = |\mathbf{S}_3|$.

Choice (2) is correct, since $P_2 = 0.022\ W$, $P_3 = 0.011\ W$, and $P_2 = 2P_3$.

Choice (3) is correct, since $Q_1 = 0.022\ W$, $Q_2 = -0.011\ W$, and $Q_1 = -2Q_2$.

Choice (4) is incorrect, since $Q_3 = 0.022\ W$, $Q_1 = 0.022\ W$, and $Q_3 \neq 4Q_1$.

Choice (4) is the answer.

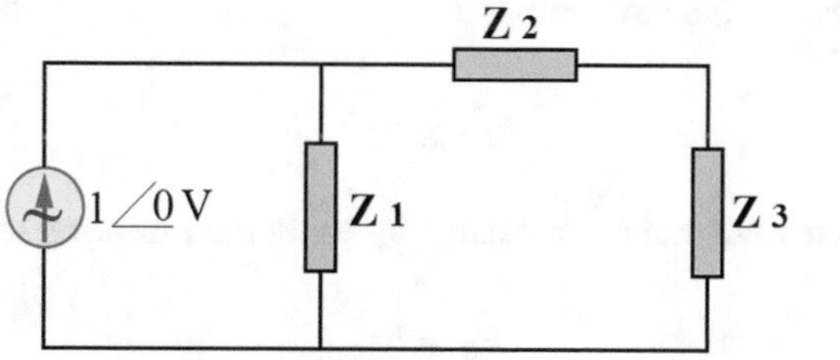

Figure 2.83 The circuit of solution of problem 2.85

2.86. The circuit is shown in frequency domain in Figure 2.84.2 for the angular frequency of $\omega = 1\ rad/sec$. The impedances of the components are presented in the following:

$$\mathbf{Z_R} = R\,\Omega \tag{1}$$

$$\mathbf{Z_{1\,\Omega}} = 1\,\Omega \tag{2}$$

$$\mathbf{Z_{2\,H}} = j\omega L = j \times 1 \times 2 = j2\,\Omega \tag{3}$$

$$\mathbf{Z_{1\,F}} = \frac{1}{j\omega C} = \frac{1}{j \times 1 \times 1} = -j\,\Omega \tag{4}$$

To transfer the maximum average power to R, the relation below must be held, where $\mathbf{Z_{Th}}$ is the Thevenin impedance seen by R.

$$R = |\mathbf{Z_{Th}}| \tag{5}$$

To calculate the Thevenin impedance seen by R, we need to replace it with a test source and determine the value of $\frac{\mathbf{V_t}}{\mathbf{I_t}}$, as can be seen in the following and in Figure 2.84.3:

Defining $\mathbf{I_1}$ based on the mesh currents:

$$\mathbf{I_1} = \mathbf{I_t} - \mathbf{I_2} \tag{6}$$

Applying KVL in the right-side mesh:

$$(-j)(\mathbf{I_2} - \mathbf{I_t}) + (j2)\mathbf{I_2} - \mathbf{I_1} = 0 \Rightarrow \mathbf{I_2} = \frac{\mathbf{I_1} - j\mathbf{I_t}}{j} = -j\mathbf{I_1} - \mathbf{I_t} \xrightarrow{Using\ (6)} \mathbf{I_2} = -j(\mathbf{I_t} - \mathbf{I_2}) - \mathbf{I_t}$$

$$\Rightarrow (1-j)\mathbf{I_2} = (-1-j)\mathbf{I_t} \Rightarrow \mathbf{I_2} = \frac{(-1-j)}{1-j}\mathbf{I_t} \tag{7}$$

Applying KVL in the left-side mesh:

$$-\mathbf{V_t} + \mathbf{I_t} + (-j)(\mathbf{I_t} - \mathbf{I_2}) = 0 \Rightarrow -\mathbf{V_t} + (1-j)\mathbf{I_t} + j\mathbf{I_2} = 0 \tag{8}$$

$$\xrightarrow{Using\ (7)} -\mathbf{V_t} + (1-j)\mathbf{I_t} + j\frac{(-1-j)}{1-j}\mathbf{I_t} = 0 \Rightarrow -\mathbf{V_t} + \left(1 - j + \frac{1-j}{1-j}\right)\mathbf{I_t} = 0$$

$$\Rightarrow -\mathbf{V_t} + (2-j)\mathbf{I_t} = 0 \Rightarrow \frac{\mathbf{V_t}}{\mathbf{I_t}} = 2 - j \Rightarrow \mathbf{Z_{Th}} = (2-j)\,\Omega \tag{9}$$

Solving (5) and (9):

$$R = |\mathbf{Z_{Th}}| = |2-j| = \sqrt{5} \tag{10}$$

Based on the information given in the problem:

$$P_{R,max} = 2\sqrt{5} \tag{11}$$

As we know, the power of a resistor can be calculated by using the following relation:

$$P_R = RI_{R,rms}^2 \tag{12}$$

Solving (10)–(12):

$$2\sqrt{5} = \sqrt{5}I_{R,rms}^2 \Rightarrow I_{R,rms} = \sqrt{2} \tag{13}$$

Now, we can calculate the total average power consumed in the circuit, which is equal to the total average power wasted in the resistors:

$$P_{Consumed} = (R+1)I_{R,rms}^2 \xrightarrow{Using\ (10),(13)} P_{Consumed} = \left(\sqrt{5}+1\right)\left(\sqrt{2}\right)^2 = 2\left(\sqrt{5}+1\right)\ W$$

Choice (1) is the answer.

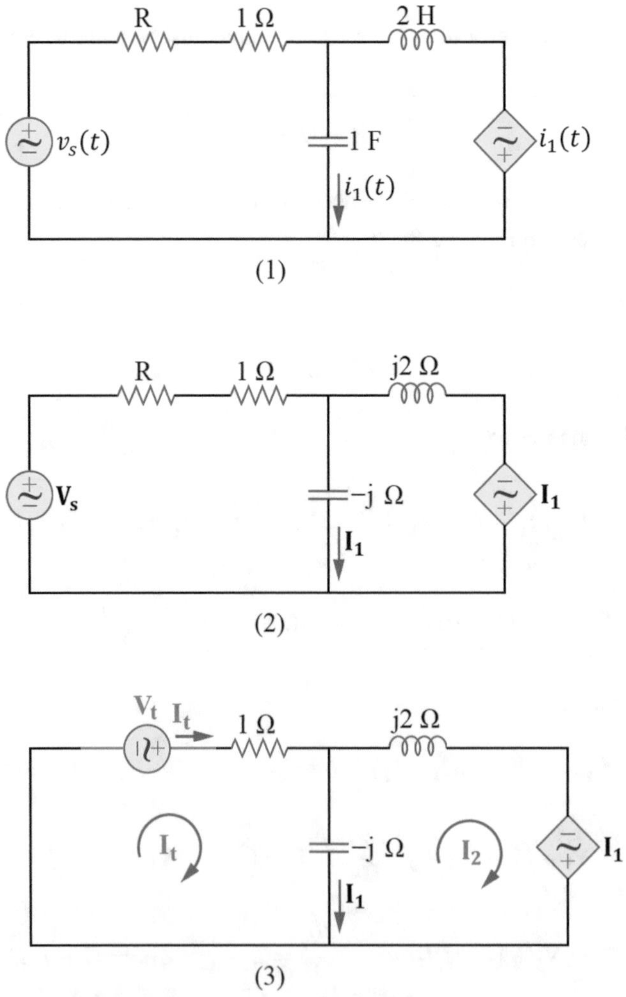

Figure 2.84 The circuit of solution of problem 2.86

Reference

1. Mehdi Rahmani-Andebili. (2020). DC electrical circuit analysis: Practice problems, methods, and solutions, *Springer Nature*.

Problems: Sinusoidal Steady-State Analysis of Circuits Including Transformers and Magnetically Coupled Inductors

3

Abstract

This chapter helps both groups of underprepared and knowledgeable students taking courses in AC electrical circuit analysis. In this chapter, the basic and advanced problems of another important subject of AC circuit analysis, that is, sinusoidal steady-state analysis of circuits including transformers and magnetically coupled inductors, are presented. Like the first chapter, the problems are categorized in different levels based on their difficulty levels (easy, normal, and hard) and calculation amounts (small, normal, and large). Additionally, the problems are ordered from the easiest problem with the smallest computations to the most difficult problems with the largest calculations.

3.1. In the circuit of Figure 3.1, what must be the angular frequency of the voltage source to transfer the maximum average power [1] to the 12 Ω resistor in sinusoidal steady state?

Difficulty level ● Easy ○ Normal ○ Hard
Calculation amount ● Small ○ Normal ○ Large

1) 3 *rad/sec*
2) 6 *rad/sec*
3) 9 *rad/sec*
4) 12 *rad/sec*

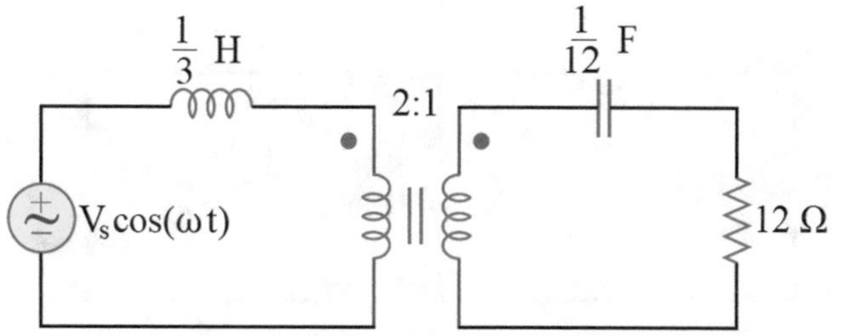

Figure 3.1 The circuit of problem 3.1

3.2. The inductance matrix of a circuit including three magnetically coupled inductors is $[L] = \begin{bmatrix} 2 & 1 & -1 \\ 1 & 4 & -1 \\ -1 & -1 & 3 \end{bmatrix}$ H, and their currents are $i_1 = 3\ A$, $i_2 = 1\ A$, and $i_3 = 2\ A$. Calculate the magnetic energy stored in the circuit.

Difficulty level ● Easy ○ Normal ○ Hard
Calculation amount ● Small ○ Normal ○ Large

1) 17 *J*
2) 12 *J*
3) 28 *J*
4) 24 *J*

M. Rahmani-Andebili, *AC Electrical Circuit Analysis*, https://doi.org/10.1007/978-3-030-60986-3_3

3.3. In the circuit of Figure 3.2, parametrically determine the value of R_L so that it can absorb the maximum average power in sinusoidal steady state.

Difficulty level ● Easy ○ Normal ○ Hard

Calculation amount ● Small ○ Normal ○ Large

1) $\sqrt{(R_0)^2 + \left(L_0\omega - \frac{1}{\omega C_0}\right)^2}\ \Omega$
2) $\sqrt{\left(\frac{R_0}{n^2}\right)^2 + \left(n^2 L_0\omega - \frac{1}{\omega n^2 C_0}\right)^2}\ \Omega$
3) $\left(\frac{R_0}{n^2}\right)^2\ \Omega$
4) $\sqrt{\left(\frac{R_0}{n^2}\right)^2 + \left(L_0\omega - \frac{1}{\omega n^2 C_0}\right)^2}\ \Omega$

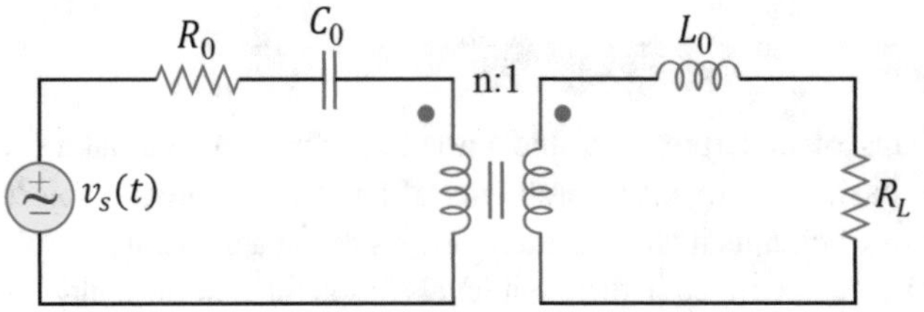

Figure 3.2 The circuit of problem 3.3

3.4. In the circuit of Figure 3.3, determine the number of turns of the primary coils of the transformer to transfer the maximum average power to the 4 Ω resistor in sinusoidal steady state.

Difficulty level ● Easy ○ Normal ○ Hard

Calculation amount ● Small ○ Normal ○ Large

1) 100
2) 4
3) 25
4) 5

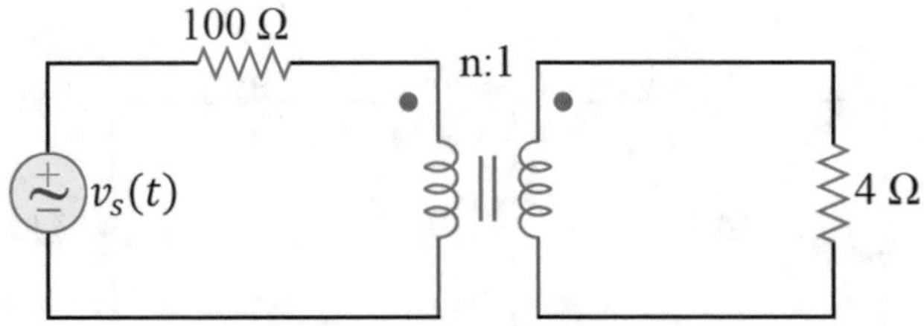

Figure 3.3 The circuit of problem 3.4

3.5. In the circuit of Figure 3.4, calculate the primary voltage of the transformer in sinusoidal steady state ($v_1(t)$).

Difficulty level ● Easy ○ Normal ○ Hard

Calculation amount ● Small ○ Normal ○ Large

1) $100\cos(10t)\ V$
2) $-200\cos(10t)\ V$
3) $-100\cos(10t)\ V$
4) $-30\cos(10t)\ V$

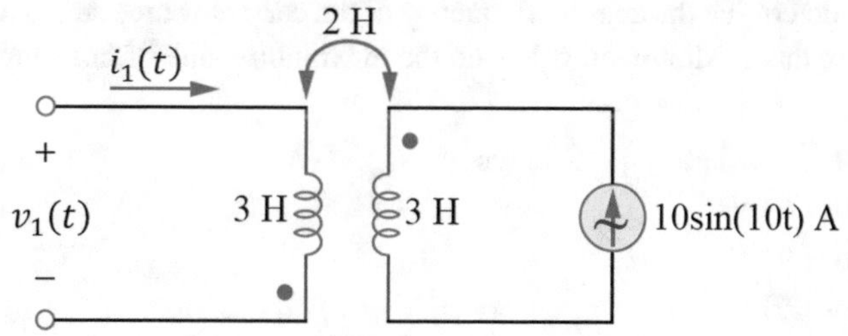

Figure 3.4 The circuit of problem 3.5

3.6. In the circuit of Figure 3.5, determine the resistance of R so that it can absorb the maximum average power in sinusoidal steady state.

Difficulty level ● Easy ○ Normal ○ Hard
Calculation amount ○ Small ● Normal ○ Large

1) 3 Ω
2) 0.75 Ω
3) 12 Ω
4) 1.25 Ω

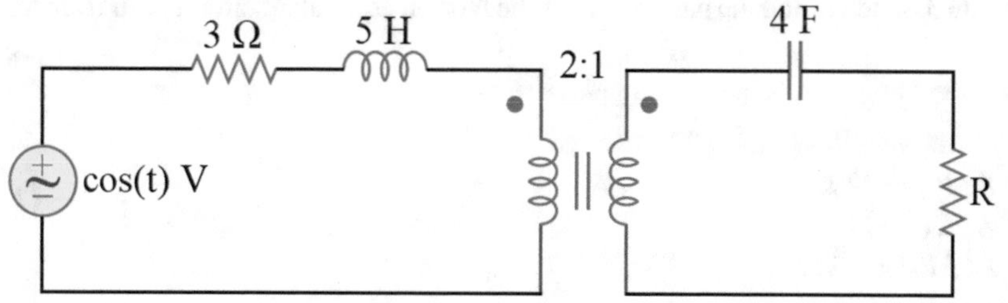

Figure 3.5 The circuit of problem 3.6

3.7. In the circuit of Figure 3.6, calculate the load impedance ($\mathbf{Z_L}$) so that it can absorb the maximum average power in sinusoidal steady state.

Difficulty level ● Easy ○ Normal ○ Hard
Calculation amount ○ Small ● Normal ○ Large

1) $(5 - j5)$ Ω
2) $(5 + j5)$ Ω
3) $(20 - j20)$ Ω
4) $(20 + j20)$ Ω

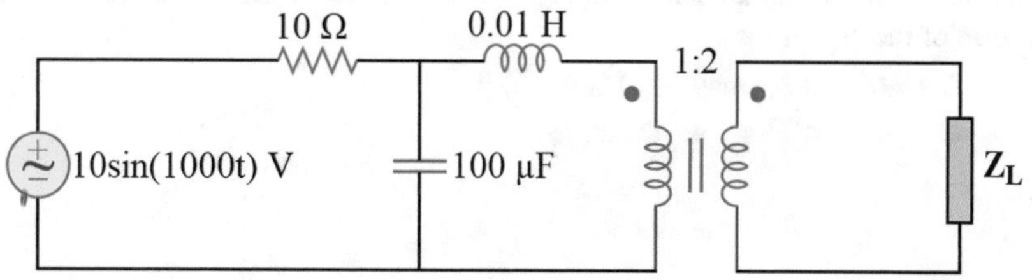

Figure 3.6 The circuit of problem 3.7

3.8. In the circuit of Figure 3.7, determine the angular frequency of the current source so that the sinusoidal steady-state value of the output voltage is maximum. Moreover, calculate the maximum value of this output voltage.

Difficulty level ○ Easy ● Normal ○ Hard

Calculation amount ● Small ○ Normal ○ Large

1) $\omega_0 = 2\,\frac{rad}{sec}, v_o(t) = \frac{1}{2}\cos(2t)\ V$
2) $\omega_0 = 4\,\frac{rad}{sec}, v_o(t) = \frac{1}{2}\cos(4t)\ V$
3) $\omega_0 = 2\,\frac{rad}{sec}, v_o(t) = \frac{1}{4}\cos(2t)\ V$
4) $\omega_0 = 4\,\frac{rad}{sec}, v_o(t) = \frac{1}{4}\cos(4t)\ V$

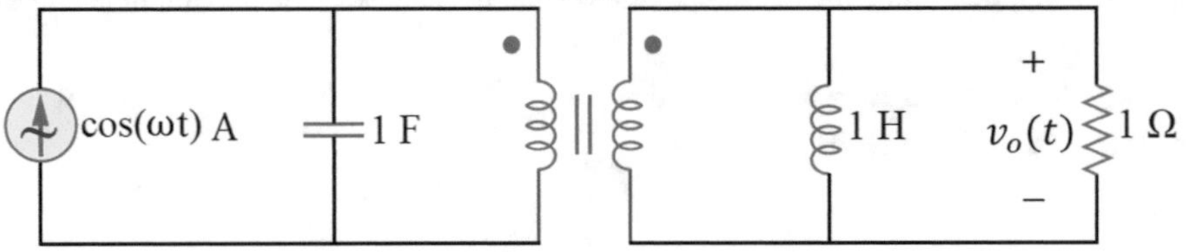

Figure 3.7 The circuit of problem 3.8

3.9. In the circuit of Figure 3.8, determine the parameters of the Norton equivalent circuit seen from terminal a–b in sinusoidal steady state.

Difficulty level ○ Easy ● Normal ○ Hard

Calculation amount ● Small ○ Normal ○ Large

1) $i_N(t) = \sin(t)\ A,\ R_N = 12\ \Omega$
2) $i_N(t) = \sin(t)\ A,\ R_N = 11\ \Omega$
3) $i_N(t) = 3\sin(t)\ A,\ R_N = 12\ \Omega$
4) $i_N(t) = 3\sin(t)\ A,\ R_N = 11\ \Omega$

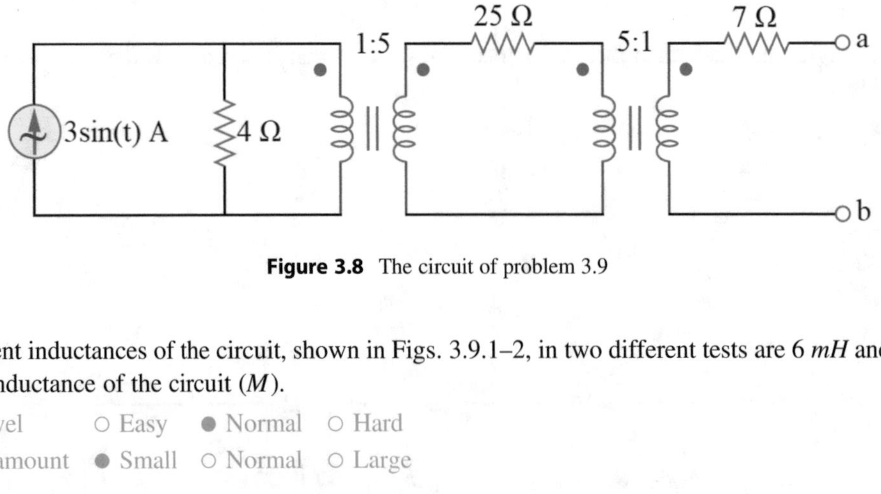

Figure 3.8 The circuit of problem 3.9

3.10. The equivalent inductances of the circuit, shown in Figs. 3.9.1–2, in two different tests are 6 mH and 2 mH. Determine the mutual inductance of the circuit (M).

Difficulty level ○ Easy ● Normal ○ Hard

Calculation amount ● Small ○ Normal ○ Large

1) 6 mH
2) 4 mH
3) 2 mh
4) 1 mH

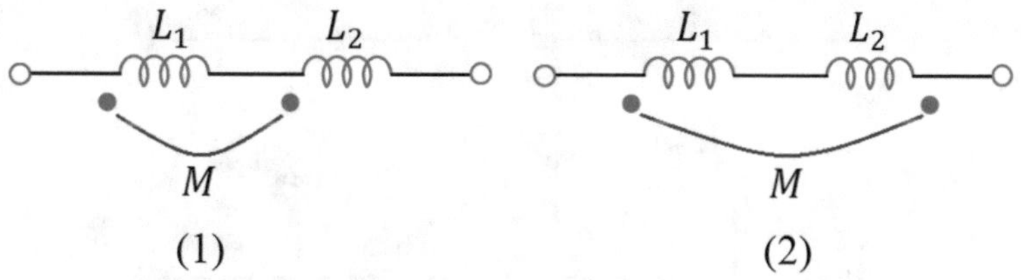

Figure 3.9 The circuit of problem 3.10

3.11. In the circuit of Figure 3.10, determine the impedance seen from terminal a–b in sinusoidal steady state with the angular frequency of 1 rad/sec?

Difficulty level ○ Easy ● Normal ○ Hard
Calculation amount ● Small ○ Normal ○ Large

1) $(2 + j4)\ \Omega$
2) $(2 + j3)\ \Omega$
3) $(2 + j4.5)\ \Omega$
4) $(2 + j3.5)\ \Omega$

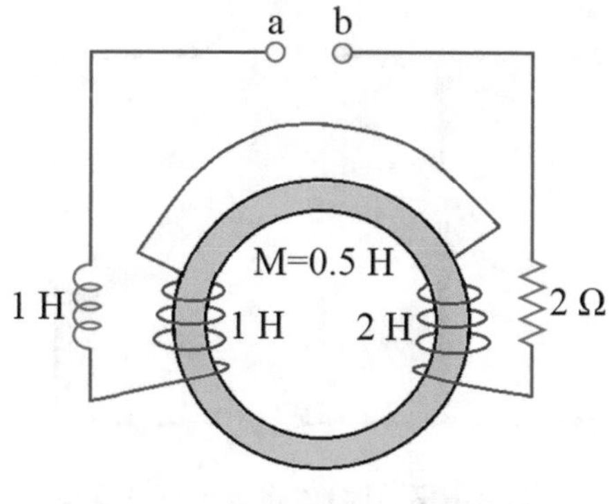

Figure 3.10 The circuit of problem 3.11

3.12. In the circuit of Figure 3.11, determine the inductance matrix of the circuit for $n_1 = 1$ and $n_2 = 1$.

Difficulty level ○ Easy ● Normal ○ Hard
Calculation amount ○ Small ● Normal ○ Large

1) $\begin{bmatrix} \frac{3}{2} & \frac{3}{2} \\ \frac{3}{2} & \frac{3}{2} \end{bmatrix} H$

2) $\begin{bmatrix} \frac{1}{2} & \frac{3}{2} \\ \frac{3}{2} & \frac{1}{2} \end{bmatrix} H$

3) $\begin{bmatrix} \frac{3}{2} & \frac{1}{2} \\ \frac{1}{2} & \frac{3}{2} \end{bmatrix} H$

4) $\begin{bmatrix} \frac{1}{2} & \frac{1}{2} \\ \frac{1}{2} & \frac{1}{2} \end{bmatrix} H$

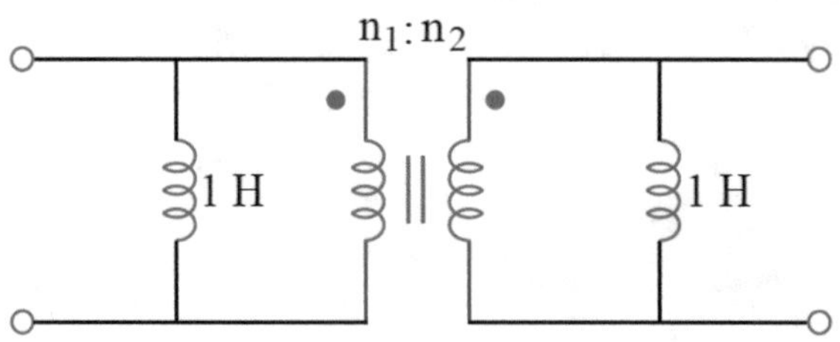

Figure 3.11 The circuit of problem 3.12

3.13. In the circuit of Figure 3.12, determine the equivalent inductance seen from terminal a–b.

Difficulty level ○ Easy ● Normal ○ Hard

Calculation amount ○ Small ● Normal ○ Large

1) $\frac{23}{16}\ H$
2) $\frac{5}{16}\ H$
3) $\frac{16}{23}\ H$
4) $\frac{16}{5}\ H$

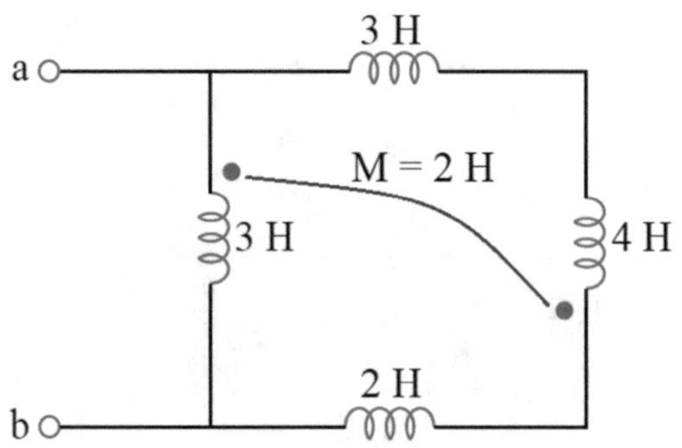

Figure 3.12 The circuit of problem 3.13

3.14. In the circuit of Figure 3.13, determine the equivalent inductance seen from the terminal.

Difficulty level ○ Easy ● Normal ○ Hard

Calculation amount ○ Small ● Normal ○ Large

1) 1 H
2) 0.5 H
3) 1 H
4) 2.5 H

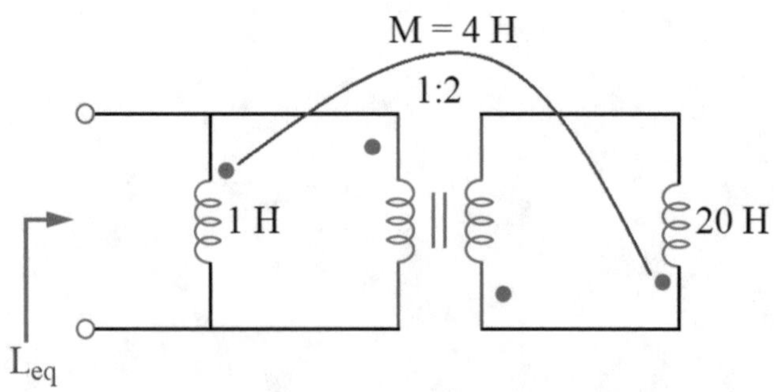

Figure 3.13 The circuit of problem 3.14

3.15. Determine the resonance frequency of the circuit of Figure 3.14 that can be seen from the terminal for $L_1 = \frac{3}{5}\ H, L_2 = \frac{2}{5}\ H$, $|M| = \frac{1}{5}\ H$, and $C = \frac{1}{7}\ F$.

Difficulty level ○ Easy ● Normal ○ Hard
Calculation amount ○ Small ● Normal ○ Large

1) $7\ rad/sec$
2) $\sqrt{21}\ rad/sec$
3) $1.6\ rad/sec$
4) $\frac{5}{\sqrt{21}}\ rad/sec$

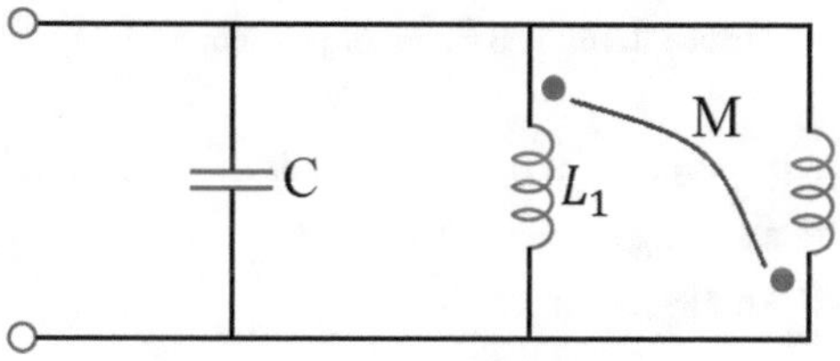

Figure 3.14 The circuit of problem 3.15

3.16. In the circuit of Figure 3.15, determine the value of $\frac{n_1}{n_2}$ and C so that the maximum average power is transferred to the secondary side of the transformer in sinusoidal steady state.

Difficulty level ○ Easy ● Normal ○ Hard
Calculation amount ○ Small ● Normal ○ Large

1) $\frac{n_1}{n_2} = 2, C = \frac{1}{20}\ F$
2) $\frac{n_1}{n_2} = \frac{1}{2}, C = \frac{1}{5}\ F$
3) $\frac{n_1}{n_2} = 2, C = 5\ F$
4) $\frac{n_1}{n_2} = \frac{1}{2}, C = 20\ F$

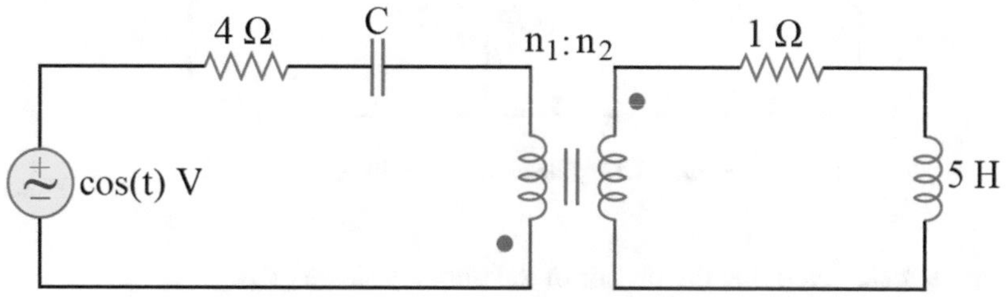

Figure 3.15 The circuit of problem 3.16

3.17. In the circuit of Figure 3.16, determine the equivalent inductance of the circuit seen from terminal a–b (L_{ab}). k is the coupling coefficient.

Difficulty level ○ Easy ● Normal ○ Hard
Calculation amount ○ Small ● Normal ○ Large

1) L_1
2) $L_1\sqrt{1-k^2}$
3) $L_1(1-k^2)$
4) $L_1\left(1+\frac{M}{L_2^2}\right)$

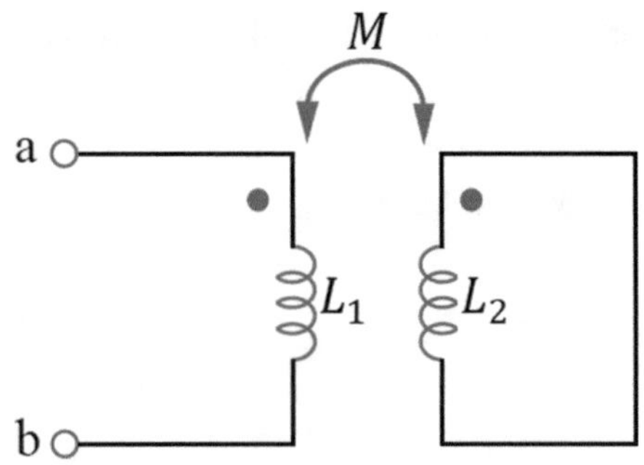

Figure 3.16 The circuit of problem 3.17

3.18. Calculate the sinusoidal steady-state response of $v(t)$ in the circuit of Figure 3.17.

Difficulty level ○ Easy ● Normal ○ Hard
Calculation amount ○ Small ● Normal ○ Large

1) $-3\sin(2t) - \cos(2t)\ V$
2) $3\sin(2t) + \cos(2t)\ V$
3) $-\sin(2t) - 3\cos(2t)\ V$
4) $\sin(2t) + 3\cos(2t)\ V$

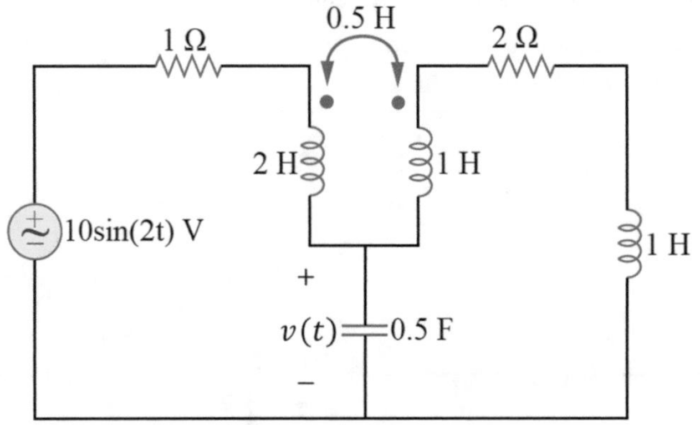

Figure 3.17 The circuit of problem 3.18

3.19. In the circuit of Figure 3.18, determine the phasor of the output voltage ($\mathbf{V_o}$).

Difficulty level ○ Easy ● Normal ○ Hard
Calculation amount ○ Small ● Normal ○ Large

1) $36e^{j53}\ V$
2) $36e^{-j53}\ V$
3) $72e^{-j53}\ V$
4) $72e^{j53}\ V$

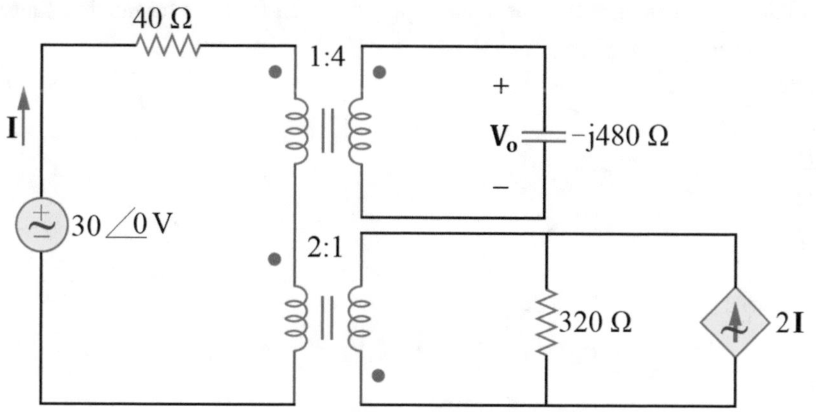

Figure 3.18 The circuit of problem 3.19

3.20. In the circuit of Figure 3.19, what must be the value of $\frac{n_1}{n_2}$ and C to transfer the maximum average power to 2 Ω resistor in sinusoidal steady state?

Difficulty level ○ Easy ● Normal ○ Hard
Calculation amount ○ Small ● Normal ○ Large

1) $\frac{n_1}{n_2} = 0.5, C = 0.02\ F$
2) $\frac{n_1}{n_2} = 5, C = 0.02\ F$
3) $\frac{n_1}{n_2} = 0.25, C = 0.01\ F$
4) $\frac{n_1}{n_2} = 5, C = 0.2\ F$

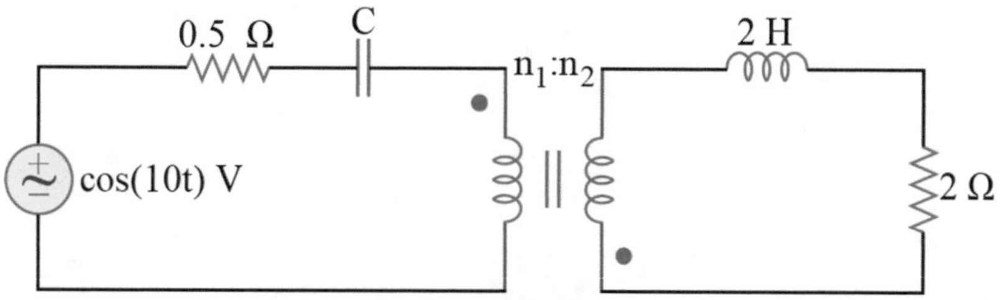

Figure 3.19 The circuit of problem 3.20

3.21. In the circuit of Figure 3.20, calculate the power of 2 Ω resistor.

Difficulty level ○ Easy ● Normal ○ Hard
Calculation amount ○ Small ● Normal ○ Large

1) 25 mW
2) 50 mW
3) 80 mW
4) 100 mW

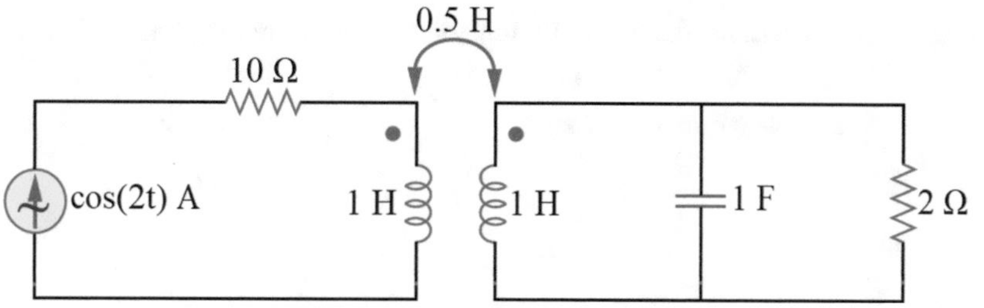

Figure 3.20 The circuit of problem 3.21

3.22. In the circuit of Figure 3.21, calculate the Thevenin impedance seen from the terminal in sinusoidal steady state.

Difficulty level ○ Easy ● Normal ○ Hard
Calculation amount ○ Small ● Normal ○ Large

1) $\left(\frac{162}{221} - j\frac{100}{221}\right)\ \Omega$
2) $\left(\frac{50}{221} - j\frac{140}{221}\right)\ \Omega$
3) $\left(\frac{320}{221} + j\frac{430}{221}\right)\ \Omega$
4) $\left(\frac{112}{221} + j\frac{40}{221}\right)\ \Omega$

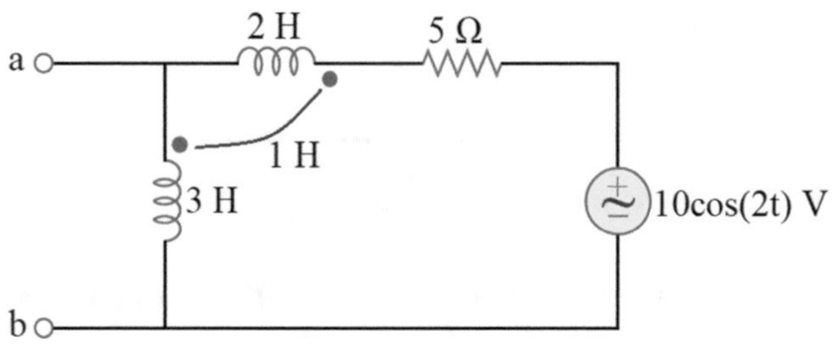

Figure 3.21 The circuit of problem 3.22

3.23. In the circuit of Figure 3.22, determine the magnitude of the Thevenin impedance seen from terminal a–b in sinusoidal steady state with the angular frequency of 2 rad/sec.

Difficulty level ○ Easy ● Normal ○ Hard
Calculation amount ○ Small ● Normal ○ Large

1) 5 Ω
2) 10 Ω
3) 2 Ω
4) 1 Ω

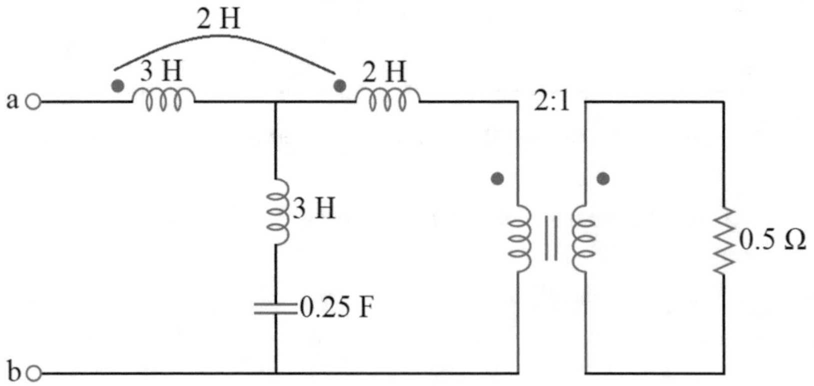

Figure 3.22 The circuit of problem 3.23

3.24. In the circuit of Figure 3.23, determine the Thevenin impedance seen from the terminal in sinusoidal steady state.

Difficulty level ○ Easy ● Normal ○ Hard
Calculation amount ○ Small ● Normal ○ Large

1) $\frac{4}{3}\ \Omega$
2) $\frac{3}{4}\ \Omega$
3) $\frac{7}{5}\ \Omega$
4) $\frac{5}{7}\ \Omega$

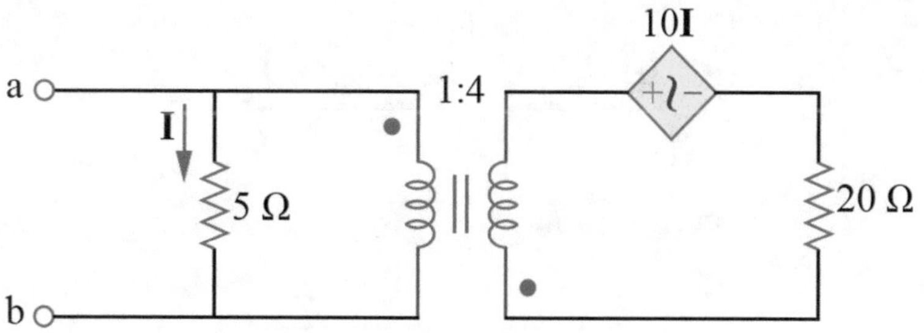

Figure 3.23 The circuit of problem 3.24

3.25. In the circuit of Figure 3.24, calculate the root mean square (rms) value of the primary and secondary currents of the transformer.

Difficulty level ○ Easy ● Normal ○ Hard
Calculation amount ○ Small ● Normal ○ Large

1) $\begin{cases} I_1 = 10\,A \\ I_2 = -3\,A \end{cases}$

2) $\begin{cases} I_1 = 12\,A \\ I_2 = -3\,A \end{cases}$

3) $\begin{cases} I_1 = 10\,A \\ I_2 = -2.5\,A \end{cases}$

4) $\begin{cases} I_1 = -10\,A \\ I_2 = -2\,A \end{cases}$

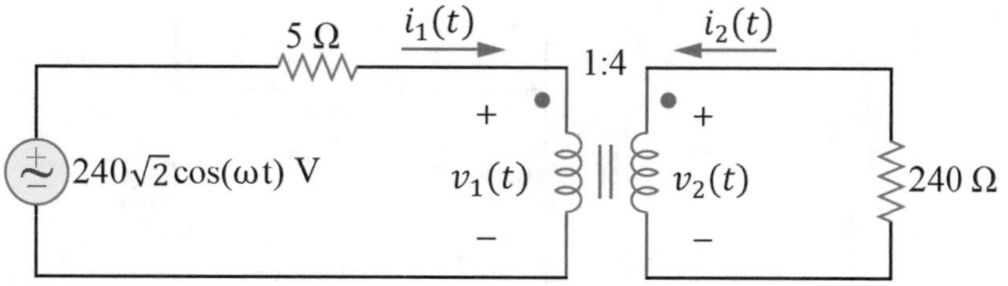

Figure 3.24 The circuit of problem 3.25

3.26. In the circuit of Figure 3.25, determine the inductance matrix based on the given information.
Test 1: 5 H inductance is measured in the left-side terminal while the right-side terminal is open.
Test 2: 16 V rms voltage is measured in the right-side terminal (open circuit voltage) by applying a sinusoidal voltage source with the rms voltage of 5 V to the left-side terminal.
Test 3: 4 V rms voltage is measured in the left-side terminal (open circuit voltage) by applying a sinusoidal voltage source with the rms voltage of 20 V to the right-side terminal.

Difficulty level ○ Easy ● Normal ○ Hard
Calculation amount ○ Small ○ Normal ● Large

1) $\begin{bmatrix} 5 & 15 \\ 15 & 80 \end{bmatrix} H$

2) $\begin{bmatrix} 5 & 15 \\ 10 & 60 \end{bmatrix} H$

3) $\begin{bmatrix} 5 & 16 \\ 16 & 60 \end{bmatrix} H$

4) $\begin{bmatrix} 5 & 16 \\ 16 & 80 \end{bmatrix} H$

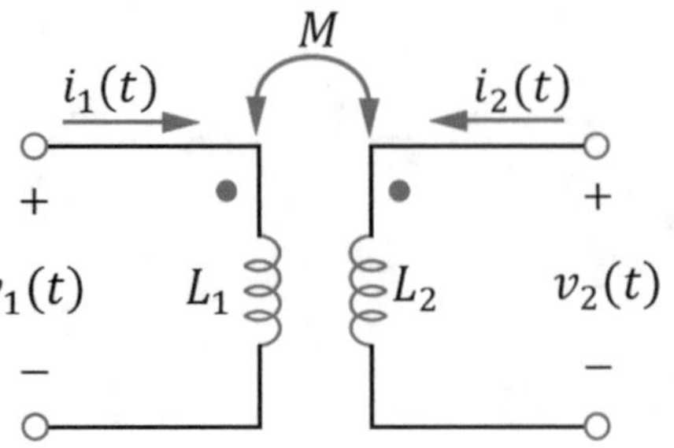

Figure 3.25 The circuit of problem 3.26

3.27. In the circuit of Figure 3.26, determine the mutual inductance (M) so that no current flows from node "a" to node "b" if it is short-circuited in sinusoidal steady state.

Difficulty level ○ Easy ○ Normal ● Hard
Calculation amount ● Small ○ Normal ○ Large

1) $\frac{R_1L_2-R_2L_1}{R_1+R_2}\ H$
2) $\frac{R_1L_2-R_2L_1}{R_1-R_2}\ H$
3) $\frac{R_2L_1-R_1L_2}{R_1+R_2}\ H$
4) $\frac{R_2L_1-R_1L_2}{R_1-R_2}\ H$

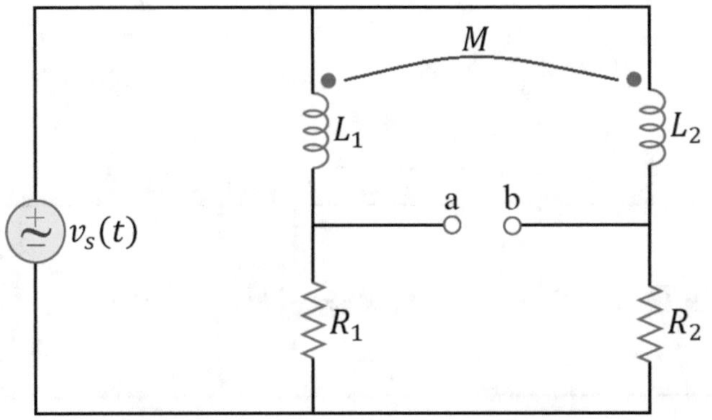

Figure 3.26 The circuit of problem 3.27

3.28. In the circuit of Figure 3.27, calculate $\mathbf{I}_1$ if $\mathbf{I}_2 = 20\sqrt{2}e^{-j45}\ A$ and $\mathbf{I}_3 = 24\ A$.

Difficulty level ○ Easy ○ Normal ● Hard
Calculation amount ● Small ○ Normal ○ Large

1) $(2\sqrt{89}\angle{-22°})\ A$
2) $(2\sqrt{89}\angle{-32°})\ A$
3) $(10\sqrt{2}\angle{-45°})\ A$
4) $(6\angle{0°})\ A$

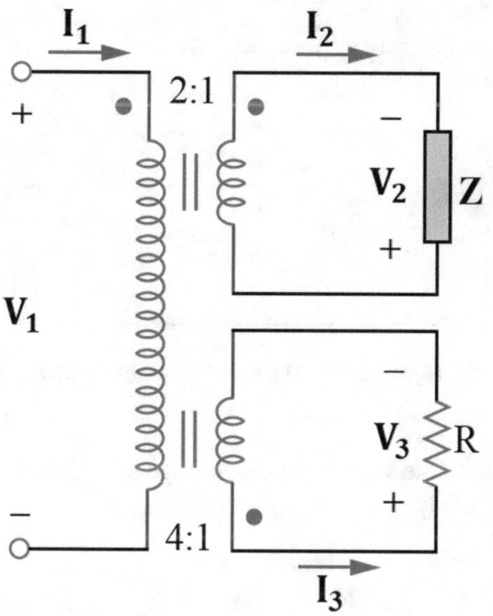

Figure 3.27 The circuit of problem 3.28

3.29. The circuit illustrated in Figure 3.28 is in sinusoidal steady state. The inverse of the inductance matrix of the circuit is $[L]^{-1} = \Gamma = \begin{bmatrix} 5 & -2 \\ -2 & 2 \end{bmatrix}$. Determine the capacitance of the capacitor so that the voltage of the resistor is maximum.

Difficulty level ○ Easy ○ Normal ● Hard
Calculation amount ● Small ○ Normal ○ Large

1) 10 *F*
2) 50 *F*
3) 100 *F*
4) 200 *F*

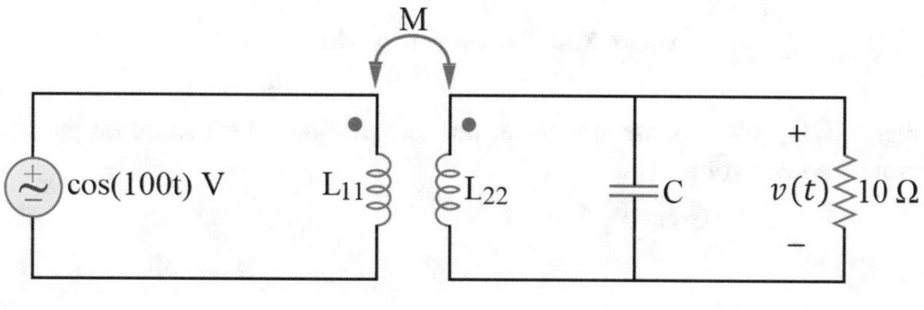

Figure 3.28 The circuit of problem 3.29

3.30. In the circuit of Figure 3.29, determine the impedance of the load ($\mathbf{Z}_L$) so that it can absorb the maximum average power in sinusoidal steady state.

Difficulty level ○ Easy ○ Normal ● Hard
Calculation amount ● Small ○ Normal ○ Large

1) $(1.25 - j3.75)\ \Omega$
2) $(0.8 - j2.4)\ \Omega$
3) $(1.25 + j3.75)\ \Omega$
4) $(0.8 + j2.4)\ \Omega$

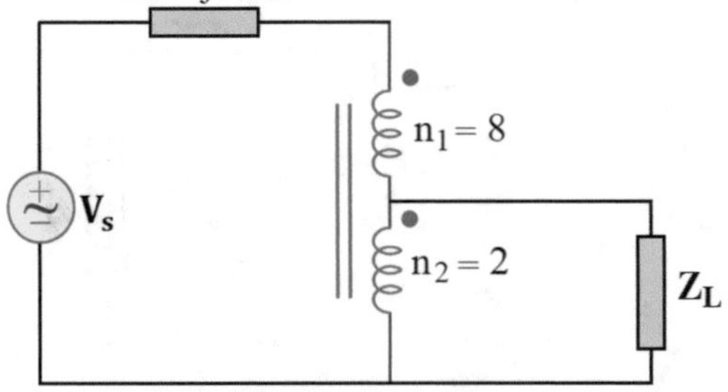

Figure 3.29 The circuit of problem 3.30

3.31. In the circuit of Figure 3.30, calculate the voltage of the voltage source in sinusoidal steady state.

Difficulty level ○ Easy ○ Normal ● Hard
Calculation amount ● Small ○ Normal ○ Large

1) $2\sqrt{5}\sin\left(2t - 26^\circ\right)\ V$
2) $2\sin\left(2t - 26^\circ\right)\ V$
3) $2\sin\left(2t + 26^\circ\right)\ V$
4) $\sqrt{5}\sin\left(2t - 26^\circ\right)\ V$

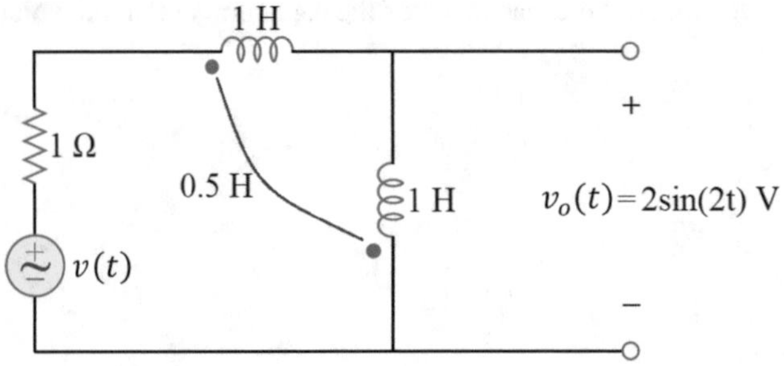

Figure 3.30 The circuit of problem 3.31

3.32. In the circuit of Figure 3.31, calculate the amount of mutual inductance (M) based on the given parameters and the information read from the ideal voltmeters.

Difficulty level ○ Easy ○ Normal ● Hard
Calculation amount ● Small ○ Normal ○ Large

1) $\frac{R}{2\omega} \times \left|\frac{\mathbf{V_2}}{\mathbf{V_R}}\right|\ H$
2) $\frac{2R}{\omega} \times \left|\frac{\mathbf{V_R}}{\mathbf{V_2}}\right|\ H$
3) $\frac{R}{\omega} \times \left|\frac{\mathbf{V_2}}{\mathbf{V_R}}\right|\ H$
4) $\frac{R}{\omega} \times \left|\frac{\mathbf{V_R}}{\mathbf{V_2}}\right|\ H$

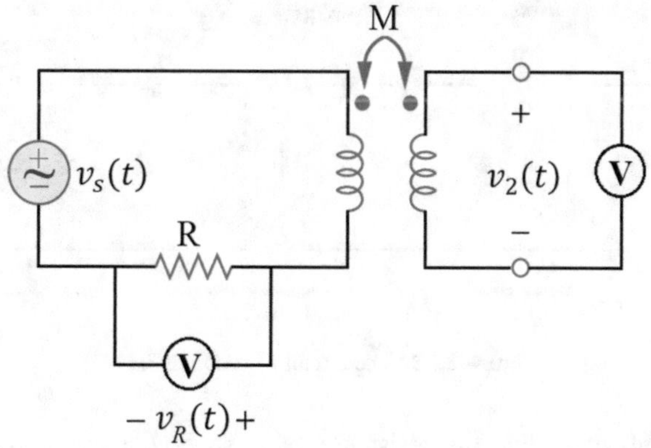

Figure 3.31 The circuit of problem 3.32

3.33. In the circuit of Figure 3.32, calculate the sinusoidal steady-state voltage of the circuit for the given inductance matrix.

$$[L] = \begin{bmatrix} 1 & -1 & 0 \\ -1 & 1 & 1 \\ 0 & 1 & 1 \end{bmatrix} H$$

Difficulty level ○ Easy ○ Normal ● Hard
Calculation amount ○ Small ● Normal ○ Large
1) $\frac{1}{3}\sin(t)\ V$
2) $\cos(t)\ V$
3) $\sin(t)\ V$
4) $\frac{1}{3}\cos(t)\ V$

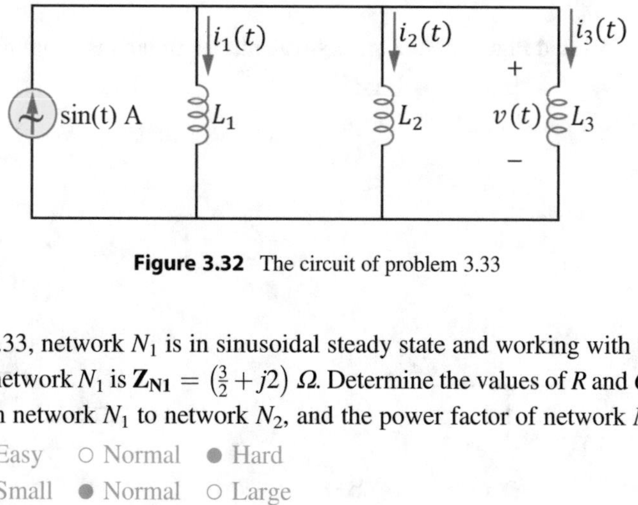

Figure 3.32 The circuit of problem 3.33

3.34. In the circuit of Figure 3.33, network N_1 is in sinusoidal steady state and working with the angular frequency of 1 Hz. The input impedance of network N_1 is $\mathbf{Z_{N1}} = \left(\frac{3}{2} + j2\right)\ \Omega$. Determine the values of R and C so that the maximum average power is transferred from network N_1 to network N_2, and the power factor of network N_2 is unity.

Difficulty level ○ Easy ○ Normal ● Hard
Calculation amount ○ Small ● Normal ○ Large
1) $R = \frac{3}{2}\ \Omega, C = \frac{1}{4}\ F$
2) $R = \frac{5}{2}\ \Omega, C = \frac{4}{3}\ F$
3) $R = \frac{5}{2}\ \Omega, C = 1\ F$
4) For no value of C, unity power factor is achieved for network N_2

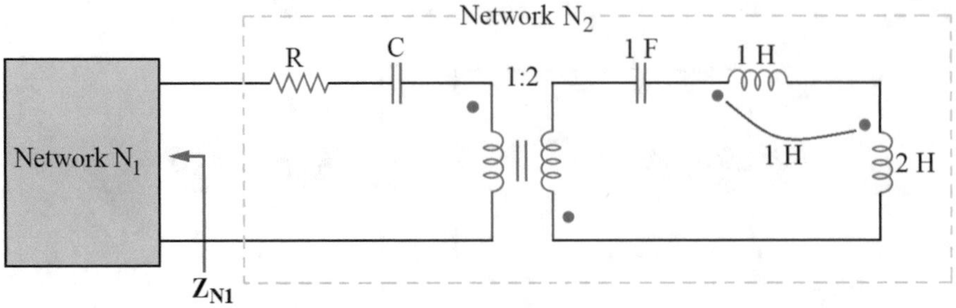

Figure 3.33 The circuit of problem 3.34

3.35. In the circuit of Figure 3.34, determine the resistance of R so that it can absorb the maximum average power. Additionally, calculate the maximum average power.

Difficulty level ○ Easy ○ Normal ● Hard

Calculation amount ○ Small ● Normal ○ Large

1) $R = \frac{2}{7}\ \Omega, P_{max} = \frac{8}{7}\ W$
2) $R = \frac{2}{7}\ \Omega, P_{max} = 1\ W$
3) $R = \frac{4}{7}\ \Omega, P_{max} = \frac{16}{7}\ W$
4) $R = \frac{4}{7}\ \Omega, P_{max} = \frac{6}{7}\ W$

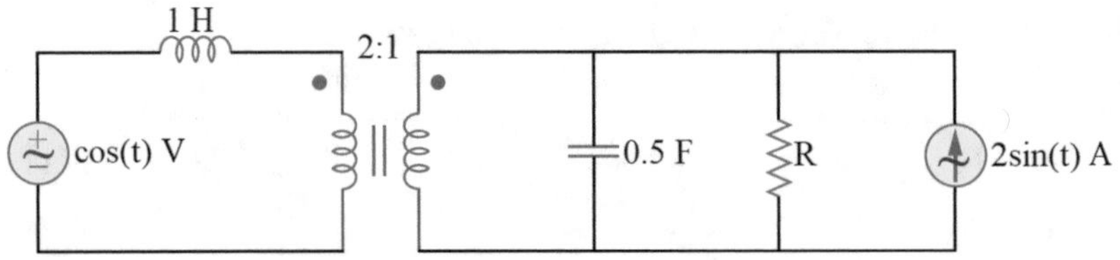

Figure 3.34 The circuit of problem 3.35

3.36. If the circuits of Figure 3.35.1 and Figure 3.35.2 are equivalent, determine the coupling coefficient (k).

Difficulty level ○ Easy ○ Normal ● Hard

Calculation amount ○ Small ● Normal ○ Large

1) $\frac{2}{9}$
2) $\frac{2}{3}$
3) $\frac{3}{2}$
4) $\frac{9}{2}$

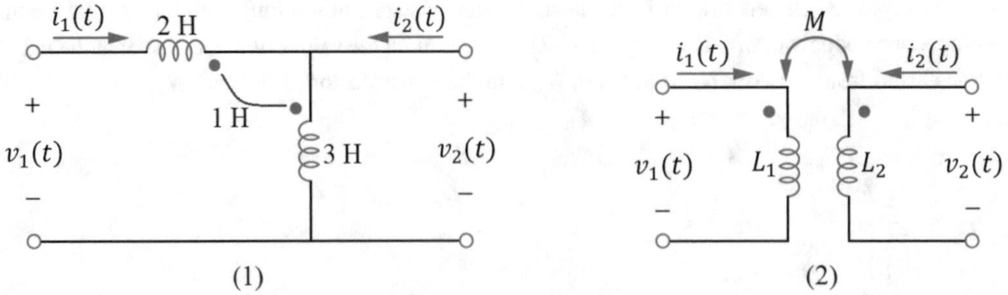

Figure 3.35 The circuit of problem 3.36

3.37. In the circuit of Figure 3.36, calculate the Thevenin resistor seen from terminal a–b.

Difficulty level ○ Easy ○ Normal ● Hard

Calculation amount ○ Small ● Normal ○ Large

1) $\left(\frac{n+1}{n}\right)^2(R_1+R_2)$
2) $\left(\frac{n}{n+1}\right)^2(R_1+R_2)$
3) $\left(\frac{n-1}{n}\right)^2(R_1+R_2)$
4) $\left(\frac{n}{n-1}\right)^2(R_1+R_2)$

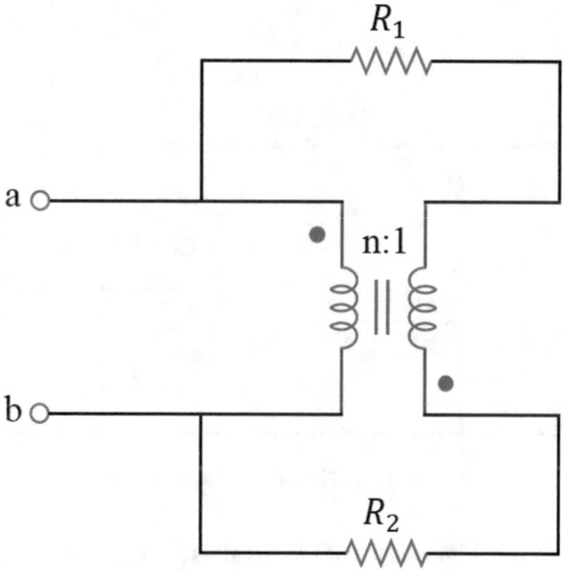

Figure 3.36 The circuit of problem 3.37

3.38. In the circuit of Figure 3.37, the sinusoidal steady-state current flowing through the load is zero. Calculate the amount of mutual inductance (M).

Difficulty level ○ Easy ○ Normal ● Hard
Calculation amount ○ Small ● Normal ○ Large

1) R^2C and $\frac{L_1}{2}$
2) $\frac{L_1+L_2}{2}$
3) $\frac{L_1L_2}{R^2C}$
4) $\frac{L_2R^2C}{L_1}$

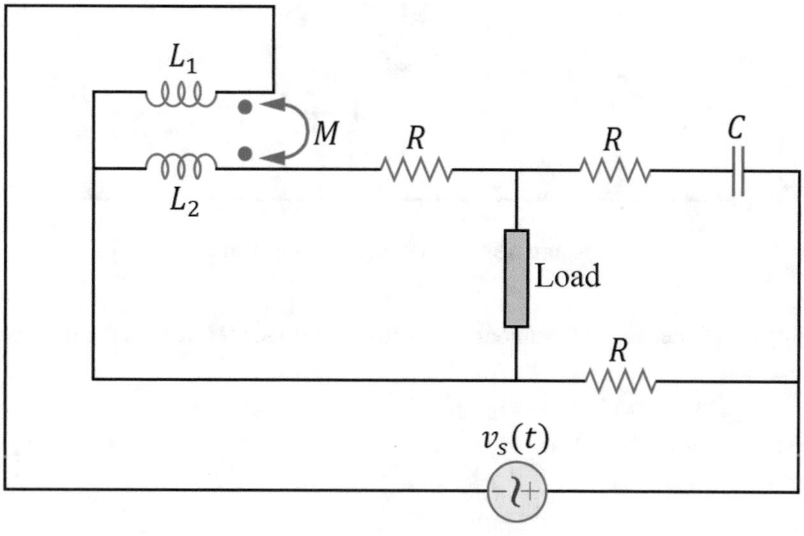

Figure 3.37 The circuit of problem 3.38

3.39. In the circuit of Figure 3.38, determine the value of $\frac{n_1}{n_2}$ so that the maximum average power is transferred to the 400 Ω resistor in sinusoidal steady state.

Difficulty level ○ Easy ○ Normal ● Hard
Calculation amount ○ Small ● Normal ○ Large

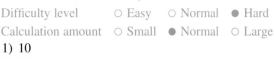

1) 10
2) 12
3) 14
4) 16

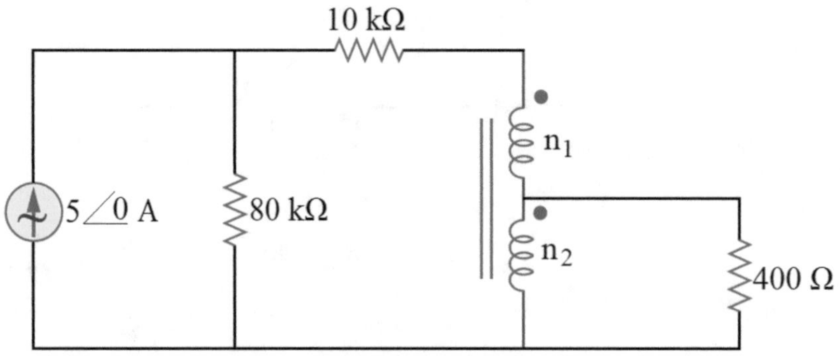

Figure 3.38 The circuit of problem 3.39

3.40. In the circuit of Figure 3.39 the input impedance of the circuit is $\frac{R}{2}$. Determine the parameter of "*a*."

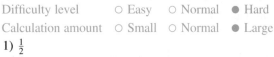

Difficulty level ○ Easy ○ Normal ● Hard
Calculation amount ○ Small ○ Normal ● Large

1) $\frac{1}{2}$
2) $\frac{1}{2}C$
3) 1
4) $2C$

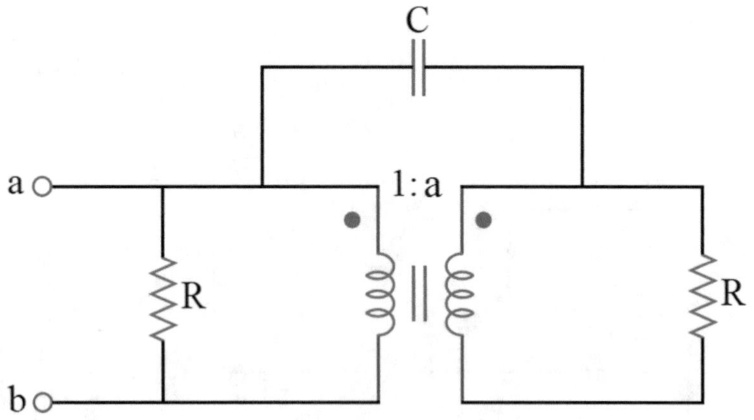

Figure 3.39 The circuit of problem 3.40

3.41. In the circuit of Figure 3.40, determine the amount of mutual inductance (M) so that the impedance seen from terminal a–b is purely inductive.

Difficulty level ○ Easy ○ Normal ● Hard
Calculation amount ○ Small ○ Normal ● Large

1) $\frac{1}{3}\,H$
2) $\frac{1}{2}\,H$
3) $\frac{2}{3}\,H$
4) No value of mutual inductance can result in a purely inductive impedance

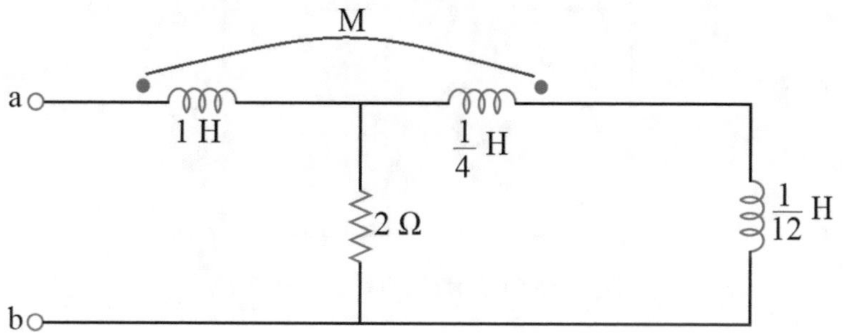

Figure 3.40 The circuit of problem 3.41

3.42. In the circuit of Figure 3.41, determine the resistance of R so that it can absorb the maximum average power.

Difficulty level ○ Easy ○ Normal ● Hard
Calculation amount ○ Small ○ Normal ● Large

1) $\frac{1810}{33}\,\Omega$
2) $\frac{1810}{11}\,\Omega$
3) $\frac{1810}{99}\,\Omega$
4) $\frac{905}{99}\,\Omega$

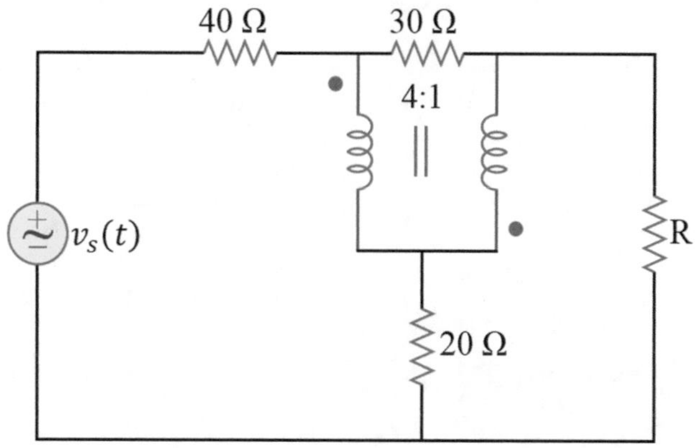

Figure 3.41 The circuit of problem 3.42

3.43. In the circuit of Figure 3.42, calculate the resonance frequency seen from terminal a–b.

Difficulty level ○ Easy ○ Normal ● Hard
Calculation amount ○ Small ○ Normal ● Large

1) $\frac{1}{\sqrt{LC}}\ rad/sec$
2) $\frac{n}{\sqrt{LC}}\ rad/sec$
3) $\frac{1}{n\sqrt{LC}}\ rad/sec$
4) This circuit does not have any resonance frequency from the terminal

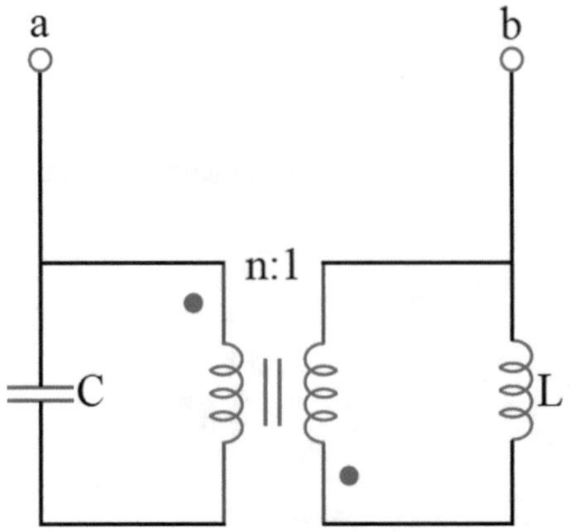

Figure 3.42 The circuit of problem 3.43

References

1. Rahmani-Andebili, M. (2020). DC electrical circuit analysis: Practice problems, methods, and solutions, *Springer Nature*.

Solutions of Problems: Sinusoidal Steady-State Analysis of Circuits Including Transformers and Magnetically Coupled Inductors

4

Abstract

In this chapter, the problems of the third chapter are fully solved, in detail, step-by-step, and with different methods.

4.1. By transferring the whole circuit to the left side of the transformer, we can simplify the problem solution [1], as can be seen in Figure 4.1.2. The updated parameters are as follows:

$$R'_{12\ \Omega} = (12\ \Omega) \times \left(\frac{2}{1}\right)^2 = 48\ \Omega \tag{1}$$

$$C'_{\frac{1}{12}F} = \left(\frac{1}{12}\ F\right) \times \left(\frac{1}{2}\right)^2 = \frac{1}{48}\ F \tag{2}$$

As can be noticed from Figure 4.1.2, if the capacitor and the inductor create a resonance state, the impedance of their series connection will be zero (i.e., a short circuit branch). Hence, the whole voltage of the voltage source will be applied on the resistor, and consequently it will absorb the maximum average power.

The resonance frequency of the series connection of a capacitor and an inductor is calculated as follows:

$$\omega_0 = \frac{1}{\sqrt{LC}} = \frac{1}{\sqrt{\frac{1}{3} \times \frac{1}{48}}} = 12\ rad/sec \tag{3}$$

Therefore, the frequency of the voltage source must be equal to the resonance frequency.

$$\omega = \omega_0 = 12\ rad/sec$$

Choice (4) is the answer.

M. Rahmani-Andebili, *AC Electrical Circuit Analysis*, https://doi.org/10.1007/978-3-030-60986-3_4

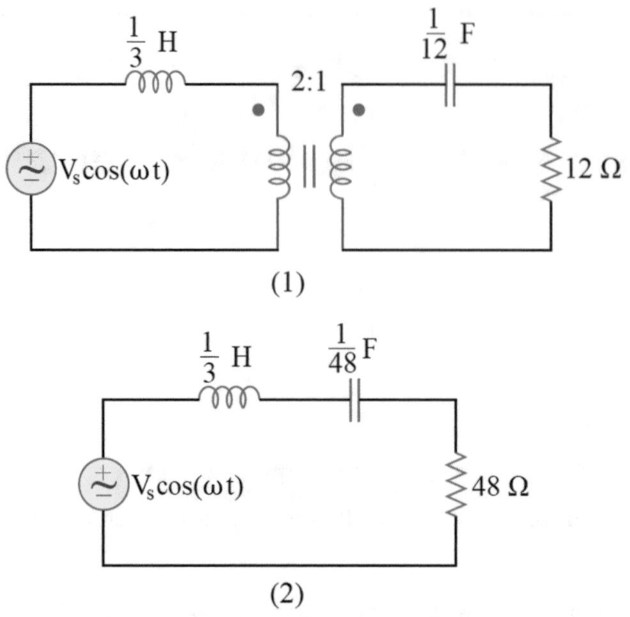

Figure 4.1 The circuit of solution of problem 4.1

4.2. The magnetic energy stored in the magnetically coupled inductors can be calculated by using the following relation:

$$W = \frac{1}{2}[i]^T[L][i] \ J \tag{1}$$

where $[L]$ and $[I]$ are the inductance matrix of the circuit and the currents of the inductors, respectively, as can be seen in (2) and (3).

$$[L] = \begin{bmatrix} 2 & 1 & -1 \\ 1 & 4 & -1 \\ -1 & -1 & 3 \end{bmatrix} H \tag{2}$$

$$[i] = \begin{bmatrix} 3 \\ 1 \\ 2 \end{bmatrix} A \tag{3}$$

Solving (1), (2), and (3):

$$W = \frac{1}{2}\begin{bmatrix} 3 \\ 1 \\ 2 \end{bmatrix}^T \begin{bmatrix} 2 & 1 & -1 \\ 1 & 4 & -1 \\ -1 & -1 & 3 \end{bmatrix}\begin{bmatrix} 3 \\ 1 \\ 2 \end{bmatrix} = 12 \ J$$

Choice (2) is the answer.

4.3. To simplify the problem solution, we can eliminate the effect of the transformer by transferring the circuit of one side of the transformer to its other side. Figure 4.2.2 shows the circuit, where the circuit of the left side of the transformer has been transferred to its right side. The updated values of R_O and C_O can be calculated as follows:

$$R'_O = \frac{R_O}{n^2} \tag{1}$$

$$C'_O = n^2 C_O \tag{2}$$

Figure 4.2.3 illustrates the circuit of Figure 4.2.2 in frequency domain. The impedances of the components are as follows:

$$\mathbf{Z}_{\mathbf{R}_0'} = \frac{R_0}{n^2}\ \Omega \tag{3}$$

$$\mathbf{Z}_{\mathbf{C}_0'} = \frac{1}{j\omega n^2 C_0}\ \Omega \tag{4}$$

$$\mathbf{Z}_{\mathbf{L}_0} = j\omega L_0\ \Omega \tag{5}$$

In Figure 4.2.3, the voltage source has been turned off (short-circuited), since we want to calculate the Thevenin impedance ($\mathbf{Z}_{\mathbf{Th}}$) seen by the load. After that, by using the following relation, we can calculate R_L so that it can absorb the maximum average power based on maximum average power transfer theorem.

$$R_L = |\mathbf{Z}_{\mathbf{Th}}| \tag{6}$$

The Thevenin impedance, seen by the load, can be calculated as follows:

$$\mathbf{Z}_{\mathbf{Th}} = \frac{R_0}{n^2} + \frac{1}{j\omega n^2 C_0} + jL_0\omega = \frac{R_0}{n^2} + j\left(L_0\omega - \frac{1}{\omega n^2 C_0}\right)\ \Omega \tag{7}$$

Solving (6) and (7):

$$R_L = \left|\frac{R_0}{n^2} + j\left(L_0\omega - \frac{1}{\omega n^2 C_0}\right)\right| = \sqrt{\left(\frac{R_0}{n^2}\right)^2 + \left(L_0\omega - \frac{1}{\omega n^2 C_0}\right)^2}\ \Omega$$

Choice (4) is the answer.

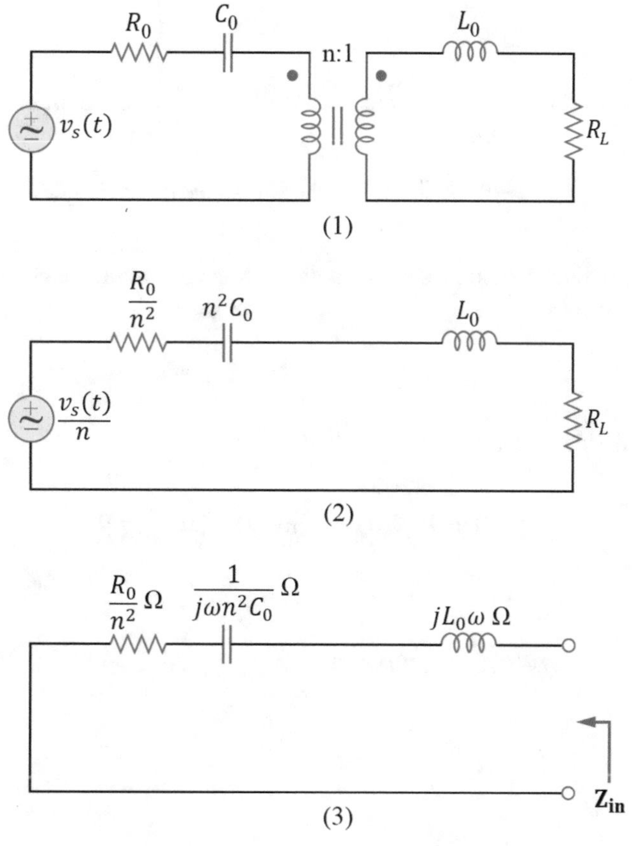

Figure 4.2 The circuit of solution of problem 4.3

4.4. This problem can be solved in time domain, since the circuit is purely resistive. Based on maximum average power transfer theorem, to transfer the maximum average power to the load, the resistance of the load (4 Ω) must be equal to the Thevenin resistance seen by the load. In other words:

$$R_L = R_{Th} \Rightarrow R_{Th} = 4\ \Omega \tag{1}$$

To calculate the Thevenin resistance, we must turn off the voltage source. In addition, to simplify the problem, we can transfer the 100 Ω resistor from the left side of the transformer to its right side, as is shown in Figure 4.3.2. The updated resistance of the resistor is as follows:

$$R'_{100\ \Omega} = (100\ \Omega) \times \left(\frac{1}{n}\right)^2 = \frac{100}{n^2}\ \Omega \tag{2}$$

From Figure 4.3.2, we see that:

$$R_{Th} = \frac{100}{n^2}\ \Omega \tag{3}$$

Solving (1) and (3):

$$\frac{100}{n^2} = 4 \Rightarrow n^2 = 25 \Rightarrow n = 5$$

Choice (4) is the answer.

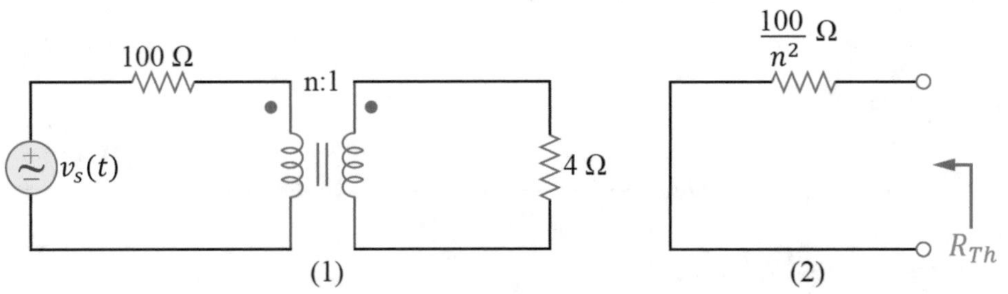

Figure 4.3 The circuit of solution of problem 4.4

4.5. This problem can be solved in time domain. As can be noticed from the circuit, the current of the left-side mesh is zero, since the circuit is open. Therefore:

$$i_1(t) = 0 \tag{1}$$

By applying KVL in the left-side mesh, we can write:

$$-v_1(t) + 3\frac{d}{dt}i_1(t) - 2\frac{d}{dt}(10\sin(10t)) = 0 \tag{2}$$

Solving (1) and (2):

$$-v_1(t) - 200\cos(10t) = 0 \Rightarrow v_1(t) = -200\cos(10t)\ V$$

Choice (2) is the answer.

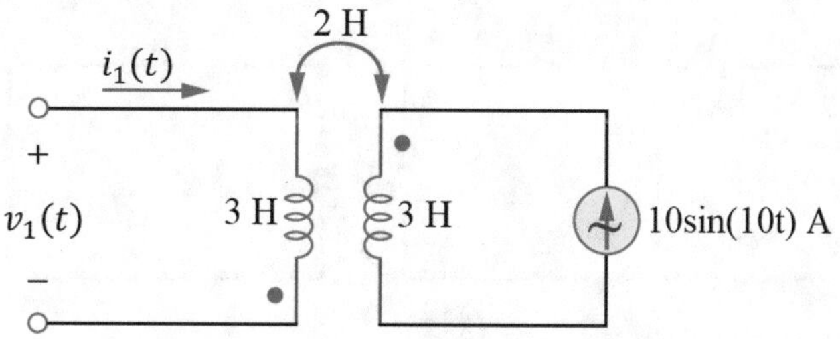

Figure 4.4 The circuit of solution of problem 4.5

4.6. Based on maximum average power transfer theorem, to transfer the maximum average power to a resistor, its resistance must be equal to the magnitude of the Thevenin impedance seen by the resistor. In other words:

$$R = |\mathbf{Z_{Th}}| \tag{1}$$

To simplify the circuit, we should transfer the whole circuit to the right side of the transformer, as is shown in Figure 4.5.2. The updated parameters are as follows:

$$R'_{3\ \Omega} = (3\ \Omega) \times \left(\frac{1}{2}\right)^2 = \frac{3}{4}\ \Omega \tag{2}$$

$$L'_{5\ H} = (5\ H) \times \left(\frac{1}{2}\right)^2 = \frac{5}{4}\ H \tag{3}$$

$$v'_s(t) = \cos(t) \times \left(\frac{1}{2}\right) = \frac{1}{2}\cos(t) V \tag{4}$$

Figure 4.5.3 shows the circuit in frequency domain, while the voltage source has been turned off. The impedances of the components are as follows:

$$\mathbf{Z}_{\frac{3}{4}\,\mathbf{\Omega}} = \frac{3}{4}\ \Omega \tag{5}$$

$$\mathbf{Z}_{\frac{5}{4}\,\mathbf{H}} = j\omega L = j \times 1 \times \frac{5}{4} = j\frac{5}{4}\ \Omega \tag{6}$$

$$\mathbf{Z_{4\ F}} = \frac{1}{j\omega C} = \frac{1}{j \times 1 \times 4} = -j\frac{1}{4}\ \Omega \tag{7}$$

The Thevenin impedance can be calculated as follows:

$$\mathbf{Z_{Th}} = \frac{3}{4} + j\frac{5}{4} + \left(-j\frac{1}{4}\right) = \left(\frac{3}{4} + j\right)\ \Omega \tag{8}$$

Solving (1) and (8):

$$R = \left|\frac{3}{4} + j\right| = \sqrt{\left(\frac{3}{4}\right)^2 + 1^2} = \frac{5}{4}\ \Omega$$

Choice (4) is the answer.

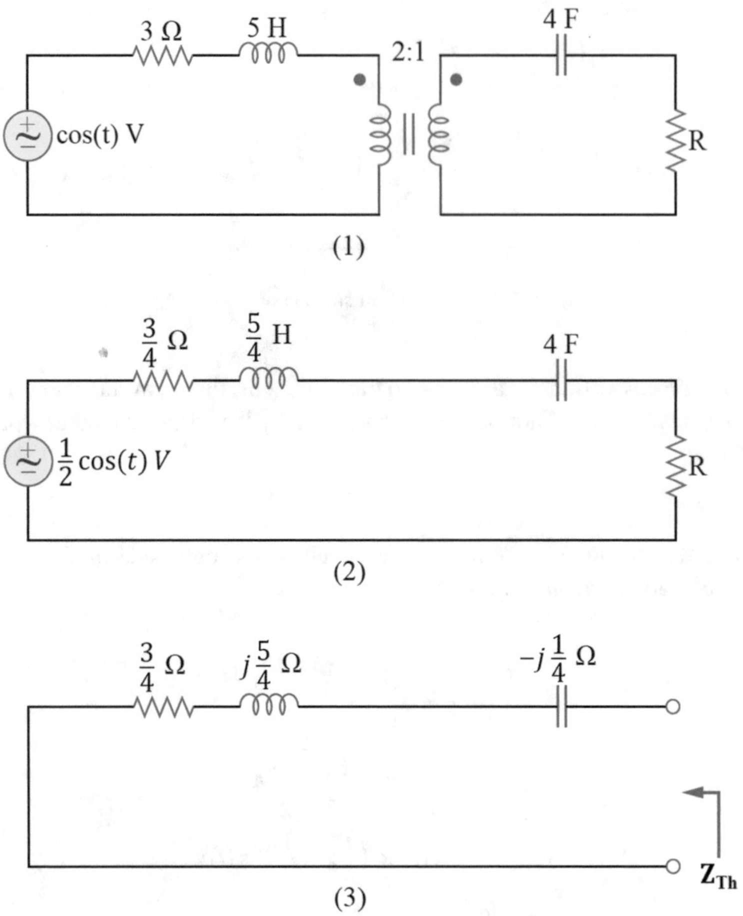

Figure 4.5 The circuit of solution of problem 4.6

4.7. To transfer the maximum average power to the load, the impedance of the load must be equal to the complex conjugate of the Thevenin impedance seen by the load. In other words:

$$\mathbf{Z_L} = \mathbf{Z_{Th}}^* \tag{1}$$

Figure 4.6.2 shows the circuit in frequency domain. The impedances of the components are as follows:

$$\mathbf{Z_{10\ \Omega}} = 10\ \Omega \tag{2}$$

$$\mathbf{Z_{0.01\ H}} = j\omega L = j \times 1000 \times 0.01 = j10\ \Omega \tag{3}$$

$$\mathbf{Z_{100\ \mu F}} = \frac{1}{j\omega C} = \frac{1}{j \times 1000 \times 100 \times 10^{-6}} = -j10\ \Omega \tag{4}$$

Herein, to calculate the Thevenin impedance (input impedance), we need to turn off the independent voltage source, as can be seen in Figure 4.6.2.

Calculating the input impedance seen from the left side of the transformer:

$$\mathbf{Z'_{in}} = 10||(-j10) + j10 = \frac{10 \times (-j10)}{10 + (-j10)} + j10 = 5 - j5 + j10 = (5 + j5)\ \Omega \tag{5}$$

Calculating the input impedance seen from the right side of the transformer:

$$\mathbf{Z_{in}} = \mathbf{Z'_{in}} \times \left(\frac{2}{1}\right)^2 = (20 + j20)\ \Omega \tag{6}$$

Solving (1) and (6):

$$Z_L = (20 + j20)^* = (20 - j20)\ \Omega$$

Choice (3) is the answer.

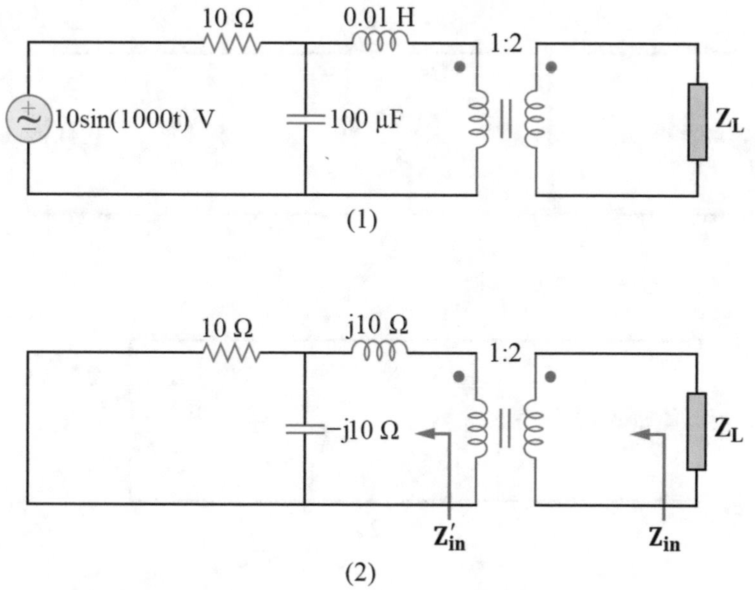

Figure 4.6 The circuit of solution of problem 4.7

4.8. To simplify the circuit, we can transfer the whole circuit to the left side of the transformer. The updated parameters are as follows:

$$R'_{1\ \Omega} = (1\ \Omega) \times \left(\frac{1}{2}\right)^2 = \frac{1}{4}\ \Omega \tag{1}$$

$$L'_{1\ H} = (1\ H) \times \left(\frac{1}{2}\right)^2 = \frac{1}{4}\ H \tag{2}$$

$$v'_o(t) = v_o(t) \times \left(\frac{1}{2}\right) = \frac{v_o(t)}{2} \tag{3}$$

By studying the circuit of Figure 4.7.2, it is realized that if the capacitor and the inductor create a resonance state, the impedance of their parallel connection will be infinite (i.e., an open circuit branch). Therefore, the whole current of the current source will flow through the resistor, and consequently the output voltage will be maximum.

As we know, the resonance frequency of the parallel connection of a capacitor and an inductor can be calculated by using the following relation:

$$\omega_0 = \frac{1}{\sqrt{LC}} = \frac{1}{\sqrt{\frac{1}{4} \times 1}} = 2\ rad/sec \tag{4}$$

Figure 4.7.3 shows the updated circuit of Figure 4.7.2 in frequency domain. The phasor of cos(ωt), that is, $1\angle 0^\circ$ is defined as the reference phasor, where "$\angle$" is the symbol of phase angle. Thus, the phasor of the current of the current source is $1\angle 0^\circ$ or 1 A.

By using Ohm's law for the resistor, we have:

$$\frac{\mathbf{V_o}}{2} = 1 \times \frac{1}{4} \Rightarrow \mathbf{V_o} = \frac{1}{2}\ \text{V} \tag{5}$$

Transferring to time domain:

$$v_o(t) = \frac{1}{2}\cos(2t)\ V$$

Choice (1) is the answer.

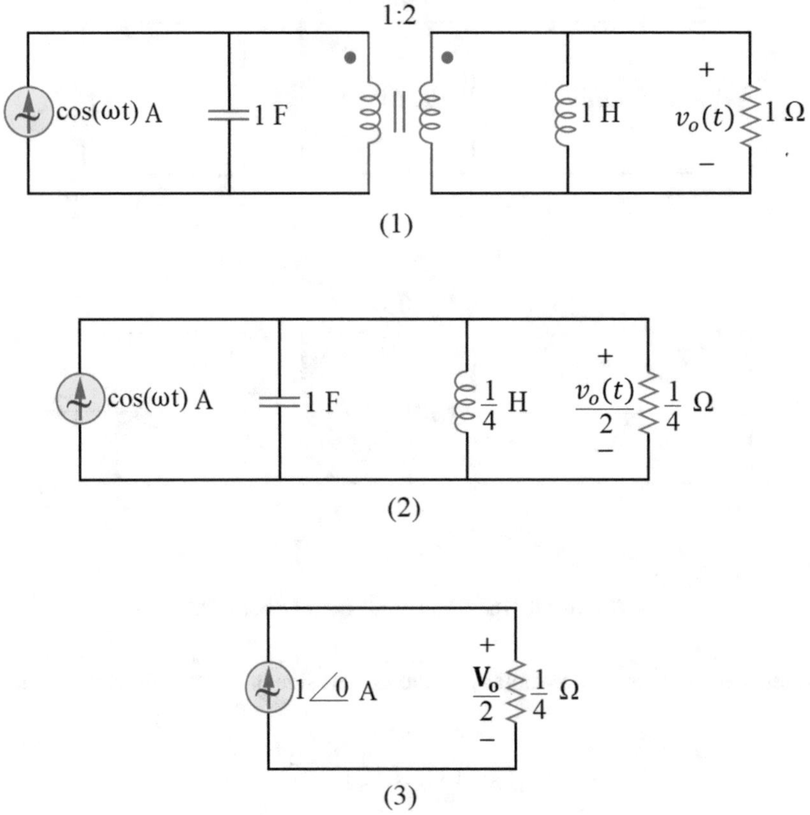

Figure 4.7 The circuit of solution of problem 4.8

4.9. This problem can be solved in time domain, since the circuit does not include any capacitor or inductor.

To simplify the problem solution, we should transfer all parts of the circuit to one side. Before that, we need to apply source transformation theorem for the parallel connection of the current source and the 4 Ω resistor, as can be seen in Figure 4.8.2 and in the following:

$$v_s(t) = 4 \times 3\sin(t) = 12\sin(t)\ V \tag{1}$$

Transferring the left-side circuit of the left-side transformer to its right side, as is shown in Figure 4.8.3, the updated parameters are as follows:

$$R'_{4\ \Omega} = (4\ \Omega) \times \left(\frac{5}{1}\right)^2 = 100\ \Omega \tag{2}$$

$$v'_s(t) = 12\sin(t) \times \left(\frac{5}{1}\right) = 60\sin(t)\ V \tag{3}$$

Combining the 100 Ω and 25 Ω resistors and transferring the left-side circuit of the right-side transformer to its right side, as is shown in Figure 4.8.4:

$$R''_{100+25\ \Omega} = (100\ \Omega + 25\ \Omega) \times \left(\frac{1}{5}\right)^2 = 5\ \Omega \tag{4}$$

$$v'_s(t) = 60 \sin(t) \times \left(\frac{1}{5}\right) = 12 \sin(t)\ V \tag{5}$$

The circuit of Figure 4.8.4 is simplified and shown in the form of Norton equivalent circuit in Figure 4.8.5 by using source transformation theorem as follows:

$$i_s(t) = \frac{12 \sin(t)}{5+7} = \sin(t)\ A \tag{6}$$

Based on Norton theorem, we can conclude that:

$$R_N = 12\ \Omega$$

$$i_N(t) = \sin(t)\ A$$

Choice (1) is the answer.

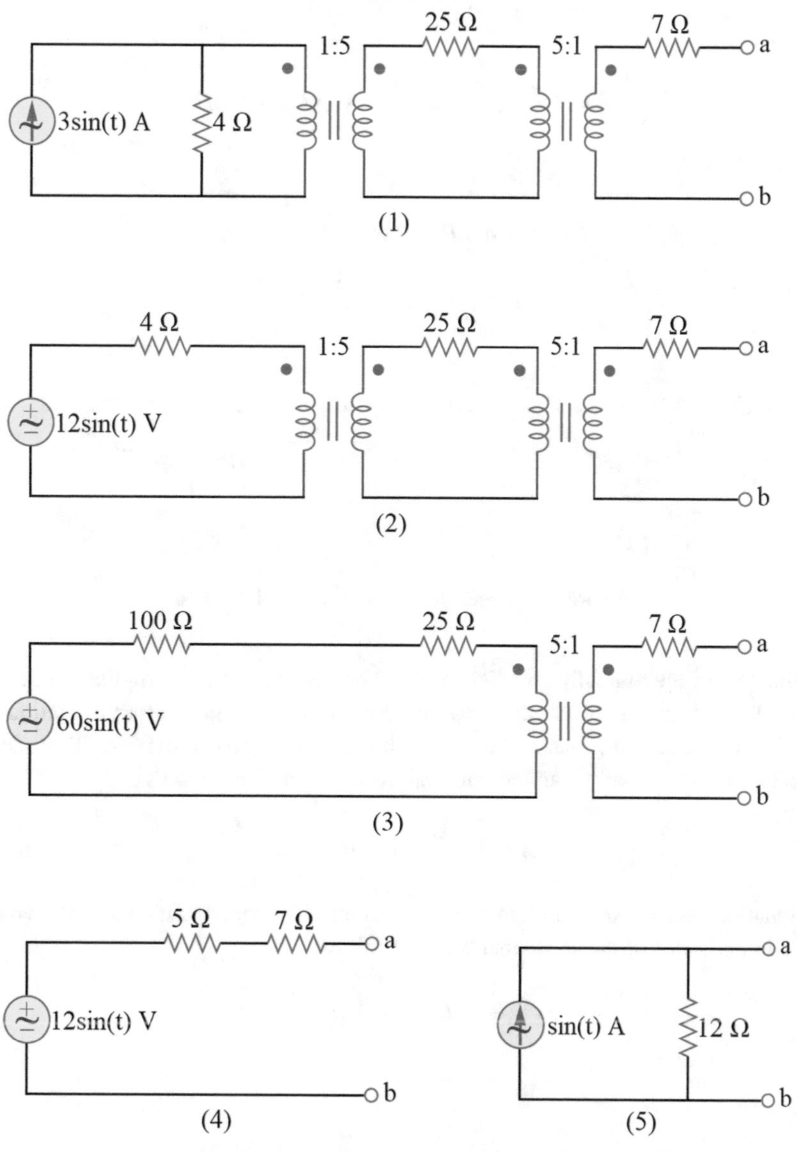

Figure 4.8 The circuit of solution of problem 4.9

4.10. Based on the information given in the problem, we know that:

$$L_{eq-1} = 6\ mH \quad (1)$$

$$L_{eq-2} = 2\ mH \quad (2)$$

The equivalent inductance of the circuit, in the first test, can be calculated as follows:

$$L_{eq-1} = L_1 + L_2 + 2M \quad (3)$$

Herein, plus sign is applied for the mutual inductance, due to the direct position of the dots of the inductors. The equivalent inductance of the circuit, in the second test, is:

$$L_{eq-2} = L_1 + L_2 - 2M \quad (4)$$

Herein, minus sign is applied for the mutual inductance because of the inverse position of the dots of the inductors.

Solving (3) and (4):

$$L_{eq-1} + L_{eq-2} = 2(L_1 + L_2) \quad (5)$$

Solving (1), (2), and (5):

$$8\ mH = 2(L_1 + L_2) \Rightarrow L_1 + L_2 = 4\ mH \quad (6)$$

Solving (1), (3), and (6):

$$6\ mH = 4\ mH + 2M \Rightarrow M = 1\ mH$$

Choice (4) is the answer.

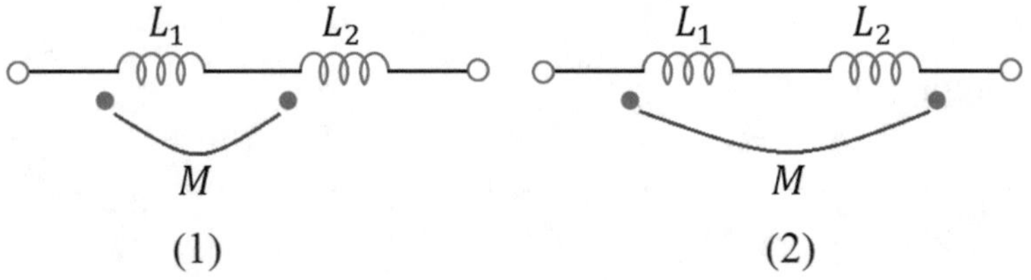

Figure 4.9 The circuit of solution of problem 4.10

4.11. The positions of the dots of the mutually coupled coils can be determined by using the right-hand rule. As is shown in Figure 4.10.1, the magnetic fluxes of the coils oppose each other, for an arbitrary direction of the current flowing through the coils. Therefore, the positions of the dots of the coils must be like the ones illustrated in Figure 4.10.2. The equivalent inductance of the mutually coupled coils can be calculated as follows:

$$L_{eq} = 1 + 2 - 2 \times 0.5 = 2\ H \quad (1)$$

Based on the information given in the problem, $\omega = 1\ rad/sec$. The simplified circuit is shown in frequency domain in Figure 4.10.3. The impedances of the components are as follows:

$$\mathbf{Z}_{1\ \mathbf{H}} = j\omega L = j \times 1 \times 1 = j\ \Omega \quad (2)$$

$$\mathbf{Z}_{2\,\mathbf{H}} = j\omega L_{eq} = j \times 1 \times 2 = j2\,\Omega \tag{3}$$

$$\mathbf{Z}_{2\,\Omega} = 2\,\Omega \tag{4}$$

Therefore:

$$\mathbf{Z}_{\mathrm{ab}} = j + j2 + 2 = (2 + j3)\,\Omega$$

Choice (2) is the answer.

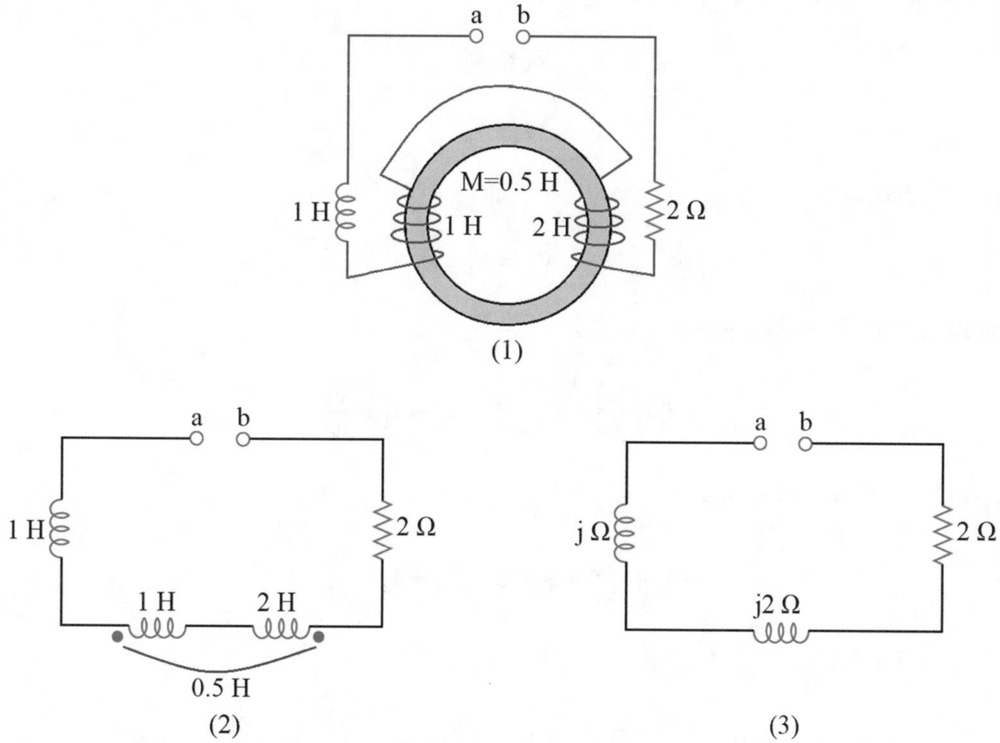

Figure 4.10 The circuit of solution of problem 4.11

4.12. The circuit of Figure 4.11.1 illustrates the primary circuit in frequency domain. The impedance of each inductor is as follows:

$$\mathbf{Z}_{1\,\mathbf{H}} = j\omega L = j\omega \times 1 = j\omega\,\Omega \tag{1}$$

To determine the inductance matrix of the circuit, we need to find a relation between the phasors of the primary and secondary voltages and currents of the circuit in the following matrix form:

$$\begin{bmatrix} \mathbf{V}'_1 \\ \mathbf{V}'_2 \end{bmatrix} = j\omega \begin{bmatrix} L_{11} & L_{12} \\ L_{21} & L_{22} \end{bmatrix} \begin{bmatrix} \mathbf{I}'_1 \\ \mathbf{I}'_2 \end{bmatrix} \tag{2}$$

As we know, the relation below exists between the primary and secondary voltages, currents, and the numbers of turns of a transformer.

$$\frac{\mathbf{V}_1}{\mathbf{V}_2} = -\frac{\mathbf{I}_2}{\mathbf{I}_1} = \frac{n_1}{n_2} \tag{3}$$

Based on the information given in the problem:

$$n_1 = n_2 = 1 \tag{4}$$

Solving (3) and (4):

$$\frac{\mathbf{V}_1}{\mathbf{V}_2} = -\frac{\mathbf{I}_2}{\mathbf{I}_1} = 1 \Rightarrow \begin{cases} \mathbf{V}_1 = \mathbf{V}_2 & (5) \\ \mathbf{I}_2 = -\mathbf{I}_1 & (6) \end{cases}$$

From the circuit of Figure 4.11.2, it is seen that:

$$\mathbf{V}_1 = \mathbf{V}_1' \tag{7}$$

$$\mathbf{V}_2 = \mathbf{V}_2' \tag{8}$$

Solving (5), (7), and (8):

$$\mathbf{V}_1' = \mathbf{V}_2' \tag{9}$$

Applying KCL in the left-side node:

$$-\mathbf{I}_1' + \frac{\mathbf{V}_1'}{j\omega} + \mathbf{I}_1 = 0 \Rightarrow \mathbf{I}_1 = \mathbf{I}_1' - \frac{\mathbf{V}_1'}{j\omega} \tag{10}$$

Applying KCL in the right-side node:

$$-\mathbf{I}_2' + \frac{\mathbf{V}_2'}{j\omega} + \mathbf{I}_2 = 0 \Rightarrow \mathbf{I}_2 = \mathbf{I}_2' - \frac{\mathbf{V}_2'}{j\omega} \tag{11}$$

Solving (6), (10), and (11):

$$\mathbf{I}_2' - \frac{\mathbf{V}_2'}{j\omega} = -\left(\mathbf{I}_1' - \frac{\mathbf{V}_1'}{j\omega}\right) \xrightarrow{Using\ (9)} \mathbf{I}_1' - \frac{2\mathbf{V}_1'}{j\omega} + \mathbf{I}_2' = 0 \Rightarrow \mathbf{V}_1' = \frac{j\omega}{2}\left(\mathbf{I}_1' + \mathbf{I}_2'\right) \tag{12}$$

Solving (9) and (12):

$$\mathbf{V}_2' = \frac{j\omega}{2}\left(\mathbf{I}_1' + \mathbf{I}_2'\right) \tag{13}$$

Solving (12), (13), and (2):

$$\begin{bmatrix} \mathbf{V}_1' \\ \mathbf{V}_2' \end{bmatrix} = j\omega \begin{bmatrix} \frac{1}{2} & \frac{1}{2} \\ \frac{1}{2} & \frac{1}{2} \end{bmatrix} \begin{bmatrix} \mathbf{I}_1' \\ \mathbf{I}_2' \end{bmatrix}$$

Choice (4) is the answer.

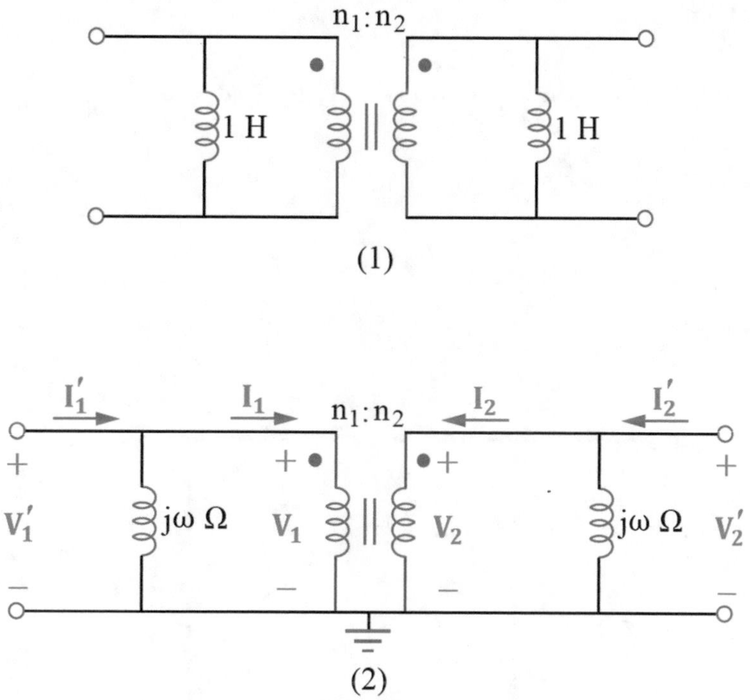

Figure 4.11 The circuit of solution of problem 4.12

4.13. The main circuit is shown in Figure 4.12.2 in frequency domain. The impedances of the components are as follows:

$$\mathbf{Z}_{3\ \mathrm{H}} = j\omega L = j\omega \times 3\ \Omega \tag{1}$$

$$\mathbf{Z}_{4\ \mathrm{H}} = j\omega L = j\omega \times 4\ \Omega \tag{2}$$

$$\mathbf{Z}_{2\ \mathrm{H}} = j\omega L = j\omega \times 2\ \Omega \tag{3}$$

$$\mathbf{Z}_{2\ \mathrm{H}} = j\omega M = j\omega \times 2\ \Omega \tag{4}$$

To determine the equivalent inductance of the circuit in frequency domain, we need to connect a test source (e.g., test voltage source) to the terminal (see Figure 4.12.2), analyze the circuit, and determine the value of $\frac{\mathbf{V_t}}{j\omega \mathbf{I_t}}$. Then:

$$L_{eq} = \frac{\mathbf{V_t}}{j\omega \mathbf{I_t}} \tag{5}$$

In this problem, mesh analysis is the best approach to analyze the circuit.

Applying KVL in the left-side mesh:

$$-\mathbf{V_t} + j\omega \times 3 \times (\mathbf{I_t} - \mathbf{I}) - j\omega \times 2 \times \mathbf{I} = 0 \Rightarrow \ -\mathbf{V_t} + j3\omega \mathbf{I_t} - j5\omega \mathbf{I} = 0 \tag{6}$$

Applying KVL in the right-side mesh:

$$j\omega \times 3 \times \mathbf{I} + j\omega \times 4 \times \mathbf{I} + j\omega \times 2 \times (\mathbf{I} - \mathbf{I_t}) + j\omega \times 2 \times \mathbf{I} + j\omega \times 3 \times (\mathbf{I} - \mathbf{I_t}) + j\omega \times 2 \times \mathbf{I} = 0$$

$$\Rightarrow j16\omega \mathbf{I} - j5\omega \mathbf{I_t} = 0 \Rightarrow \mathbf{I} = \frac{5}{16}\mathbf{I_t} \tag{7}$$

Solving (6) and (7):

$$-\mathbf{V_t} + j3\omega\mathbf{I_t} - j5\omega \times \frac{5}{16}\mathbf{I_t} = 0 \Rightarrow -\mathbf{V_t} + j\frac{23}{16}\omega\mathbf{I_t} = 0 \Rightarrow \frac{\mathbf{V_t}}{j\omega\mathbf{I_t}} = \frac{23}{16} \quad (8)$$

Solving (5) and (8):

$$L_{eq} = \frac{23}{16}\,H$$

Choice (1) is the answer.

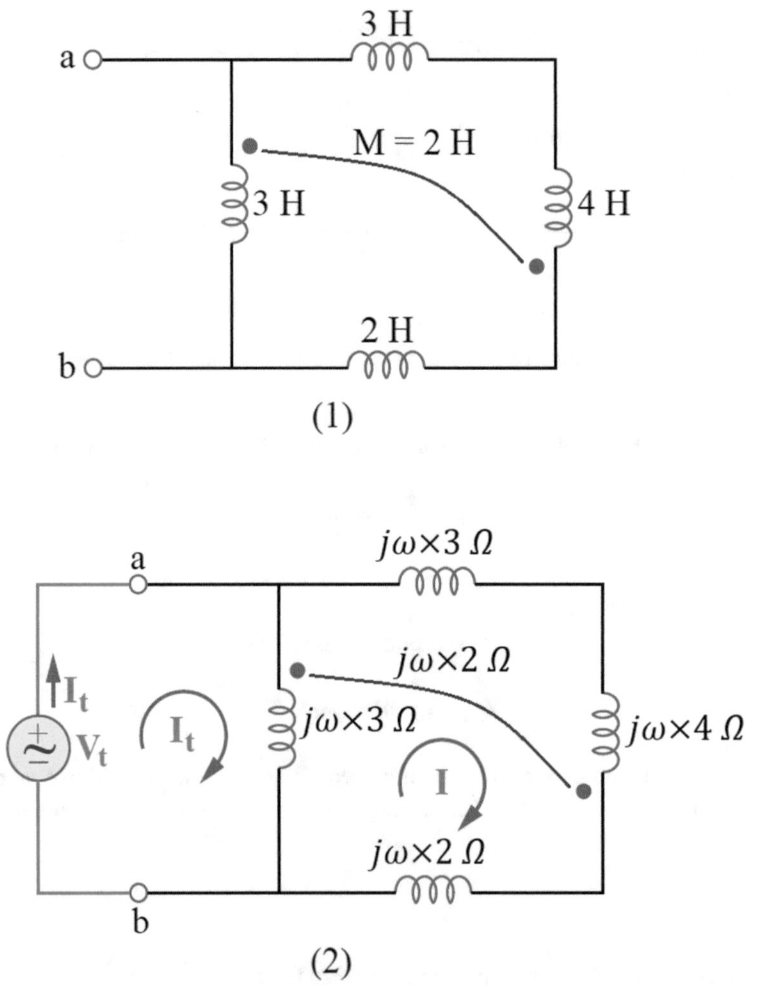

Figure 4.12 The circuit of solution of problem 4.13

4.14. To simplify the circuit, we should transfer the circuit of the right side of the transformer to its left side. The values of inductance and the mutual inductance are updated, as can be seen in the following.

$$M'_{4\,H} = (4\,H) \times \left(\frac{1}{2}\right) = 2\,H \quad (1)$$

$$L'_{20\,H} = (20\,H) \times \left(\frac{1}{2}\right)^2 = 5\,H \quad (2)$$

The updated circuit is shown in Figure 4.13.2. As can be seen, due to the inverse positions of the dots of the transformer's coils, the position of one of the dots of the coils of the inductors needs to change.

Since the circuit includes mutually coupled inductors, we need to apply a test source to determine the equivalent inductance of the circuit. Then, we need to analyze the circuit and determine the value of $\frac{\mathbf{V_t}}{j\omega\mathbf{I_t}}$ that will give us the equivalent inductance of the circuit (L_{eq}).

Figure 4.13.3 shows the circuit of Figure 4.13.2 in frequency domain, where a test voltage source with the voltage and current of $\mathbf{V_t}$ and $\mathbf{I_t}$, respectively, has been connected to its terminal. Herein, it is assumed the angular frequency of the test source is $\omega = 1\ rad/sec$. The impedances of the components are as follows:

$$\mathbf{Z_{1\,H}} = j\omega L = j \times 1 \times 1 = j\,\Omega \tag{3}$$

$$\mathbf{Z_{5\,H}} = j\omega L = j \times 1 \times 5 = j5\,\Omega \tag{4}$$

$$\mathbf{Z_{2\,H}} = j\omega M = j \times 1 \times 2 = j2\,\Omega \tag{5}$$

KVL in the right-side mesh (Figure 4.13.3):

$$-j\mathbf{I_1} - j2\mathbf{I_2} + j5\mathbf{I_2} + j2\mathbf{I_1} = 0 \Rightarrow j\mathbf{I_1} + j3\mathbf{I_2} = 0 \Rightarrow \mathbf{I_1} = -3\mathbf{I_2} \tag{6}$$

Applying KCL in the top node (Figure 4.13.3):

$$\mathbf{I_t} = \mathbf{I_1} + \mathbf{I_2} \xRightarrow{Using\ (6)} \mathbf{I_t} = -3\mathbf{I_2} + \mathbf{I_2} = -2\mathbf{I_2} \Rightarrow \mathbf{I_2} = -\tfrac{1}{2}\mathbf{I_t}$$

KVL in the left-side mesh:

$$-\mathbf{V_t} + j\mathbf{I_1} + j2\mathbf{I_2} = 0 \xRightarrow{Using\ (6)} -\mathbf{V_t} + j(-3\mathbf{I_2}) + j2\mathbf{I_2} = 0 \Rightarrow -\mathbf{V_t} - j\mathbf{I_2} = 0$$

$$\xRightarrow{Using\ (7)} -\mathbf{V_t} - j\left(-\frac{1}{2}\mathbf{I_t}\right) = 0 \Rightarrow \frac{\mathbf{V_t}}{j\mathbf{I_t}} = \frac{1}{2} \Rightarrow L_{eq} = \frac{1}{2}\ H$$

Choice (2) is the answer.

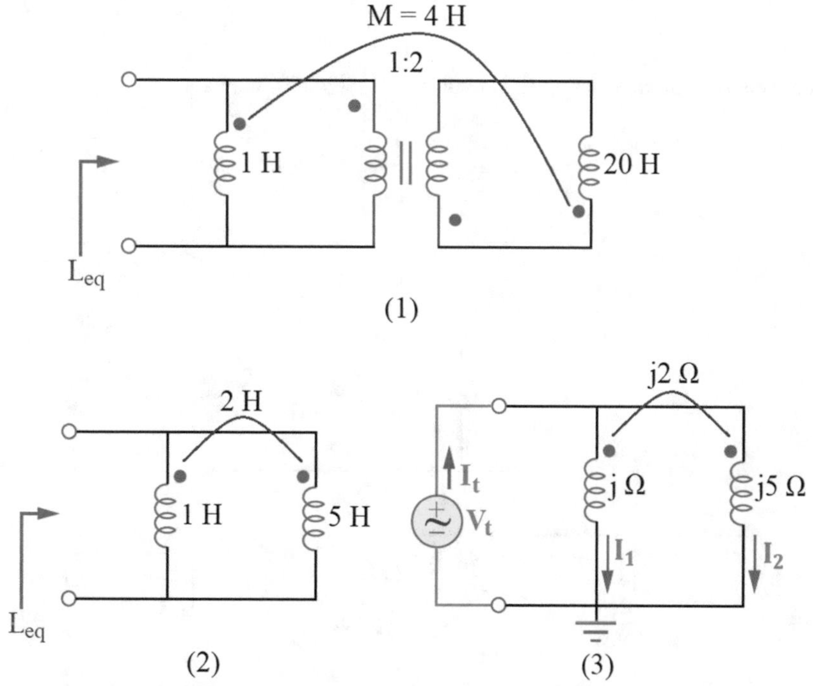

Figure 4.13 The circuit of solution of problem 4.14

4.15. The resonance frequency of the parallel connection of the single capacitor and the mutually coupled inductors can be calculated by using the following relation, where L_{eq} is the equivalent inductance of the mutually coupled inductors.

$$\omega_0 = \frac{1}{\sqrt{L_{eq}C}} \tag{1}$$

Therefore, we need to determine the inductance of L_{eq}. Since the circuit includes mutually coupled inductors, we need to apply a test source (with the voltage and current of $\mathbf{V_t}$ and $\mathbf{I_t}$), analyze the circuit, and determine the value of $\frac{\mathbf{V_t}}{j\omega \mathbf{I_t}}$, which will give us the equivalent inductance of the circuit (L_{eq}). The circuit of Figure 4.14.2 shows the mutually coupled inductors in frequency domain. Herein, it is assumed the angular frequency of the test source is $\omega = 1\ rad/sec$. The impedances of the components are as follows:

$$\mathbf{Z}_{\frac{3}{5}\mathbf{H}} = j\omega L = j \times 1 \times \frac{3}{5} = j\frac{3}{5}\ \Omega \tag{2}$$

$$\mathbf{Z}_{\frac{2}{5}\mathbf{H}} = j\omega L = j \times 1 \times \frac{2}{5} = j\frac{2}{5}\ \Omega \tag{3}$$

$$\mathbf{Z}_{\frac{1}{5}\mathbf{H}} = j\omega M = j \times 1 \times \frac{1}{5} = j\frac{1}{5}\ \Omega \tag{4}$$

KVL in the left-side mesh of the circuit of Figure 4.14.2:

$$-\mathbf{V_t} + j\frac{3}{5}(\mathbf{I_t} - \mathbf{I}) - j\frac{1}{5}\mathbf{I} = 0 \Rightarrow -\mathbf{V_t} + j\frac{3}{5}\mathbf{I_t} - j\frac{4}{5}\mathbf{I} = 0 \tag{5}$$

KVL in the right-side mesh of the circuit of Figure 4.14.2:

$$-\left(j\frac{3}{5}(\mathbf{I_t} - \mathbf{I}) - j\frac{1}{5}\mathbf{I}\right) + j\frac{2}{5}\mathbf{I} - j\frac{1}{5}(\mathbf{I_t} - \mathbf{I}) = 0 \Rightarrow -j\frac{4}{5}\mathbf{I_t} + j\frac{7}{5}\mathbf{I} = 0 \Rightarrow \mathbf{I} = \frac{4}{7}\mathbf{I_t} \tag{6}$$

Solving (5) and (6):

$$-\mathbf{V_t} + j\frac{3}{5}\mathbf{I_t} - j\frac{4}{5} \times \frac{4}{7}\mathbf{I_t} = 0 \Rightarrow -\mathbf{V_t} + j\frac{1}{7}\mathbf{I_t} = 0 \Rightarrow \frac{\mathbf{V_t}}{j\mathbf{I_t}} = \frac{1}{7} \Rightarrow L_{eq} = \frac{1}{7}\ H \tag{7}$$

Figure 4.14.3 illustrates the simplified circuit. By solving (1) and (7), we have:

$$\omega_0 = \frac{1}{\sqrt{L_{eq}C}} = \frac{1}{\sqrt{\frac{1}{7} \times \frac{1}{7}}} = 7\ rad/sec$$

Choice (1) is the answer.

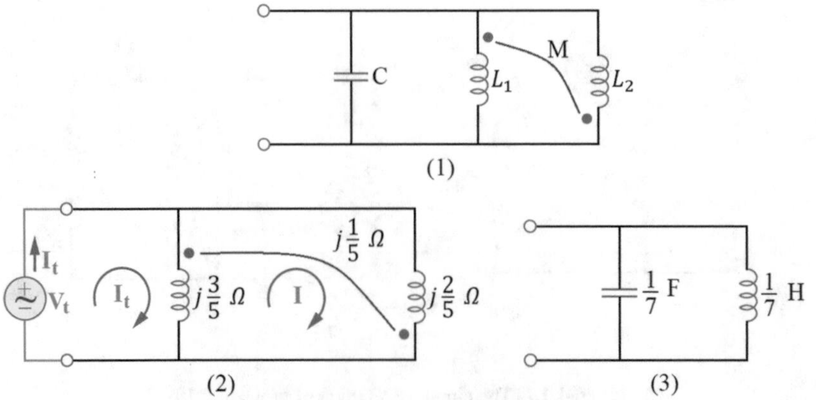

Figure 4.14 The circuit of solution of problem 4.15

4.16. To transfer the maximum average power to the right-side circuit (RSC) of the transformer, the impedance of the right-side circuit of the transformer must be equal to the complex conjugate of the Thevenin impedance seen by itself. In other words:

$$\mathbf{Z_{RSC}} = \mathbf{Z_{Th}}^* \tag{1}$$

To calculate the $\mathbf{Z_{Th}}$, we must turn off the independent voltage source, as is shown in Figure 4.15.2. Moreover, to simplify the problem, we should transfer the left-side circuit of the transformer to another side of the transformer. The updated parameters are as follows:

$$R'_{4\,\Omega} = 4 \times \left(\frac{n_2}{n_1}\right)^2 \Omega \tag{2}$$

$$C' = C \times \left(\frac{n_1}{n_2}\right)^2 F \tag{3}$$

Calculating $\mathbf{Z_{Th}}$ by using (2) and (3) (see Figure 4.15.2):

$$\mathbf{Z_{Th}} = 4 \times \left(\frac{n_2}{n_1}\right)^2 + \frac{1}{j \times 1 \times C \times \left(\frac{n_1}{n_2}\right)^2} = \left(\frac{n_2}{n_1}\right)^2 \left(4 - j\frac{1}{C}\right) \Omega \tag{4}$$

From the circuit of Figure 4.15.2:

$$\mathbf{Z_{RSC}} = 1 + j \times 1 \times 5 = (1 + j5)\,\Omega \tag{5}$$

Solving (1), (4), and (5):

$$1 + j5 = \left(\frac{n_2}{n_1}\right)^2 \left(4 - j\frac{1}{C}\right)^* \Rightarrow 1 + j5 = \left(\frac{n_2}{n_1}\right)^2 \left(4 + j\frac{1}{C}\right) \Rightarrow \begin{cases} 1 = \left(\frac{n_2}{n_1}\right)^2 \times 4 & (6) \\ 5 = \left(\frac{n_2}{n_1}\right)^2 \times \frac{1}{C} & (7) \end{cases}$$

From (6), we have:

$$\left(\frac{n_2}{n_1}\right)^2 = \frac{1}{4} \Rightarrow \left(\frac{n_2}{n_1}\right) = \frac{1}{2} \Rightarrow \left(\frac{n_1}{n_2}\right) = 2 \tag{8}$$

Solving (7) and (8):

$$5 = \frac{1}{4} \times \frac{1}{C} \Rightarrow C = \frac{1}{20} F$$

Choice (1) is the answer.

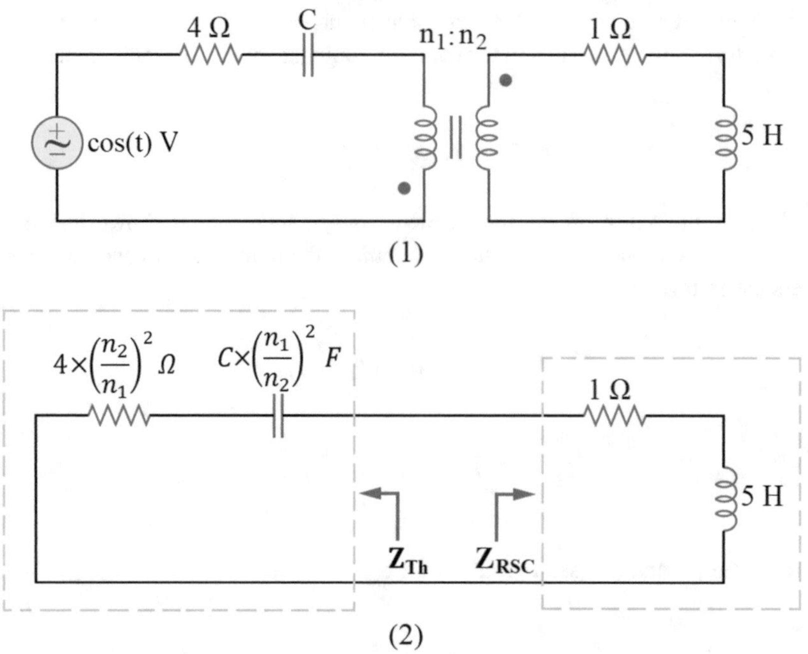

Figure 4.15 The circuit of solution of problem 4.16

4.17. This problem can be solved either in time domain or in frequency domain. To calculate the equivalent inductance of the circuit in time domain, we need to connect a test source (e.g., test voltage source) to the terminal, analyze the circuit, and determine the value of $\frac{V_t}{\frac{d}{dt}I_t}$. Then:

$$L_{eq} = \frac{V_t}{\frac{d}{dt}I_t} \tag{1}$$

Applying KVL in the right-side mesh of the circuit of Figure 4.16.2:

$$L_2\frac{d}{dt}I_2 + M\frac{d}{dt}I_t = 0 \Rightarrow I_2 = -\frac{M}{L_2}I_t \tag{2}$$

Applying KVL in the left-side mesh of the circuit of Figure 4.16.2:

$$-V_t + L_1\frac{d}{dt}I_t + M\frac{d}{dt}I_2 = 0 \xrightarrow{\text{Using (2)}} -V_t + L_1\frac{d}{dt}I_t + M\frac{d}{dt}\left(-\frac{M}{L_2}I_t\right) = 0$$

$$\Rightarrow -V_t + \left(L_1 - \frac{M^2}{L_2}\right)\frac{d}{dt}I_t = 0 \Rightarrow \frac{V_t}{\frac{d}{dt}I_t} = L_1 - \frac{M^2}{L_2} \tag{3}$$

Solving (1) and (3):

$$L_{eq} = L_1 - \frac{M^2}{L_2} \tag{4}$$

Based on the definition, the coupling coefficient (k) can be determined as follows:

$$k = \frac{M}{\sqrt{L_1L_2}} \tag{5}$$

Solving (4) and (5):

$$L_{eq} = L_1 - \frac{k^2 L_1 L_2}{L_2} = L_1(1 - k^2)$$

Choice (3) is the answer.

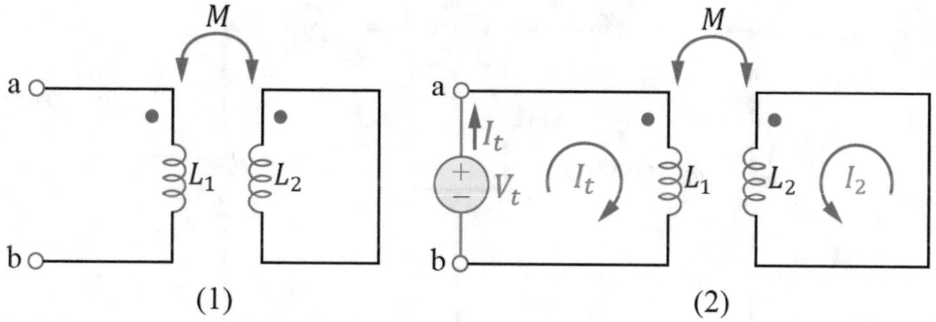

Figure 4.16 The circuit of solution of problem 4.17

4.18. Figure 4.17.2 illustrates the main circuit in frequency domain. The phasor of sin(2t), that is, $1\angle 0^\circ$ is defined as the reference phasor, where "$\angle$" is the symbol of phase angle. Thus, the phasor of the voltage of the voltage source is $10\angle 0^\circ$ or 10 V. The impedances of the components are as follows:

$$\mathbf{Z}_{1\,\Omega} = 1\,\Omega \quad (1)$$

$$\mathbf{Z}_{2\,\mathrm{H}} = j\omega L = j \times 2 \times 2 = j4\,\Omega \quad (2)$$

$$\mathbf{Z}_{0.5\,\mathrm{H}} = j\omega M = j \times 2 \times 0.5 = j\,\Omega \quad (3)$$

$$\mathbf{Z}_{1\,\mathrm{H}} = j\omega L = j \times 2 \times 1 = j2\,\Omega \quad (4)$$

$$\mathbf{Z}_{0.5\,\mathrm{F}} = \frac{1}{j\omega C} = \frac{1}{j \times 2 \times 0.5} = -j\,\Omega \quad (5)$$

$$\mathbf{Z}_{2\,\Omega} = 2\,\Omega \quad (6)$$

The problem can be solved by using mesh analysis as is presented in the following.

KVL in the left-side mesh:

$$-10 + \mathbf{I}_1 + j4\mathbf{I}_1 - j\mathbf{I}_2 - j(\mathbf{I}_1 - \mathbf{I}_2) = 0 \Rightarrow -10 + (1 + j3)\mathbf{I}_1 = 0 \Rightarrow \mathbf{I}_1 = \frac{10}{1 + j3} \quad (7)$$

KVL in the right-side mesh:

$$-j(\mathbf{I}_2 - \mathbf{I}_1) + j2\mathbf{I}_2 - j\mathbf{I}_1 + 2\mathbf{I}_2 + j2\mathbf{I}_2 = 0 \Rightarrow (2 + j3)\mathbf{I}_2 = 0 \Rightarrow \mathbf{I}_2 = 0 \quad (8)$$

Using Ohm's law for the impedance of the capacitor:

$$\mathbf{V} = -j(\mathbf{I}_1 - \mathbf{I}_2) \quad (9)$$

Solving (7), (8), and (9):

$$\mathbf{V} = -j\left(\frac{10}{1 + j3} - 0\right) = \frac{-10j}{1 + j3} = (-3 - j)\,V \quad (10)$$

Transferring to time domain:

$$v(t) = -3\sin(2t) - \sin\left(2t + 90^{\circ}\right) = -3\sin(2t) - \cos(2t)\ V$$

Choice (1) is the answer.

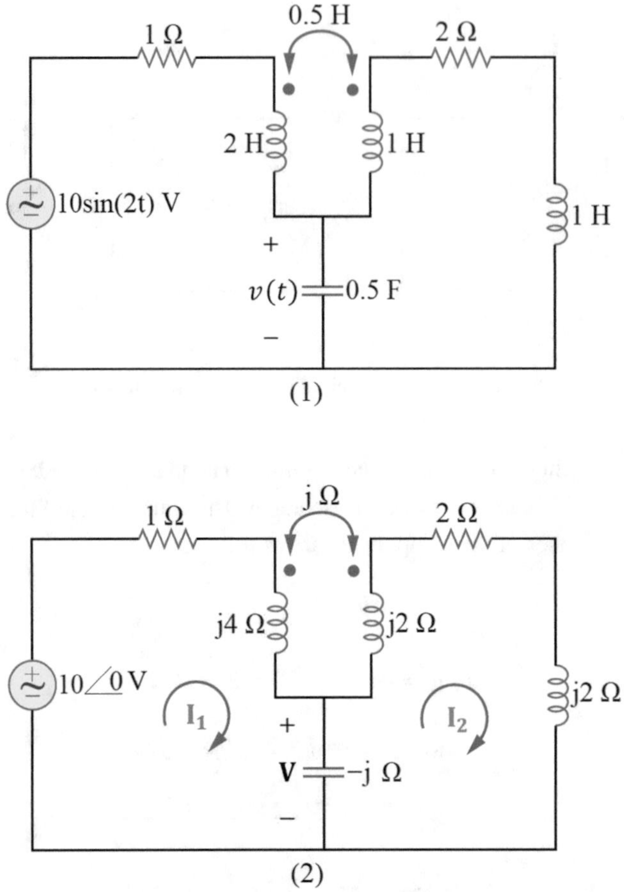

Figure 4.17 The circuit of solution of problem 4.18

4.19. First, we should apply source transformation theorem for the parallel connection of the dependent current source (DCS) and the 320 Ω resistor, as can be seen in Figure 4.18.2.

$$\mathbf{V}_{\mathrm{D.C.S.}} = 320 \times 2\mathbf{I} = 640\mathbf{I} \tag{1}$$

To simplify the circuit, we can transfer the whole circuit to the left-side mesh, as is shown in Figure 4.18.3. The updated parameters are as follows:

$$\mathbf{V}_{\mathbf{o}}' = (\mathbf{V}_{\mathbf{o}}) \times \left(\frac{1}{4}\right) = \frac{\mathbf{V}_{\mathbf{o}}}{4} \tag{2}$$

$$\mathbf{Z}_{\mathbf{C}}' = (-j480\ \Omega) \times \left(\frac{1}{4}\right)^2 = -j30\ \Omega \tag{3}$$

$$\mathbf{Z}_{\mathbf{R}}' = (320\ \Omega) \times \left(\frac{2}{1}\right)^2 = 1280\ \Omega \tag{4}$$

$$\mathbf{V}_{\mathrm{D.V.S.}}' = -(640\mathbf{I}) \times \left(\frac{2}{1}\right) = -1280\mathbf{I} \tag{5}$$

In (5), due to the inverse position of the dots of the bottom transformer, minus sign needs to be applied for the voltage of the dependent voltage source (DVS).

Applying KVL in the only mesh of the circuit of Figure 4.18.3:

$$-30 + 40\mathbf{I} - j30\mathbf{I} + 1280\mathbf{I} - 1280\mathbf{I} = 0 \Rightarrow -30 + (40 - j30)\mathbf{I} = 0 \Rightarrow \mathbf{I} = \frac{30}{40 - j30} \quad (6)$$

Using Ohm's law for the impedance of the capacitor:

$$\frac{\mathbf{V_o}}{4} = \frac{30}{40 - j30} \times (-j30) \Rightarrow \mathbf{V_o} = \frac{-j3600}{40 - j30} \Rightarrow \mathbf{V_o} = (72\angle{-53^\circ})\ V = 72e^{-j53}\ V$$

Choice (3) is the answer.

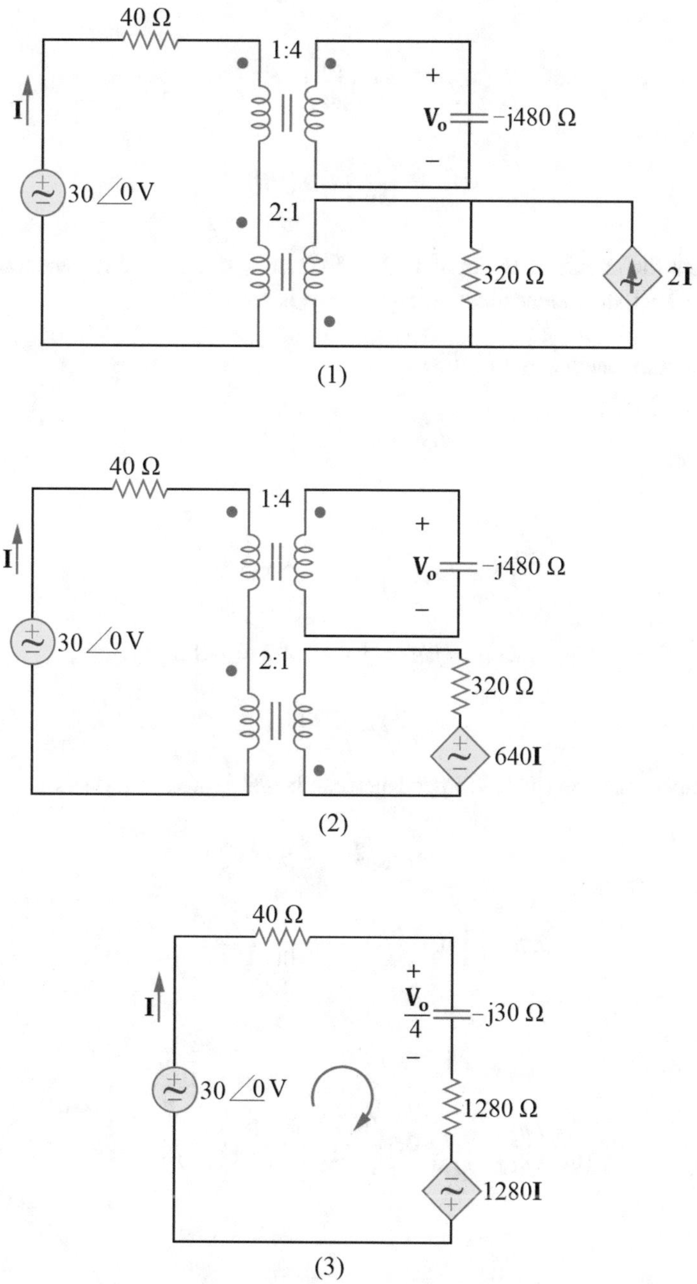

Figure 4.18 The circuit of solution of problem 4.19

4.20. This problem needs to be heuristically solved. Instead of transferring the maximum average power to 2 Ω resistor, we can assume that we want to transfer the maximum average power to the series connection of the 2 Ω resistor and the 2 H inductor. This assumption is right, since the inductor does not consume any average power.

Based on maximum average power transfer theorem, to transfer the maximum average power to the load (herein, the series connection of the resistor and inductor), the impedance of the load must be equal to the complex conjugate of the Thevenin impedance seen by the load. In other words:

$$\mathbf{Z_L} = \mathbf{Z_{Th}}^* \tag{1}$$

To simplify the problem, we should have the whole circuit in one side of the transformer, as is shown in Figure 4.19.2. Herein, the circuit of the left side of the transformer has been transferred to its right side. The value of parameters is updated as follows:

$$R'_{0.5\,\Omega} = 0.5\left(\frac{n_2}{n_1}\right)^2\ \Omega \tag{2}$$

$$C' = C\left(\frac{n_1}{n_2}\right)^2\ F \tag{3}$$

$$v(t)' = \left(\frac{n_2}{n_1}\right)\cos(10t)\ V \tag{4}$$

Figure 4.19.3 illustrates the circuit of Figure 4.19.2 in frequency domain. Since we want to calculate the Thevenin impedance seen by the load, the independent voltage source is turned off.

The impedances of the components are as follows:

$$\mathbf{Z_{R'}} = 0.5\left(\frac{n_2}{n_1}\right)^2\ \Omega \tag{5}$$

$$\mathbf{Z_{C'}} = \frac{1}{j\omega C\left(\frac{n_1}{n_2}\right)^2} = -j\frac{1}{10C}\left(\frac{n_2}{n_1}\right)^2\ \Omega \tag{6}$$

$$\mathbf{Z_{2\,H}} = j\omega L = j \times 10 \times 2 = j20\ \Omega \tag{7}$$

$$\mathbf{Z_{2\,\Omega}} = 2\ \Omega \tag{8}$$

Now, the Thevenin impedance and the load impedance can be calculated as follows:

$$\mathbf{Z_L} = (2 + j20)\ \Omega \tag{9}$$

$$\mathbf{Z_{Th}} = \left(0.5\left(\frac{n_2}{n_1}\right)^2 - j\frac{1}{10C}\left(\frac{n_2}{n_1}\right)^2\right)\ \Omega \tag{10}$$

Solving (1), (9), and (10):

$$2 + j20 = \left(0.5\left(\frac{n_2}{n_1}\right)^2 - j\frac{1}{10C}\left(\frac{n_2}{n_1}\right)^2\right)^* = 0.5\left(\frac{n_2}{n_1}\right)^2 + j\frac{1}{10C}\left(\frac{n_2}{n_1}\right)^2 \Rightarrow \begin{cases} 2 = 0.5\left(\frac{n_2}{n_1}\right)^2 & (11) \\ 20 = \frac{1}{10C}\left(\frac{n_2}{n_1}\right)^2 & (12) \end{cases}$$

$$\stackrel{(11)}{\Rightarrow} \left(\frac{n_2}{n_1}\right)^2 = 4 \Rightarrow \frac{n_2}{n_1} = 2 \Rightarrow \frac{n_1}{n_2} = 0.5 \tag{13}$$

Solving (12) and (13):

$$20 = \frac{1}{10C} \times (2)^2 \Rightarrow C = 0.02\ F$$

Choice (1) is the answer.

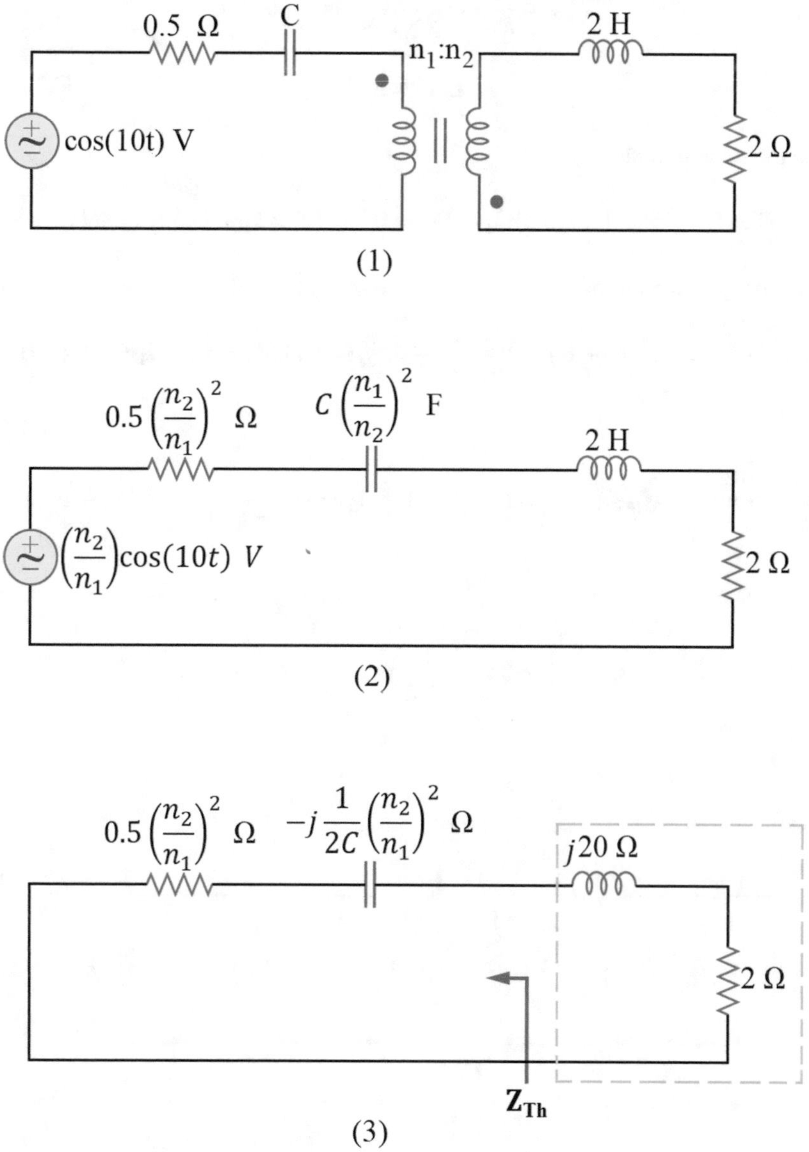

Figure 4.19 The circuit of solution of problem 4.20

4.21. The average power of a resistor can be calculated by using the relation below:

$$P = \frac{1}{2}R|\mathbf{I}|^2 \tag{1}$$

Figure 4.20.2 shows the circuit in frequency domain. The phasor of cos(2*t*), that is, $1\angle 0^\circ$ is defined as the reference phasor, where "$\angle$" is the symbol of phase angle. Thus, the phasor of the current of the current source is $1\angle 0^\circ$ or 1 *A*. The impedances of the components are as follows:

$$\mathbf{Z}_{10\ \Omega} = 10\ \Omega \tag{2}$$

$$\mathbf{Z}_{1\ \mathrm{H}} = j\omega L = j \times 2 \times 1 = j2\ \Omega \tag{3}$$

$$\mathbf{Z}_{0.5\ \mathrm{H}} = j\omega M = j \times 2 \times 0.5 = j\ \Omega \tag{4}$$

$$\mathbf{Z}_{1\ \mathrm{F}} = \frac{1}{j\omega C} = \frac{1}{j \times 2 \times 1} = -j0.5\ \Omega \tag{5}$$

$$\mathbf{Z}_{2\ \Omega} = 2\ \Omega \tag{6}$$

Applying KVL in the middle mesh:

$$j2 \times \mathbf{I}_1 + j \times 1 + (-j0.5)(\mathbf{I}_1 - \mathbf{I}_2) = 0 \Rightarrow j1.5 \times \mathbf{I}_1 + j0.5 \times \mathbf{I}_2 + j = 0 \tag{7}$$

Applying KVL in the right-side mesh:

$$2 \times \mathbf{I}_2 + (-j0.5)(\mathbf{I}_2 - \mathbf{I}_1) = 0 \Rightarrow (2 - 0.5j)\mathbf{I}_2 + j0.5\mathbf{I}_1 = 0 \Rightarrow \mathbf{I}_1 = (1 + 4j)\mathbf{I}_2 \tag{8}$$

Solving (7) and (8):

$$j1.5 \times (1 + 4j)\mathbf{I}_2 + j0.5 \times \mathbf{I}_2 + j = 0 \Rightarrow \mathbf{I}_2 = \frac{-j}{-6 + j2}\ A \Rightarrow |\mathbf{I}_2| = \frac{1}{\sqrt{40}}\ A \tag{9}$$

Solving (1) and (9):

$$P_{2\ \Omega} = \frac{1}{2} \times 2 \times \left(\frac{1}{\sqrt{40}}\right)^2 = \frac{1}{40}\ W = 25\ mW$$

Choice (1) is the answer.

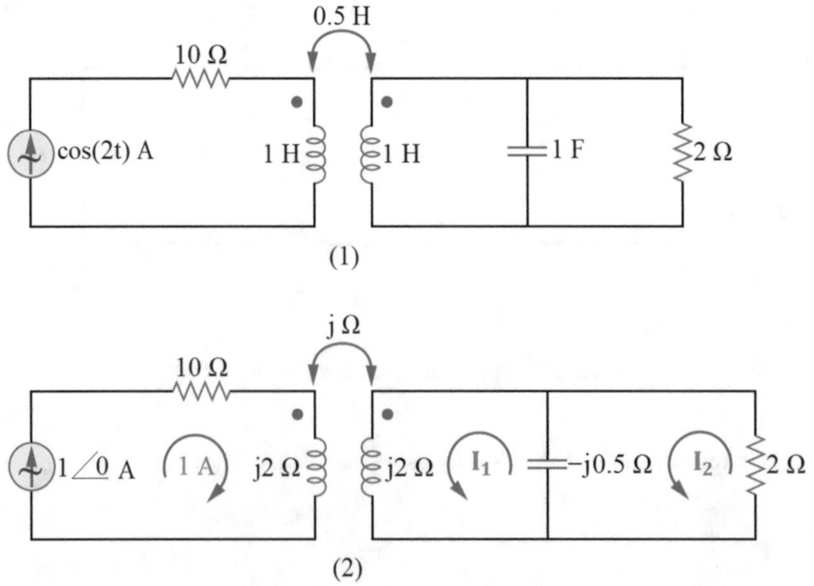

Figure 4.20 The circuit of solution of problem 4.21

4.22. To determine the Thevenin impedance or the input impedance seen from terminal a–b, we need to connect a test voltage source to the terminal, analyze the circuit, and determine the value of $\frac{\mathbf{V_t}}{\mathbf{I_t}}$. Herein, the independent voltage source needs to be turned off, as is shown in Figure 4.21.2. Figure 4.21.2 shows the circuit in frequency domain. The impedances of the components are as follows:

$$\mathbf{Z}_{3\ \mathbf{H}} = j\omega L = j \times 2 \times 3 = j6\ \Omega \quad (1)$$

$$\mathbf{Z}_{2\ \mathbf{H}} = j\omega L = j \times 2 \times 2 = j4\ \Omega \quad (2)$$

$$\mathbf{Z}_{1\ \mathbf{H}} = j\omega M = j \times 2 \times 1 = j2\ \Omega \quad (3)$$

$$\mathbf{Z}_{5\ \Omega} = 5\ \Omega \quad (4)$$

This problem can be solved by using mesh analysis as is presented in the following.

Applying KVL in the right-side mesh:

$$j4 \times \mathbf{I} + j2(\mathbf{I} - \mathbf{I_t}) + 5\mathbf{I} + j6(\mathbf{I} - \mathbf{I_t}) + j2 \times \mathbf{I} = 0 \Rightarrow -j8\mathbf{I_t} + (5 + j14)\mathbf{I} = 0$$

$$\Rightarrow \mathbf{I} = \frac{j8}{5 + j14}\mathbf{I_t} \quad (5)$$

Applying KVL in the left-side mesh:

$$-\mathbf{V_t} + j6(\mathbf{I_t} - \mathbf{I}) - j2 \times \mathbf{I} = 0 \Rightarrow -\mathbf{V_t} + j6\mathbf{I_t} - j8 \times \mathbf{I} = 0 \quad (6)$$

Solving (5) and (6):

$$-\mathbf{V_t} + j6\mathbf{I_t} - j8 \times \frac{j8}{5 + j14}\mathbf{I_t} = 0 \Rightarrow -\mathbf{V_t} + \left(\frac{320 + j430}{221}\right)\mathbf{I_t} = 0 \Rightarrow \frac{\mathbf{V_t}}{\mathbf{I_t}} = \frac{320 + j430}{221}$$

$$\Rightarrow \mathbf{Z}_{ab} = \left(\frac{320}{221} + j\frac{430}{221}\right)\ \Omega$$

Choice (3) is the answer.

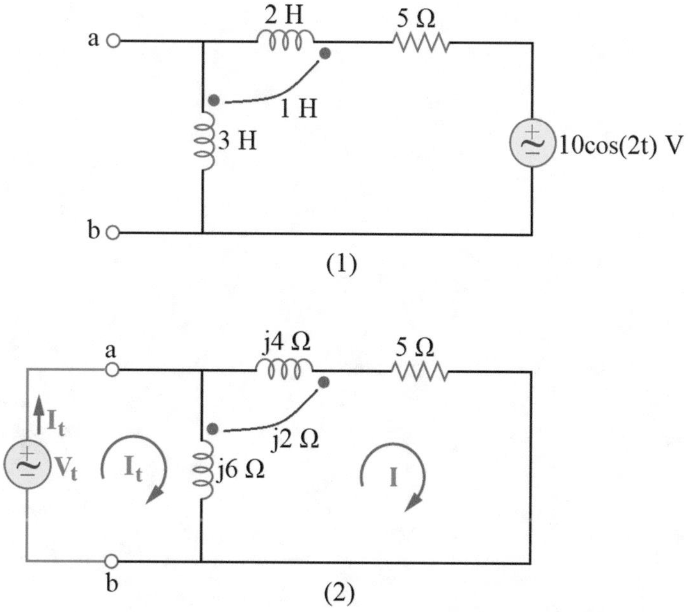

Figure 4.21 The circuit of solution of problem 4.22

4.23. To calculate the Thevenin impedance seen from a terminal, we need to apply a test source in the terminal, analyze the circuit, and calculate the value of $\frac{\mathbf{V_t}}{\mathbf{I_t}}$. To simplify the problem, we should transfer the resistor from the right side of the transformer to its left side, as is shown in Figure 4.22.2. The updated resistance of the resistor is as follows:

$$R'_{2\,\Omega} = 0.5 \times \left(\frac{2}{1}\right)^2 = 2\,\Omega \tag{1}$$

Figure 4.22.2 illustrates the circuit in frequency domain, where a test voltage source with the voltage and current of $\mathbf{V_t}$ and $\mathbf{I_t}$, respectively, has been connected to its terminal. Based on the information given in the problem:

$$\omega = 2\ rad/\,sec \tag{2}$$

The impedances of the components are as follows:

$$\mathbf{Z_{3\,H}} = j\omega L = j \times 2 \times 3 = j6\,\Omega \tag{3}$$

$$\mathbf{Z_{2\,H}} = j\omega L = j \times 2 \times 2 = j4\,\Omega \tag{4}$$

$$\mathbf{Z_{0.25\,F}} = \frac{1}{j\omega C} = \frac{1}{j \times 2 \times 0.25} = -j2\,\Omega \tag{5}$$

$$\mathbf{Z_{2\,\Omega}} = 2\,\Omega \tag{6}$$

This problem can be solved by using mesh analysis as is presented in the following.

Applying KVL in the right-side mesh of the circuit of Figure 4.22.2:

$$-\mathbf{V_t} + j6\mathbf{I_t} + j4\mathbf{I} + j6(\mathbf{I_t} - \mathbf{I}) - j2(\mathbf{I_t} - \mathbf{I}) = 0 \Rightarrow -\mathbf{V_t} + j10\mathbf{I_t} = 0 \Rightarrow \frac{\mathbf{V_t}}{\mathbf{I_t}} = j10$$

$$\Rightarrow \mathbf{Z_{in}} = j10\,\Omega \Rightarrow |\mathbf{Z_{in}}| = 10\,\Omega$$

Choice (2) is the answer.

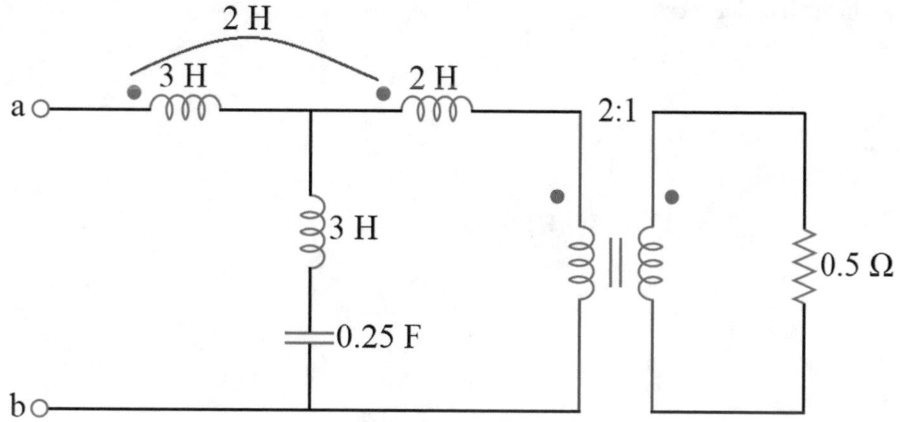

(1)

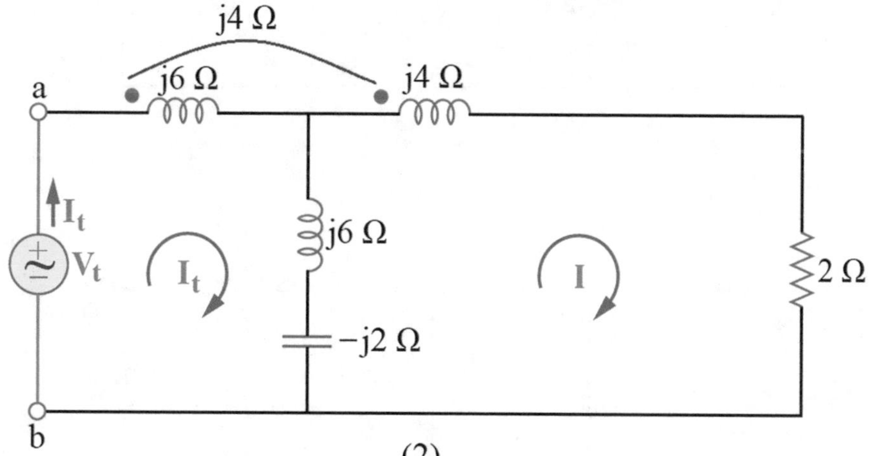

(2)

Figure 4.22 The circuit of solution of problem 4.23

4.24. As is illustrated in Figure 4.23.2, to calculate the Thevenin impedance seen from the terminal, we have connected a test voltage source to the terminal. Now, we need to analyze the circuit and calculate the value of $\frac{\mathbf{V_t}}{\mathbf{I_t}}$. To simplify the problem, we should transfer the circuit of the right side of the transformer to its left side, as is shown in Figure 4.23.2. The updated parameters are as follows:

$$R'_{20\,\Omega} = 20 \times \left(\frac{1}{4}\right)^2 = \frac{5}{4}\,\Omega \tag{1}$$

$$\mathbf{V}'_{\mathbf{D.V.S.}} = -10\mathbf{I} \times \left(\frac{1}{4}\right) = -\frac{5}{2}\mathbf{I} \tag{2}$$

In (2), minus sign must be applied for the dependent voltage source (DVS), due to the inverse position of the dots of the transformer. This problem can be solved by using mesh analysis as is presented in the following.

Simultaneously applying KCL and KVL in the right-side mesh (Figure 4.23.2):

$$-5\mathbf{I} - \frac{5}{2}\mathbf{I} + \frac{5}{4}(\mathbf{I_t} - \mathbf{I}) = 0 \Rightarrow -\frac{35}{4}\mathbf{I} + \frac{5}{4}\mathbf{I_t} = 0 \Rightarrow \mathbf{I} = \frac{1}{7}\mathbf{I_t} \tag{3}$$

Applying KVL in the left-side mesh:

$$-\mathbf{V_t} + 5\mathbf{I} = 0 \Rightarrow \mathbf{V_t} = 5\mathbf{I} \quad (4)$$

Solving (3) and (4):

$$\mathbf{V_t} = 5\left(\frac{1}{7}\mathbf{I_t}\right) \Rightarrow \frac{\mathbf{V_t}}{\mathbf{I_t}} = \frac{5}{7} \Rightarrow \mathbf{Z_{Th}} = \frac{5}{7}\ \Omega$$

Choice (4) is the answer.

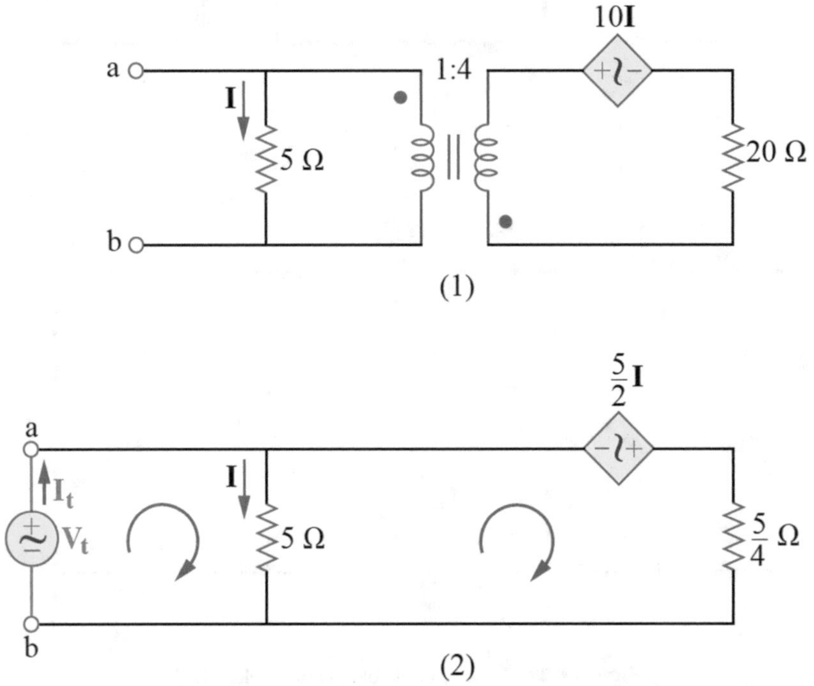

Figure 4.23 The circuit of solution of problem 4.24

4.25. This problem can be solved in time domain, since the circuit is purely resistive. In Figure 4.24.2, the circuit of the right side of the transformer has been transferred to its left side. The updated resistance of the resistor is as follows:

$$R'_{240\ \Omega} = (240\ \Omega) \times \left(\frac{1}{4}\right)^2 = 15\ \Omega \quad (1)$$

Applying KVL in the loop:

$$-240\sqrt{2}\cos(\omega t) + 5i_1(t) + 15i_1(t) = 0 \Rightarrow i_1(t) = 12\sqrt{2}\cos(\omega t)\ A \quad (2)$$

Therefore, the root mean square (rms) value of the primary current is:

$$I_{1,rms} = \frac{12\sqrt{2}}{\sqrt{2}} = 12\ A \quad (3)$$

The relation below exists between the primary and secondary currents and numbers of turns of the transformer.

$$-\frac{I_{2,rms}}{I_{1,rms}} = \frac{1}{4} \quad (4)$$

Solving (3) and (4):

$$-\frac{I_{2,rms}}{12}=\frac{1}{4}\Rightarrow I_{2,rms}=-3\ A$$

Choice (2) is the answer.

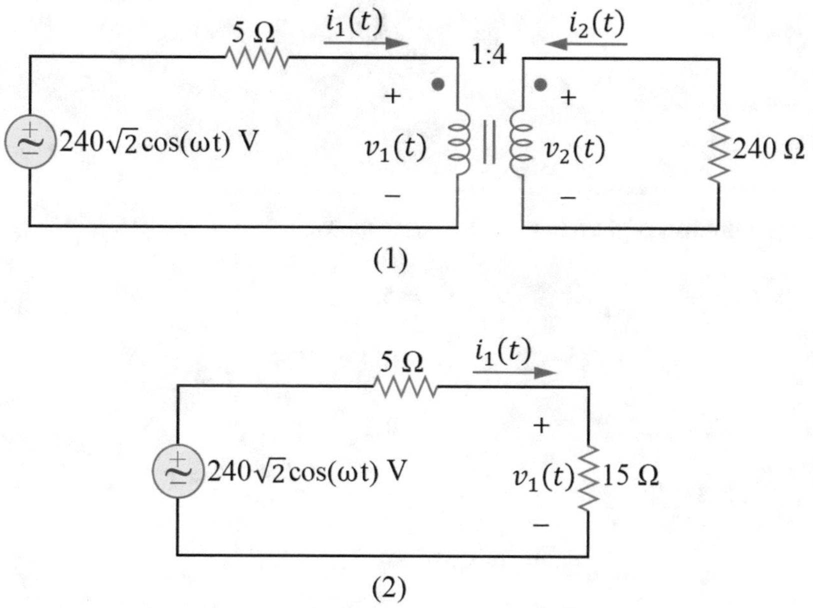

Figure 4.24 The circuit of solution of problem 4.25

4.26. Figure 4.25.2 shows the circuit in frequency domain. The impedances of the components are as follows:

$$\mathbf{Z_{L1}}=j\omega L_1 \quad (1)$$

$$\mathbf{Z_{L2}}=j\omega L_2 \quad (2)$$

$$\mathbf{Z_{M}}=j\omega M \quad (3)$$

The current–voltage relations of the mutually coupled inductors can be determined as follows:

Applying KVL in the left-side mesh:

$$\mathbf{V_1}=j\omega L_1\mathbf{I_1}+j\omega M\mathbf{I_2} \quad (4)$$

Applying KVL in the right-side mesh:

$$\mathbf{V_2}=j\omega L_2\mathbf{I_2}+j\omega M\mathbf{I_1} \quad (5)$$

Test 1: Based on the information given in the first test, we know that:

$$\mathbf{I_2}=0 \quad (6)$$

$$L_{eq1}=5\ H \quad (7)$$

Solving (4) and (6):

$$\mathbf{V_1} = j\omega L_1\mathbf{I_1} + 0 = j\omega L_1\mathbf{I_1} \Rightarrow L_1 = \frac{\mathbf{V_1}}{j\omega\mathbf{I_1}} \tag{8}$$

Based on the definition:

$$L_{eq1} \equiv \frac{\mathbf{V_1}}{j\omega\mathbf{I_1}} \tag{9}$$

Solving (7), (8), and (9):

$$L_1 = 5\ H \tag{10}$$

Test 2: Based on the information given in the second test, we know that:

$$\mathbf{V_1} = 5\ V \tag{11}$$

$$\mathbf{V_2} = 16\ V \tag{12}$$

$$\mathbf{I_2} = 0\ A \tag{13}$$

Solving (4), (10), (11), and (13):

$$5 = j\omega \times 5\mathbf{I_1} + 0 = j\omega \times 5\mathbf{I_1} \tag{14}$$

Solving (5), (12), and (13):

$$16 = 0 + j\omega M\mathbf{I_1} = j\omega M\mathbf{I_1} \tag{15}$$

Solving (14) and (15):

$$\frac{16}{5} = \frac{j\omega M\mathbf{I_1}}{j\omega \times 5\mathbf{I_1}} \Rightarrow M = 16\ H \tag{16}$$

Test 3: Based on the information given in the third test, we know that:

$$\mathbf{V_1} = 4\ V \tag{17}$$

$$\mathbf{I_1} = 0\ A \tag{18}$$

$$\mathbf{V_2} = 20\ V \tag{19}$$

Solving (4), (17), and (18):

$$4 = 0 + j\omega M\mathbf{I_2} = j\omega M\mathbf{I_2} \tag{20}$$

Solving (5), (18), and (19):

$$20 = j\omega L_2\mathbf{I_2} + 0 = j\omega L_2\mathbf{I_2} \tag{21}$$

Solving (16), (20), and (21):

$$\frac{20}{4} = \frac{j\omega L_2 \mathbf{I_2}}{j\omega \times 16\mathbf{I_2}} \Rightarrow L_2 = 80\ H \tag{22}$$

By using (10), (16), and (22), we can determine the inductance matrix of the mutually coupled inductors, as is presented in the following:

$$[L] = \begin{bmatrix} 5 & 16 \\ 16 & 80 \end{bmatrix} H$$

Choice (4) is the answer.

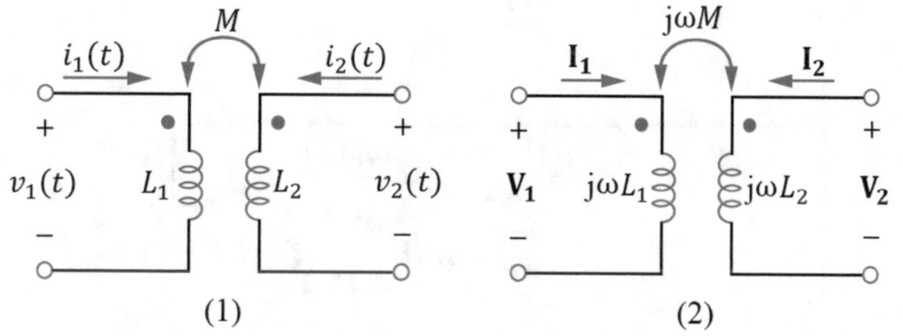

Figure 4.25 The circuit of solution of problem 4.26

4.27. Figure 4.26.2 shows the circuit of Figure 4.26.1 in frequency domain. The impedances of the components are presented in the figure. The problem can be solved by using a heuristic approach. Based on the information given in the problem, no current flows through the a–b branch. Therefore, we can define $\mathbf{I_1}$ and $\mathbf{I_2}$ for the indicated branches like the ones shown in the circuit of Figure 4.26.2.

KVL in the upper right-side mesh:

$$-(j\omega L_1 \mathbf{I_1} + j\omega M \mathbf{I_2}) + j\omega L_2 \mathbf{I_2} + j\omega M \mathbf{I_1} = 0 \Rightarrow j\omega(-L_1 + M)\mathbf{I_1} + j\omega(L_2 - M)\mathbf{I_2} = 0$$

$$\Rightarrow \mathbf{I_1} = \frac{L_2 - M}{L_1 - M}\mathbf{I_2} \tag{1}$$

KVL in the lower right-side mesh:

$$-R_1\mathbf{I_1} + R_2\mathbf{I_2} = 0 \Rightarrow \mathbf{I_1} = \frac{R_2}{R_1}\mathbf{I_2} \tag{2}$$

Solving (1) and (2):

$$\frac{L_2 - M}{L_1 - M} = \frac{R_2}{R_1} \Rightarrow L_2 - M = (L_1 - M)\frac{R_2}{R_1} \Rightarrow L_2 - \frac{R_2}{R_1}L_1 = M\left(1 - \frac{R_2}{R_1}\right) \Rightarrow M = \frac{L_2 - \frac{R_2}{R_1}L_1}{1 - \frac{R_2}{R_1}}$$

$$\Rightarrow M = \frac{R_1L_2 - R_2L_1}{R_1 - R_2}$$

Choice (2) is the answer.

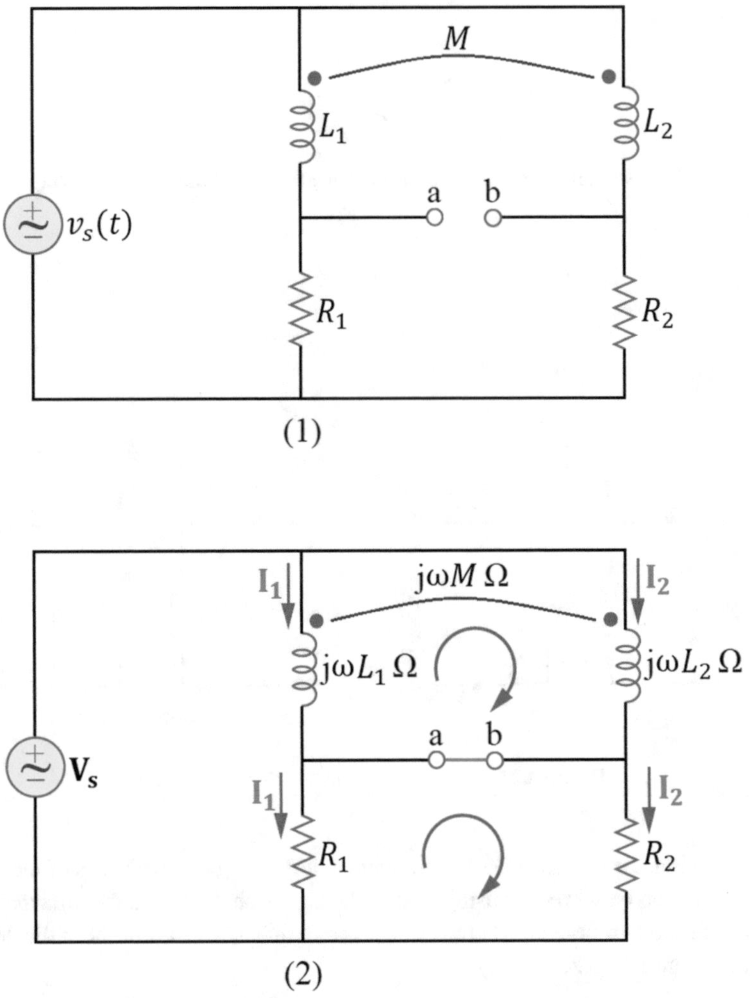

Figure 4.26 The circuit of solution of problem 4.27

4.28. Based on the information given in the problem:

$$\mathbf{I_2} = (20\sqrt{2}\angle{-45^\circ}\ A \tag{1}$$

$$\mathbf{I_3} = 24\ A \tag{2}$$

In addition, based on the information that can be extracted from the circuit:

$$\frac{n_1}{n_2} = \frac{2}{1} \tag{3}$$

$$\frac{n_1}{n_3} = \frac{4}{1} \tag{4}$$

By using (3) and (4), we can assume that:

$$n_1 = 4, n_2 = 2, n_3 = 1 \tag{5}$$

The secondary and the tertiary currents are leaving the dots of the coils. Therefore, we can write:

$$n_1\mathbf{I_1} = n_2\mathbf{I_2} + n_3\mathbf{I_3} \tag{6}$$

Solving (1), (2), (5), and (6):

$$4\mathbf{I}_1 = 2 \times (20\sqrt{2}\angle{-45^\circ}) + 1 \times 24 \Rightarrow \mathbf{I}_1 = (2\sqrt{89}\angle{-32^\circ})\ A$$

Choice (2) is the answer.

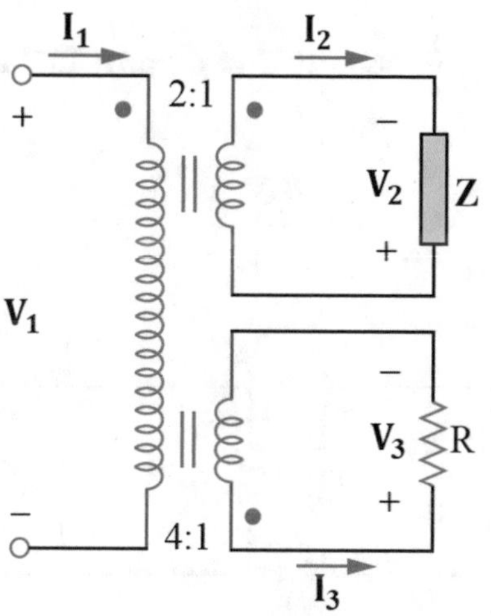

Figure 4.27 The circuit of solution of problem 4.28

4.29. If the capacitor and the mutually coupled inductors create a resonance state ($\omega = \omega_0$), the total impedance of their parallel connection will be infinite (i.e., an open circuit branch). Thus, the whole voltage of the voltage source will be applied on the resistor.

The resonance frequency of the parallel connection of the capacitor and the mutually coupled inductors is as follows:

$$\omega_0 = \frac{1}{\sqrt{L_{eq}C}} \tag{1}$$

where L_{eq} is the equivalent inductance of the mutually coupled inductors. As can be realized from the circuit of Figure 4.28.1, the angular frequency of the voltage source is:

$$\omega = 100\ rad/sec \tag{2}$$

To determine the equivalent inductance of the mutually coupled inductors, we can apply their T-equivalent circuit (see Figure 4.28.2) in the main circuit, as is illustrated in Figure 4.28.3. As can be noticed, we need to calculate the inductance matrix of the mutually coupled inductors, which is presented in the following:

$$[L] = \begin{bmatrix} 5 & -2 \\ -2 & 2 \end{bmatrix}^{-1} = \frac{1}{5 \times 2 - (-2) \times (-2)} \begin{bmatrix} 2 & 2 \\ 2 & 5 \end{bmatrix} = \begin{bmatrix} \frac{1}{3} & \frac{1}{3} \\ \frac{1}{3} & \frac{5}{6} \end{bmatrix} H$$

$$\Rightarrow L_{11} = \frac{1}{3}\ H, L_{22} = \frac{1}{3}\ H, M = \frac{1}{3}\ H \tag{3}$$

To determine the equivalent inductance, we must turn off the voltage source, as is shown in Figure 4.28.3. By using (3) and Figure 4.28.3, we can find L_{eq} as follows:

$$L_{eq} = \left(0 \| \frac{1}{3}\right) + \frac{1}{2} = \frac{1}{2}\ H \tag{4}$$

Solving (1), (2), and (4):

$$100 = \frac{1}{\sqrt{\frac{1}{2} \times C}} \Rightarrow \frac{1}{2} \times C = 10^{-4} \Rightarrow C = 2 \times 10^{-4}\ F \Rightarrow C = 200\ \mu F$$

Choice (4) is the answer.

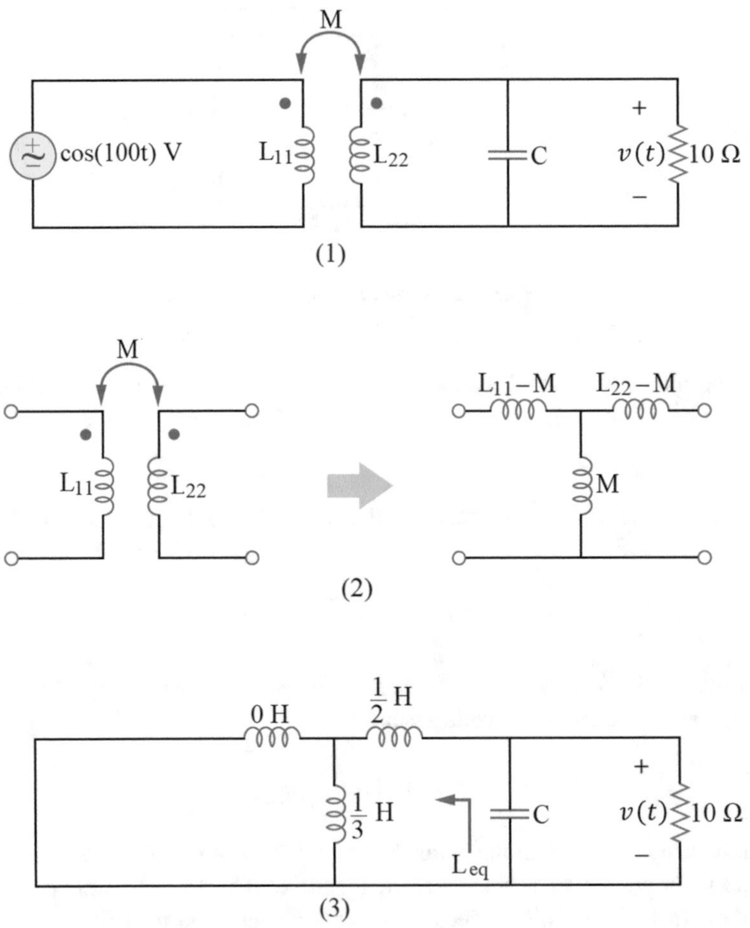

Figure 4.28 The circuit of solution of problem 4.29

4.30. Based on maximum average power transfer theorem, to transfer the maximum average power to the load, the impedance of the load must be equal to the complex conjugate of the Thevenin impedance seen by the load. In other words:

$$\mathbf{Z_L} = \mathbf{Z_{Th}}^* \tag{1}$$

To calculate the Thevenin impedance, we need to connect a test voltage source to the terminal, analyze the circuit, and determine the value of $\frac{\mathbf{V_t}}{\mathbf{I_t}}$. Herein, the independent voltage source must be turned off, as can be seen in Figure 4.29.2. The relation between the primary and secondary voltages, currents, and numbers of turns of the transformer is as follows:

$$\frac{\mathbf{V_1}}{\mathbf{V_t}} = -\frac{\mathbf{I_t}+\mathbf{I}}{\mathbf{I}} = \frac{8}{2} = 4 \Rightarrow \begin{cases} \mathbf{V_1} = 4\mathbf{V_t} & (2) \\ \mathbf{I} = -\frac{1}{5}\mathbf{I_t} & (3) \end{cases}$$

KVL in the left-side mesh:

$$(20+j60)\mathbf{I} + \mathbf{V_1} + \mathbf{V_t} = 0 \xrightarrow{Using\ (2),(3)} (20+j60) \times \left(-\frac{1}{5}\mathbf{I_t}\right) + 4\mathbf{V_t} + \mathbf{V_t} = 0$$

$$-(4+j12)\mathbf{I_t} + 5\mathbf{V_t} = 0 \Rightarrow \frac{\mathbf{V_t}}{\mathbf{I_t}} = \left(\frac{4}{5}+j\frac{12}{5}\right) \Rightarrow \mathbf{Z_{Th}} = \left(\frac{4}{5}+j\frac{12}{5}\right)\,\Omega \quad (4)$$

Solving (1) and (4):

$$Z_L = \left(\frac{4}{5}+j\frac{12}{5}\right)^* = \left(\frac{4}{5}-j\frac{12}{5}\right) = (0.8-2.4)\,\Omega$$

Choice (2) is the answer.

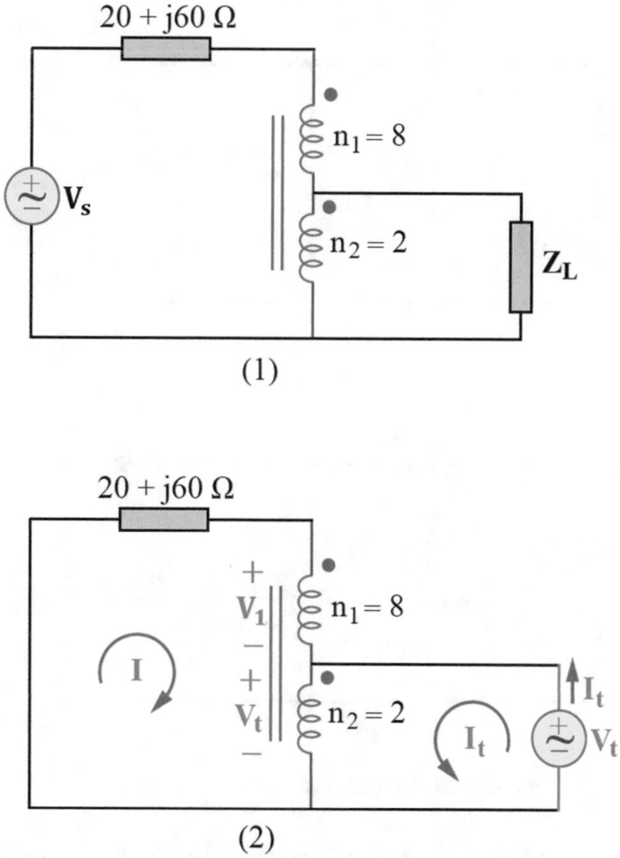

Figure 4.29 The circuit of solution of problem 4.30

4.31. The circuit is illustrated in frequency domain in Figure 4.30.2. The phasor of sin(2*t*), that is, $1\angle 0^\circ$ is defined as the reference phasor, where "$\angle$" is the symbol of phase angle. Therefore, the phasor of the output voltage is $2\angle 0^\circ$ and 2 *V*. The impedances of the components are as follows:

$$\mathbf{Z_{1\,H}} = j\omega L = j \times 2 \times 1 = j2\,\Omega \quad (1)$$

$$\mathbf{Z_{1\,\Omega}} = 1\,\Omega \quad (2)$$

Applying KVL in the right-side mesh:

$$-(j2\mathbf{I} - j\mathbf{I}) + 2 = 0 \Rightarrow \mathbf{I} = \frac{2}{j} = -j2\ V \tag{3}$$

Applying KVL in the left-side mesh:

$$-\mathbf{V} + \mathbf{I} + j2\mathbf{I} - j\mathbf{I} + j2\mathbf{I} - j\mathbf{I} = 0 \Rightarrow \mathbf{V} = (1 + j2)\mathbf{I} \tag{4}$$

Solving (3) and (4):

$$\mathbf{V} = (1 + j2) \times (-j2) = (4 - j2)\ V = (2\sqrt{5}\angle{-26.5°})\ V \tag{5}$$

Transferring to time domain:

$$v(t) = 2\sqrt{5}\sin\left(2t - 26.5^{\circ}\right) V$$

Choice (1) is the answer.

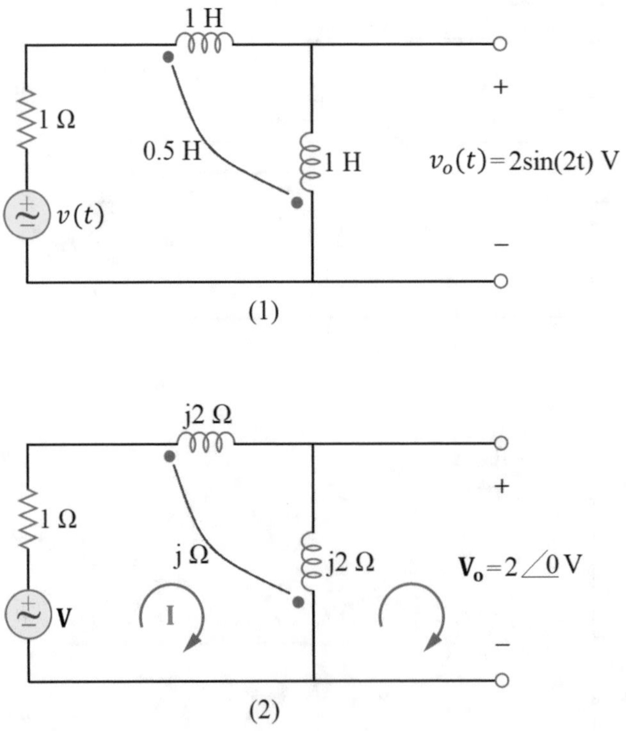

Figure 4.30 The circuit of solution of problem 4.31

4.32. Figure 4.31.2 shows the circuit in frequency domain. The impedances of the components are as follows:

$$\mathbf{Z_{L1}} = j\omega L_1 \tag{1}$$

$$\mathbf{Z_{L2}} = j\omega L_2 \tag{2}$$

$$\mathbf{Z_M} = j\omega M \tag{3}$$

$$\mathbf{Z_R} = R \tag{4}$$

Applying KVL in the right-side mesh:

$$\mathbf{V_2} = j\omega L_2\mathbf{I_2} + j\omega M\mathbf{I_1} \tag{5}$$

As we know, we can model an ideal voltmeter as an open circuit branch. Therefore:

$$\mathbf{I_2} = 0 \tag{6}$$

Solving (5) and (6):

$$\mathbf{V_2} = 0 + j\omega M\mathbf{I_1} \Rightarrow |\mathbf{V_2}| = \omega M|\mathbf{I_1}| \tag{7}$$

Applying Ohm's law for the resistor:

$$\mathbf{V_R} = R\mathbf{I_1} \Rightarrow |\mathbf{V_R}| = R|\mathbf{I_1}| \tag{8}$$

Solving (7) and (8):

$$|\mathbf{V_2}| = \omega M\frac{|\mathbf{V_R}|}{R} \Rightarrow M = \frac{R}{\omega} \times \left|\frac{\mathbf{V_2}}{\mathbf{V_R}}\right|$$

Choice (3) is the answer.

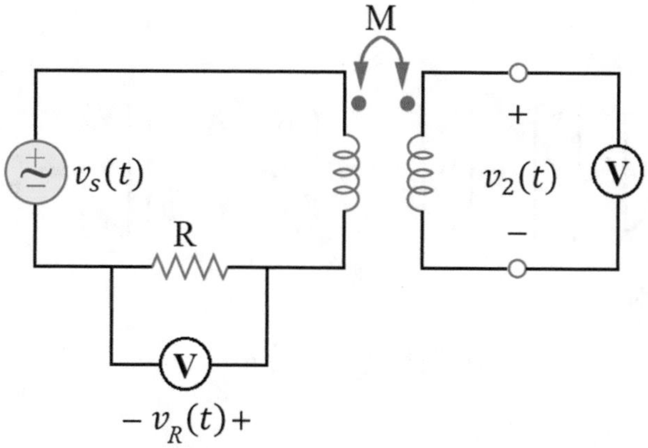

(1)

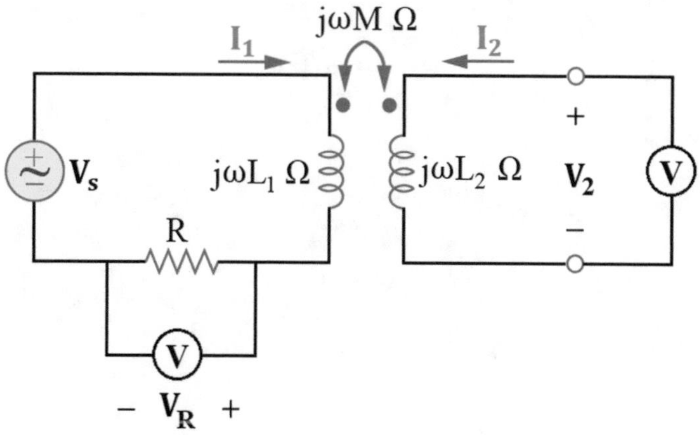

(2)

Figure 4.31 The circuit of solution of problem 4.32

4.33. Figure 4.32.2 shows the circuit of Figure 4.32.1 in frequency domain. The phasor of sin(t), that is, $1\angle 0°$ is defined as the reference phasor, where "$\angle$" is the symbol of phase angle. Thus, the phasor of the current of the current source is $1\angle 0°$ or 1 A.

The current–voltage relation matrix of the three mutually coupled inductors in phasor domain is as follows:

$$\begin{bmatrix} \mathbf{V}_1 \\ \mathbf{V}_2 \\ \mathbf{V}_3 \end{bmatrix} = j\omega \begin{bmatrix} 1 & -1 & 0 \\ -1 & 1 & 1 \\ 0 & 1 & 1 \end{bmatrix} \begin{bmatrix} \mathbf{I}_1 \\ \mathbf{I}_2 \\ \mathbf{I}_3 \end{bmatrix} \xrightarrow{\omega=1} \begin{bmatrix} \mathbf{V}_1 \\ \mathbf{V}_2 \\ \mathbf{V}_3 \end{bmatrix} = j \begin{bmatrix} 1 & -1 & 0 \\ -1 & 1 & 1 \\ 0 & 1 & 1 \end{bmatrix} \begin{bmatrix} \mathbf{I}_1 \\ \mathbf{I}_2 \\ \mathbf{I}_3 \end{bmatrix} \tag{1}$$

By determining the inverse of the inductance matrix, we can calculate the currents of the inductors based on their voltages, as can be seen in the following:

$$\begin{bmatrix} \mathbf{I}_1 \\ \mathbf{I}_2 \\ \mathbf{I}_3 \end{bmatrix} = (-j) \begin{bmatrix} 1 & -1 & 0 \\ -1 & 1 & 1 \\ 0 & 1 & 1 \end{bmatrix}^{-1} \begin{bmatrix} \mathbf{V}_1 \\ \mathbf{V}_2 \\ \mathbf{V}_3 \end{bmatrix} \tag{2}$$

From the circuit, it is clear that:

$$\mathbf{V}_1 = \mathbf{V}_2 = \mathbf{V}_3 = \mathbf{V} \tag{3}$$

By solving (2) and (3), we have:

$$\begin{bmatrix} \mathbf{I}_1 \\ \mathbf{I}_2 \\ \mathbf{I}_3 \end{bmatrix} = (-j) \begin{bmatrix} 1 & -1 & 0 \\ -1 & 1 & 1 \\ 0 & 1 & 1 \end{bmatrix}^{-1} \begin{bmatrix} \mathbf{V} \\ \mathbf{V} \\ \mathbf{V} \end{bmatrix} \Rightarrow \begin{bmatrix} \mathbf{I}_1 \\ \mathbf{I}_2 \\ \mathbf{I}_3 \end{bmatrix} = (-j) \begin{bmatrix} 0 & -1 & 1 \\ -1 & -1 & 1 \\ 1 & 1 & 0 \end{bmatrix} \begin{bmatrix} \mathbf{V} \\ \mathbf{V} \\ \mathbf{V} \end{bmatrix} \Rightarrow \begin{cases} \mathbf{I}_1 = 0 & (4) \\ \mathbf{I}_2 = j\mathbf{V} & (5) \\ \mathbf{I}_3 = -j2\mathbf{V} & (6) \end{cases}$$

Applying KCL in the supernode:

$$-1 + \mathbf{I}_1 + \mathbf{I}_2 + \mathbf{I}_3 = 0 \tag{7}$$

Solving (4)–(7):

$$-1 + 0 + j\mathbf{V} - j2\mathbf{V} = 0 \Rightarrow -1 - j\mathbf{V} = 0 \Rightarrow \mathbf{V} = j \tag{8}$$

By transferring to time domain:

$$v(t) = \cos(t)\ V$$

Choice (2) is the answer.

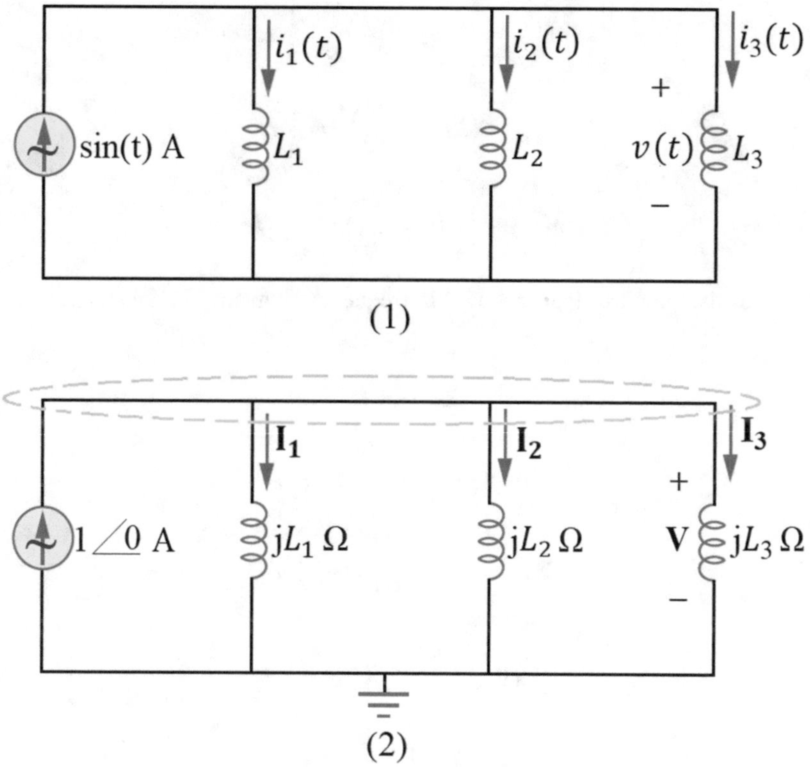

Figure 4.32 The circuit of solution of problem 4.33

4.34. The following information has been given in the problem:

$$\omega = 1\ rad/sec \tag{1}$$

$$\mathbf{Z_{N1}} = \left(\frac{3}{2} + j2\right)\ \Omega \tag{2}$$

$$PF_{N2} = 1 \Rightarrow Im\{\mathbf{Z_{N2}}\} = 0 \tag{3}$$

$$P_{N2} = P_{N2,max} \tag{4}$$

Solving (3) and (4):

$$Re\{\mathbf{Z_{N2}}\} = |\mathbf{Z_{N1}}| \xrightarrow{Using\ (2)} Re\{\mathbf{Z_{N2}}\} = \left|\frac{3}{2} + j2\right| = \sqrt{\left(\frac{3}{2}\right)^2 + 2^2} = \frac{5}{2}\ \Omega \tag{5}$$

To simplify the problem solution, we can transfer the circuit of the right side of the transformer to its left side. Figure 4.33.2 shows the updated circuit. Herein, the positions of the dots of the coils of the inductors do not need to be changed. Think about its reason. The updated value of the parameters can be calculated as follows:

$$C'_{1\ F} = (1\ F) \times \left(\frac{2}{1}\right)^2 = 4\ F \tag{6}$$

$$L'_{1\,H} = (1\,H) \times \left(\frac{1}{2}\right)^2 = \frac{1}{4}\,H \tag{7}$$

$$L'_{2\,H} = (2\,H) \times \left(\frac{1}{2}\right)^2 = \frac{1}{2}\,H \tag{8}$$

$$M'_{1\,H} = (1\,H) \times \left(\frac{1}{2}\right)^2 = \frac{1}{4}\,H \tag{9}$$

Figure 4.33.3 illustrates the circuit of Figure 4.33.2 in frequency domain. The impedances of the components are as follows:

$$\mathbf{Z_R} = R\,\Omega \tag{10}$$

$$\mathbf{Z_C} = \frac{1}{j \times 1 \times C} = -j\frac{1}{C}\,\Omega \tag{11}$$

$$\mathbf{Z_{4\,F}} = \frac{1}{j \times 1 \times 4} = -j\frac{1}{4}\,\Omega \tag{12}$$

$$\mathbf{Z_{\frac{1}{4}\,H}} = j \times 1 \times \frac{1}{4} = j\frac{1}{4}\,\Omega \tag{13}$$

$$\mathbf{Z_{\frac{1}{2}\,H}} = j \times 1 \times \frac{1}{2} = j\frac{1}{2}\,\Omega \tag{14}$$

The input impedance of the second network (N_2) is:

$$\mathbf{Z_{N2}} = R - j\frac{1}{C} - j\frac{1}{4} + j\frac{1}{4} + j\frac{1}{2} + 2 \times j\frac{1}{4} = \left(R + j\left(-\frac{1}{C} + 1\right)\right)\,\Omega \tag{15}$$

Solving (3) and (15):

$$-\frac{1}{C} + 1 = 0 \Rightarrow C = 1\,F$$

Solving (5) and (15):

$$R = \frac{5}{2}\,\Omega$$

Choice (3) is the answer.

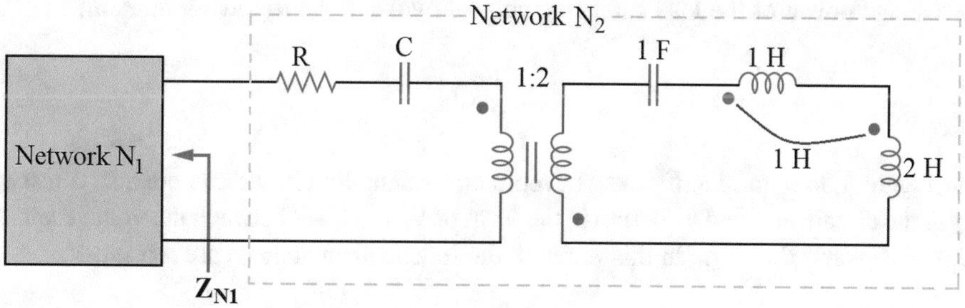

(1)

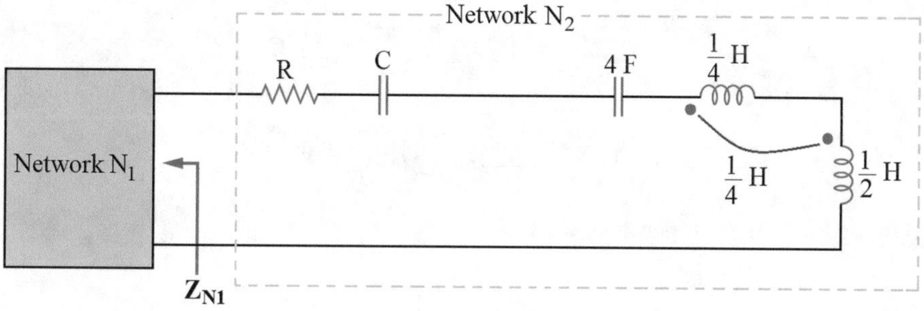

(2)

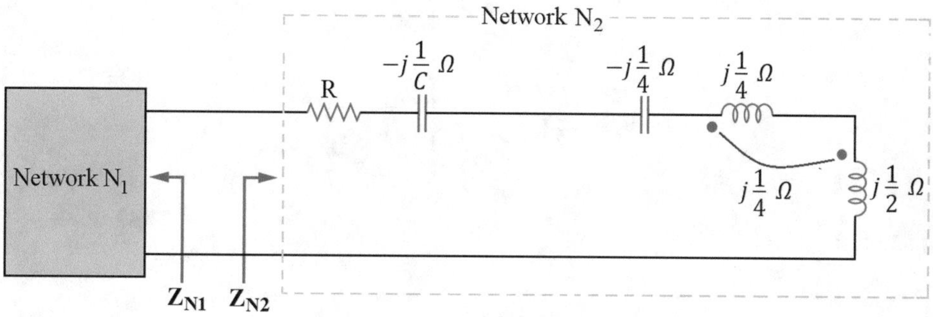

(3)

Figure 4.33 The circuit of solution of problem 4.34

4.35. Figure 4.34.2 shows the main circuit in frequency domain. The phasor of cos(t), that is, $1\angle 0^\circ$ is defined as the reference phasor, where "$\angle$" is the symbol of phase angle. Therefore, the phasor of the voltage of the voltage source and the phasor of the current of the current source are $1\angle 0^\circ\ V$ and $1\angle -90^\circ\ A$. The impedances of the components are as follows:

$$\mathbf{Z_{1\,H}} = j\omega L = j \times 1 \times 1 = j\,\Omega \tag{1}$$

$$\mathbf{Z_{0.5\,F}} = \frac{1}{j\omega C} = \frac{1}{j \times 1 \times 0.5} = -j2\,\Omega \tag{2}$$

$$\mathbf{Z_R} = R\,\Omega \tag{3}$$

Based on maximum average power transfer theorem, to transfer the maximum average power to a resistive load, the resistance of the load must be equal to the magnitude of the Thevenin impedance seen by the load. In other words:

$$R = |\mathbf{Z_{Th}}| \tag{4}$$

Moreover, the average power of the load can be calculated by using the following relation:

$$P = \frac{1}{2} R |\mathbf{I_R}|^2 \tag{5}$$

As is shown in Figure 4.34.3, to identify the Thevenin equivalent circuit, we can connect a test voltage source to the terminal, analyze the circuit, and find a relation in the form of $\mathbf{V_t} = \alpha \mathbf{I_t} + \beta$ between the voltage and the current of the test source, where $\alpha = \mathbf{Z_{Th}}$ and $\beta = \mathbf{V_{Th}}$. In this method, the independent sources are left intact.

In addition, to simplify the circuit, we can transfer the circuit of the left side of the transformer to its right side, as is shown in Figure 4.34.3. The updated parameters are as follows:

$$\mathbf{Z'_{1\,H}} = (j\,\Omega) \times \left(\frac{1}{2}\right)^2 = j\frac{1}{4}\,\Omega \tag{6}$$

$$\mathbf{V'_s} = (1\angle 0^\circ) \times \left(\frac{1}{2}\right) = \left(\frac{1}{2}\angle 0^\circ\right) V \tag{7}$$

Now, by applying KCL in the supernode, we can write:

$$-(2\angle -90^\circ) - \mathbf{I_t} + \frac{\mathbf{V_t}}{-j2} + \frac{\mathbf{V_t} - \left(\frac{1}{2}<0^\circ\right)}{j\frac{1}{4}} = 0 \Rightarrow j4 - \mathbf{I_t} - j\frac{7}{2}\mathbf{V_t} = 0$$

$$\Rightarrow \mathbf{V_t} = j\frac{2}{7}\mathbf{I_t} + \frac{8}{7} \Rightarrow \begin{cases} \mathbf{Z_{Th}} = j\frac{2}{7}\,\Omega & (8) \\ \mathbf{V_{Th}} = \frac{8}{7}\,V & (9) \end{cases}$$

Solving (4) and (8):

$$R = \left| j\frac{2}{7} \right| = \frac{2}{7}\,\Omega \tag{10}$$

Now, we have the circuit of Figure 4.34.4 with the parameters given in (8), (9), and (10).

Solving (5), (8)–(10):

$$P = \frac{1}{2} \times \frac{2}{7} \times \left| \frac{\frac{8}{7}}{\frac{2}{7} + j\frac{2}{7}} \right|^2 = \frac{1}{2} \times \frac{2}{7} \times |2 - 2i|^2 = \frac{8}{7}\,W$$

Choice (1) is the answer.

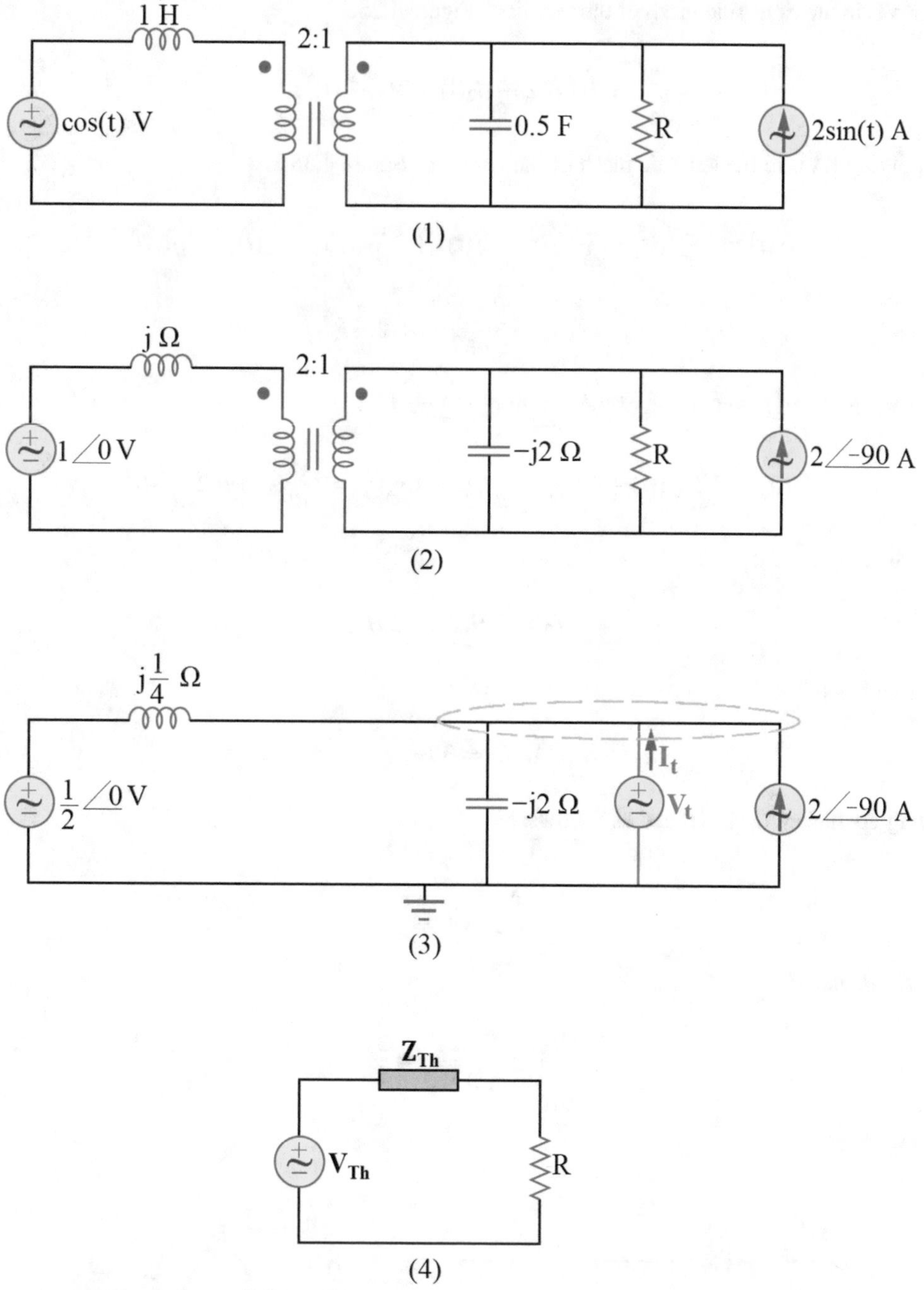

Figure 4.34 The circuit of solution of problem 4.35

4.36. This problem can be solved either in time domain or in frequency domain. In this problem, we need to write down the current–voltage equations of each circuit and equate them correspondingly.

Part 1: Applying KVL in the left-side mesh of the circuit of Figure 4.35.2:

$$v_1(t) = L_1 \frac{d}{dt} i_1(t) + M \frac{d}{dt} i_2(t) \tag{1}$$

Applying KVL in the right-side mesh of the circuit of Figure 4.35.2:

$$v_2(t) = L_2 \frac{d}{dt} i_2(t) + M \frac{d}{dt} i_1(t) \tag{2}$$

Part 2: Applying KVL in the left-side mesh of the circuit of Figure 4.35.1:

$$v_1(t) = 2\frac{d}{dt} i_1(t) - \frac{d}{dt}(i_1(t) + i_2(t)) + 3\frac{d}{dt}(i_1(t) + i_2(t)) - \frac{d}{dt} i_1(t)$$

$$\Rightarrow v_1(t) = 3\frac{d}{dt} i_1(t) + 2\frac{d}{dt} i_2(t) \tag{3}$$

Applying KVL in the right-side mesh of the circuit of Figure 4.35.1:

$$v_2(t) = 3\frac{d}{dt}(i_1(t) + i_2(t)) - \frac{d}{dt} i_1(t) \Rightarrow v_2(t) = 3\frac{d}{dt} i_2(t) + 2\frac{d}{dt} i_1(t) \tag{4}$$

Solving (1) and (3):

$$L_1 = 3\,H, M = 2\,H \tag{5}$$

Solving (2) and (4):

$$L_2 = 3\,H \tag{6}$$

The coupling coefficient (k) is defined as follows:

$$k = \frac{M}{\sqrt{L_1 L_2}} \tag{7}$$

Solving (5), (6), and (7):

$$k = \frac{2}{\sqrt{3 \times 3}} = \frac{2}{3}$$

Choice (2) is the answer.

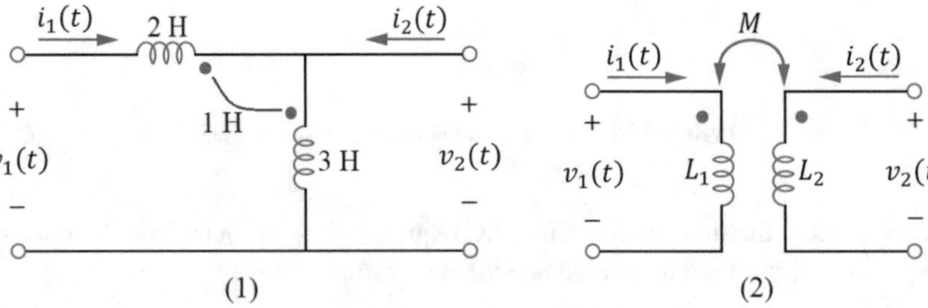

Figure 4.35 The circuit of solution of problem 4.36

4.37. This problem can be solved in time domain because the circuit is resistive. As is shown in Figure 4.36.2, to calculate the Thevenin resistance, we need to connect a test voltage source to the terminal, analyze the circuit, and determine the value of $\frac{V_t}{I_t}$ to find R_{Th}.

Applying KCL in the upper-left node:

$$-I_t + I_1 + I_2 = 0 \tag{1}$$

Based on the polarities defined for the primary and secondary voltages and currents of the transformer, we can write:

$$\frac{V_t}{V_2} = \frac{I_2}{I_1} = \frac{n}{1} \Rightarrow \begin{cases} V_2 = \frac{V_t}{n} & (2) \\ I_1 = \frac{I_2}{n} & (3) \end{cases}$$

Solving (1) and (3):

$$-I_t + \frac{I_2}{n} + I_2 = 0 \Rightarrow I_2 = \frac{1}{1+\frac{1}{n}} I_t = \frac{n}{n+1} I_t \tag{4}$$

Applying KVL in the middle mesh (clockwise):

$$-V_t + R_1 I_2 - V_2 + R_2 I_2 = 0 \xrightarrow{\text{Using (2), (4)}} -V_t + (R_1 + R_2)\frac{n}{n+1} I_t - \frac{V_t}{n} = 0$$

$$\Rightarrow -V_t\left(\frac{n+1}{n}\right) + (R_1 + R_2)\frac{n}{n+1} I_t = 0 \Rightarrow \frac{V_t}{I_t} = \left(\frac{n}{n+1}\right)^2 (R_1 + R_2)$$

$$\Rightarrow R_{Th} = \left(\frac{n}{n+1}\right)^2 (R_1 + R_2)\ \Omega$$

Choice (2) is the answer.

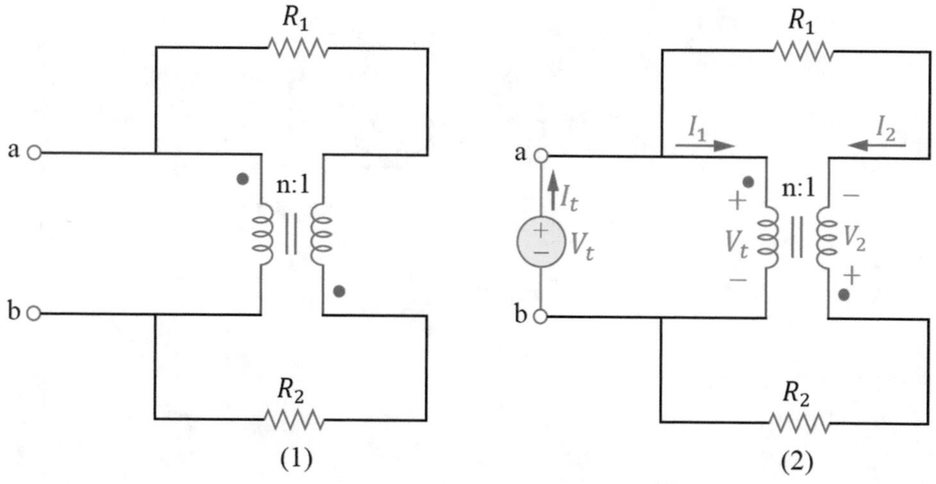

Figure 4.36 The circuit of solution of problem 4.37

4.38. Figure 4.37.2 illustrates the circuit in frequency domain. The impedances of the components are as follows:

$$\mathbf{Z_{L1}} = j\omega L_1 \ \Omega \tag{1}$$

$$\mathbf{Z_{L2}} = j\omega L_2 \ \Omega \tag{2}$$

$$\mathbf{Z_M} = j\omega M \ \Omega \tag{3}$$

$$\mathbf{Z_R} = R \ \Omega \tag{4}$$

$$\mathbf{Z_C} = \frac{1}{j\omega C} \ \Omega \tag{5}$$

Based on the information given in the problem, the current of the load is zero. Therefore, its voltage is zero as well.

Applying KVL in the left-side mesh of the circuit of Figure 4.37.2:

$$0 + R\mathbf{I_2} + j\omega L_1\mathbf{I_2} + j\omega M\mathbf{I_1} = 0 \Rightarrow \mathbf{I_1} = -\frac{R + j\omega L_1}{j\omega M}\mathbf{I_2} \tag{6}$$

Simultaneously applying KVL and KCL in the right-side mesh of the circuit of Figure 4.37.2:

$$\frac{1}{j\omega C}\mathbf{I_2} + R\mathbf{I_2} + 0 + R(\mathbf{I_1} + \mathbf{I_2}) = 0 \Rightarrow \mathbf{I_1} = -\left(\frac{1}{j\omega CR} + 2\right)\mathbf{I_2} \tag{7}$$

Solving (6) and (7):

$$-\frac{R + j\omega L_1}{j\omega M}\mathbf{I_2} = -\left(\frac{1}{j\omega CR} + 2\right)\mathbf{I_2} \Rightarrow \frac{R + j\omega L_1}{j\omega M} - \frac{1}{j\omega CR} - 2 = 0$$

$$\Rightarrow \frac{-jR}{\omega M} + \frac{L_1}{M} + \frac{j}{\omega CR} - 2 = 0 \Rightarrow j\left(-\frac{R}{\omega M} + \frac{1}{\omega CR}\right) + \frac{L_1}{M} - 2 = 0 \Rightarrow \begin{cases} -\dfrac{R}{\omega M} + \dfrac{1}{\omega CR} = 0 & (8) \\ \dfrac{L_1}{M} - 2 = 0 & (9) \end{cases}$$

$$\overset{(8)}{\Rightarrow} \frac{R}{M} = \frac{1}{CR} \Rightarrow M = R^2C \tag{10}$$

$$\overset{(9)}{\Rightarrow} \frac{L_1}{M} = 2 \Rightarrow M = \frac{L_1}{2} \tag{11}$$

Solving (10) and (11):

$$M = R^2C = \frac{L_1}{2}$$

Choice (1) is the answer.

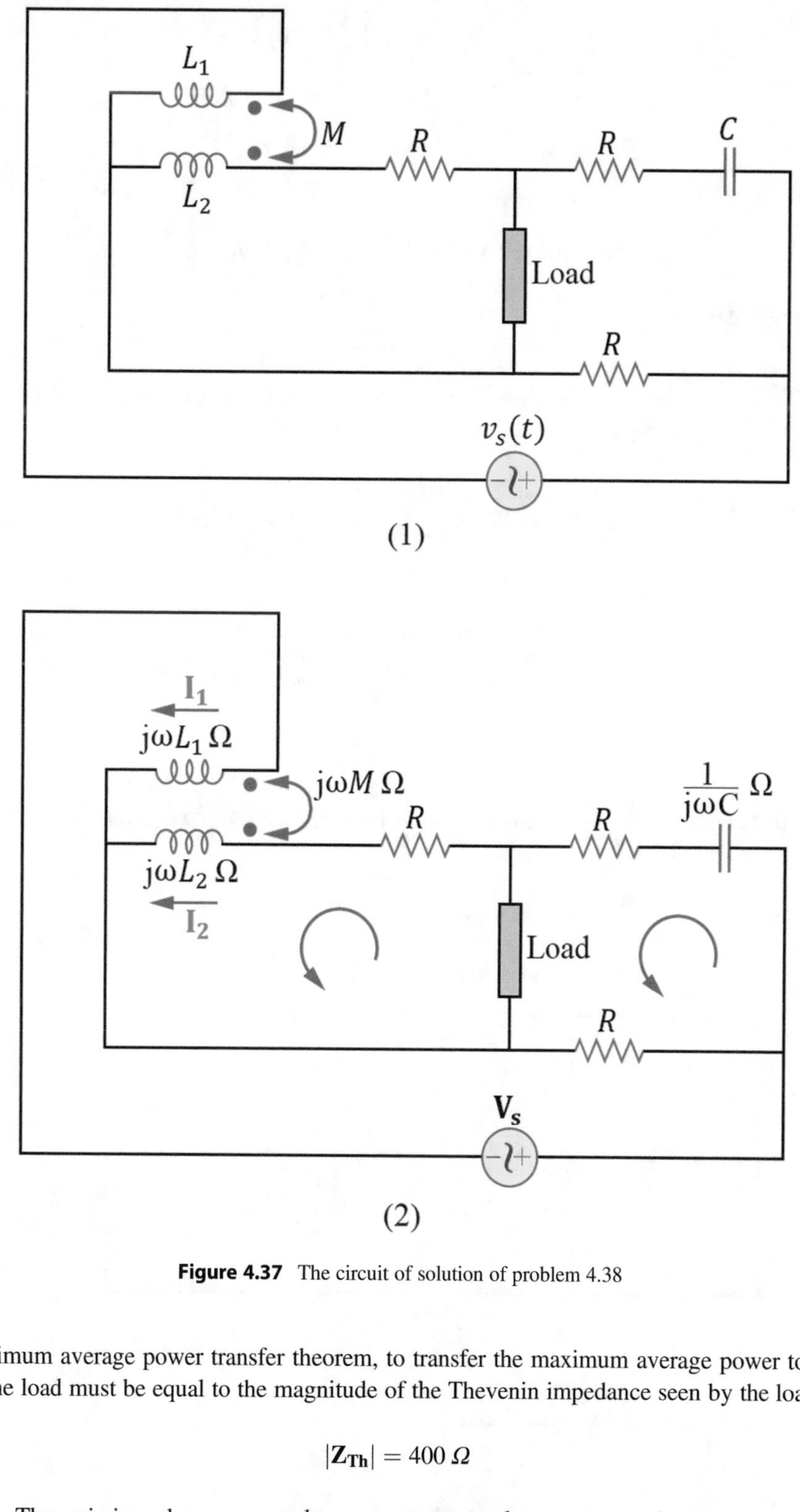

Figure 4.37 The circuit of solution of problem 4.38

4.39. Based on maximum average power transfer theorem, to transfer the maximum average power to a resistive load, the resistance of the load must be equal to the magnitude of the Thevenin impedance seen by the load. In other words:

$$|\mathbf{Z}_{\mathbf{Th}}| = 400\ \Omega \tag{1}$$

To calculate the Thevenin impedance, we need to connect a test voltage source to the terminal, analyze the circuit, and determine the value of $\frac{\mathbf{V}_t}{\mathbf{I}_t}$. Herein, the independent current source must be turned off, as can be seen in Figure 4.38.2. This problem is solved by using mesh analysis as follows:

The relation between the primary and secondary voltages, currents, and numbers of turns of the transformer is as follows:

$$\frac{\mathbf{V_1}}{\mathbf{V_t}} = -\frac{\mathbf{I_t}+\mathbf{I}}{\mathbf{I}} = \frac{n_1}{n_2} \Rightarrow \begin{cases} \mathbf{V_1} = \frac{n_1}{n_2}\mathbf{V_t} & (2) \\ \mathbf{I} = -\frac{1}{1+\frac{n_1}{n_2}}\mathbf{I_t} & (3) \end{cases}$$

KVL in the left-side mesh:

$$80 \times 10^3 \times \mathbf{I} + 10 \times 10^3 \times \mathbf{I} + \mathbf{V_1} + \mathbf{V_t} = 0 \tag{4}$$

Solving (2), (3), and (4):

$$9 \times 10^4 \times \left(-\frac{1}{1+\frac{n_1}{n_2}}\mathbf{I_t}\right) + \frac{n_1}{n_2}\mathbf{V_t} + \mathbf{V_t} = 0 \Rightarrow 9 \times 10^4 \times \left(\frac{1}{1+\frac{n_1}{n_2}}\right)\mathbf{I_t} = \left(1+\frac{n_1}{n_2}\right)\mathbf{V_t} = 0$$

$$\Rightarrow \frac{\mathbf{V_t}}{\mathbf{I_t}} = 9 \times 10^4 \times \frac{\frac{1}{1+\frac{n_1}{n_2}}}{1+\frac{n_1}{n_2}} = 9 \times 10^4 \times \frac{1}{\left(1+\frac{n_1}{n_2}\right)^2}.$$

$$\Rightarrow \mathbf{Z_{Th}} = 9 \times 10^4 \times \frac{1}{\left(1+\frac{n_1}{n_2}\right)^2}\ \Omega \tag{5}$$

Solving (1) and (5):

$$9 \times 10^4 \times \frac{1}{\left(1+\frac{n_1}{n_2}\right)^2} = 400 \Rightarrow \left(1+\frac{n_1}{n_2}\right)^2 = 225 \Rightarrow 1+\frac{n_1}{n_2} = 15 \Rightarrow \frac{n_1}{n_2} = 14$$

Choice (3) is the answer.

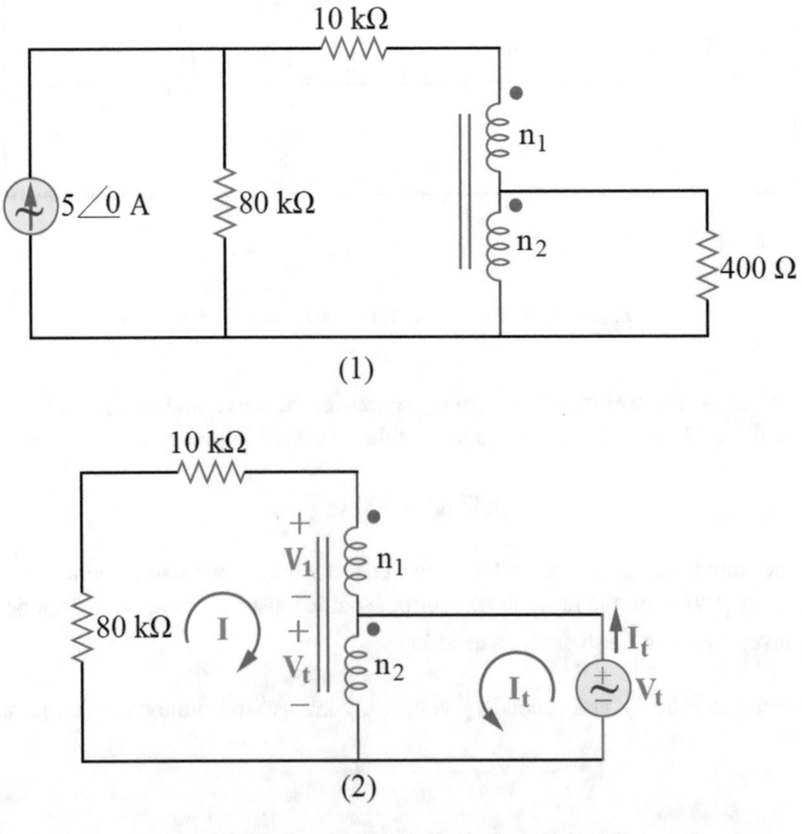

Figure 4.38 The circuit of solution of problem 4.39

4.40. To calculate the input impedance of the circuit in frequency domain, we need to connect a test source (e.g., test voltage source) to the terminal, analyze the circuit, and determine the value of $\frac{\mathbf{V_t}}{\mathbf{I_t}}$. Then:

$$\mathbf{Z_{ab}} = \frac{\mathbf{V_t}}{\mathbf{I_t}} \tag{1}$$

Figure 4.39.2 illustrates the main circuit in frequency domain. To simplify the problem solution, we can assume that the angular frequency of test source is $\omega = 1\ rad/sec$. The impedances of the components are as follows:

$$\mathbf{Z_R} = R\,\Omega \tag{2}$$

$$\mathbf{Z_C} = \frac{1}{j\omega C} = \frac{1}{j \times 1 \times C} = -j\frac{1}{C}\,\Omega \tag{3}$$

In this problem, nodal analysis is the best approach to analyze the circuit. From the circuit, it is clear that:

$$\mathbf{V_1} = \mathbf{V_t} \tag{4}$$

The primary and secondary voltages, currents, and numbers of turns of the transformer have the following relation:

$$\frac{\mathbf{V_1}}{\mathbf{V_2}} = -\frac{\mathbf{I_2}}{\mathbf{I_1}} = \frac{1}{a} \Rightarrow \begin{cases} \mathbf{V_2} = a\mathbf{V_1} & (5) \\ \mathbf{I_1} = -a\mathbf{I_2} & (6) \end{cases}$$

Solving (4) and (5):

$$\mathbf{V_2} = a\mathbf{V_t} \tag{7}$$

KCL in the left-side supernode:

$$-\mathbf{I_t} + \frac{\mathbf{V_t}}{R} + \mathbf{I_1} + \frac{\mathbf{V_t} - \mathbf{V_2}}{-j\frac{1}{C}} = 0 \Rightarrow -\mathbf{I_t} + \left(\frac{1}{R} + jC\right)\mathbf{V_t} + \mathbf{I_1} - jC\mathbf{V_2} = 0$$

$$\xrightarrow{Using\ (7)} -\mathbf{I_t} + \left(\frac{1}{R} + jC\right)\mathbf{V_t} + \mathbf{I_1} - jCa\mathbf{V_t} = 0 \Rightarrow -\mathbf{I_t} + \left(\frac{1}{R} + jC(1-a)\right)\mathbf{V_t} + \mathbf{I_1} = 0$$

$$\xrightarrow{Using\ (6)} -\mathbf{I_t} + \left(\frac{1}{R} + jC(1-a)\right)\mathbf{V_t} - a\mathbf{I_2} = 0 \tag{8}$$

KCL in the right-side node:

$$\mathbf{I_2} + \frac{\mathbf{V_2}}{R} + \frac{\mathbf{V_2} - \mathbf{V_t}}{-j\frac{1}{C}} = 0 \Rightarrow \mathbf{I_2} + \left(\frac{1}{R} + jC\right)\mathbf{V_2} - jC\mathbf{V_t} = 0$$

$$\xrightarrow{Using\ (7)} \mathbf{I_2} + \left(\frac{1}{R} + jC\right)a\mathbf{V_t} - jC\mathbf{V_t} = 0 \Rightarrow \mathbf{I_2} + \left(\frac{a}{R} + jC(a-1)\right)\mathbf{V_t} = 0 \tag{9}$$

Solving (8) and (9):

$$\frac{1}{a}\left(-\mathbf{I_t} + \left(\frac{1}{R} + jC(1-a)\right)\mathbf{V_t}\right) = -\left(\frac{a}{R} + jC(a-1)\right)\mathbf{V_t}$$

$$\Rightarrow -\frac{1}{a}\mathbf{I_t} + \left(\frac{1}{Ra} + jC\left(\frac{1}{a} - 1\right) + \frac{a}{R} + jC(a-1)\right)\mathbf{V_t} = 0$$

$$\Rightarrow -\mathbf{I_t} + \left(\frac{1}{R} + \frac{a^2}{R} + jC\left(1 - a + a^2 - a\right)\right)\mathbf{V_t} = 0$$

$$\Rightarrow \frac{\mathbf{V_t}}{\mathbf{I_t}} = \frac{1}{\frac{1}{R} + \frac{a^2}{R} + jC(a-1)^2} \Rightarrow \mathbf{Z_{ab}} = \frac{1}{\frac{1}{R} + \frac{a^2}{R} + jC(a-1)^2} \quad (10)$$

Based on the information given in the problem:

$$\mathbf{Z_{ab}} = \frac{R}{2} \quad (11)$$

Combining (10) and (11):

$$\frac{1}{\frac{1}{R} + \frac{a^2}{R} + jC(a-1)^2} = \frac{R}{2} \Rightarrow \frac{1}{R} + \frac{a^2}{R} + jC(a-1)^2 - \frac{2}{R} = 0$$

$$\Rightarrow \frac{1}{R}\left(a^2 - 1\right) + jC(a-1)^2 = 0 \Rightarrow \begin{cases} \frac{1}{R}\left(a^2 - 1\right) = 0 \Rightarrow a = \pm 1 \\ C(a-1)^2 = 0 \Rightarrow a = 1 \end{cases} \Rightarrow a = 1$$

Choice (3) is the answer.

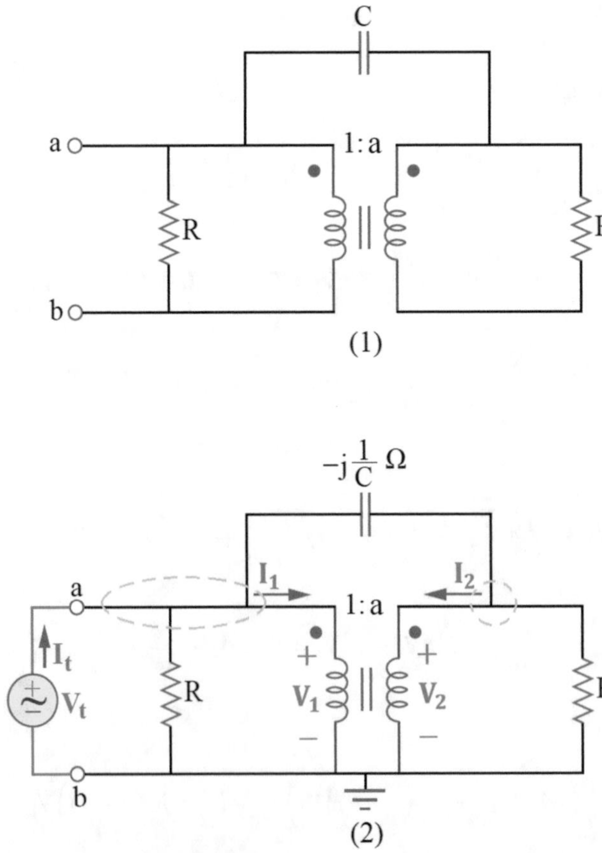

Figure 4.39 The circuit of solution of problem 4.40

4.41. As is illustrated in Figure 4.40.2, to calculate the Thevenin impedance of the circuit, we must connect a test source (e.g., test voltage source) to the terminal, analyze the circuit, and find the value of $\frac{\mathbf{V_t}}{\mathbf{I_t}}$. To simplify the problem, we can assume that the angular frequency of the circuit is 1 rad/sec ($\omega = 1\ rad/sec$).

To see a purely inductive impedance from the terminal, the real part of the Thevenin impedance must be zero. In other words:

$$Re\{\mathbf{Z_{Th}}\} = 0 \tag{1}$$

$$Im\{\mathbf{Z_{Th}}\} > 0 \tag{2}$$

Figure 4.40.2 illustrates the circuit in frequency domain. The impedances of the components are as follows:

$$\mathbf{Z_{1\,H}} = j\omega L = j \times 1 \times 1 = j\,\Omega \tag{3}$$

$$\mathbf{Z_M} = j\omega M = j \times 1 \times M = jM\,\Omega \tag{4}$$

$$\mathbf{Z_{\frac{1}{4}\,H}} = j\omega L = j \times 1 \times \frac{1}{4} = j\frac{1}{4}\,\Omega \tag{5}$$

$$\mathbf{Z_{\frac{1}{12}\,H}} = j\omega L = j \times 1 \times \frac{1}{12} = j\frac{1}{12}\,\Omega \tag{6}$$

$$\mathbf{Z_{2\,\Omega}} = 2\,\Omega \tag{7}$$

This problem is solved by using mesh analysis as follows:

KVL in the right-side mesh:

$$2(\mathbf{I} - \mathbf{I_t}) + j\frac{1}{4}\mathbf{I} - jM\mathbf{I_t} + j\frac{1}{12}\mathbf{I} = 0 \Rightarrow -(2 + jM)\mathbf{I_t} + \left(2 + j\frac{1}{3}\right)\mathbf{I} = 0$$

$$\Rightarrow \mathbf{I} = \frac{2 + jM}{2 + j\frac{1}{3}}\mathbf{I_t} \tag{8}$$

KVL in the left-side mesh:

$$-\mathbf{V_t} + j\mathbf{I_t} - jM\mathbf{I} + 2(\mathbf{I_t} - \mathbf{I}) = 0 \Rightarrow -\mathbf{V_t} + (2 + j)\mathbf{I_t} - (2 + jM)\mathbf{I} = 0 \tag{9}$$

Solving (8) and (9):

$$-\mathbf{V_t} + (2 + j)\mathbf{I_t} - (2 + jM)\frac{2 + jM}{2 + j\frac{1}{3}}\mathbf{I_t} = 0$$

$$\Rightarrow -\mathbf{V_t} + \left(\frac{4 + j\frac{2}{3} + j2 - \frac{1}{3} - (4 + j4M - M^2)}{2 + j\frac{1}{3}}\right)\mathbf{I_t} = 0$$

$$\Rightarrow -\mathbf{V_t} + \left(\frac{\left(-\frac{1}{3} + M^2\right) + j\left(\frac{8}{3} - 4M\right)}{2 + j\frac{1}{3}}\right)\mathbf{I_t} = 0 \Rightarrow \frac{\mathbf{V_t}}{\mathbf{I_t}} = \left(\frac{\left(-\frac{1}{3} + M^2\right) + j\left(\frac{8}{3} - 4M\right)}{2 + j\frac{1}{3}}\right)$$

$$\Rightarrow \frac{\mathbf{V_t}}{\mathbf{I_t}} = \frac{\left(\left(-\frac{1}{3}+M^2\right)+j\left(\frac{8}{3}-4M\right)\right)\left(2-j\frac{1}{3}\right)}{4+\frac{1}{9}}$$

$$\Rightarrow \frac{\mathbf{V_t}}{\mathbf{I_t}} = \frac{\left(-\frac{2}{3}+2M^2+\frac{8}{9}-\frac{4}{3}M\right)+j\left(\frac{16}{3}-8M+\frac{1}{9}-\frac{1}{3}M^2\right)}{\frac{37}{9}}$$

$$\Rightarrow \frac{\mathbf{V_t}}{\mathbf{I_t}} = \frac{9}{37}\left(\frac{2}{9}+2M^2-\frac{4}{3}M\right)+j\frac{9}{37}\left(\frac{49}{9}-8M-\frac{1}{3}M^2\right)$$

$$\Rightarrow \mathbf{Z_{Th}} = \left(\frac{9}{37}\left(\frac{2}{9}+2M^2-\frac{4}{3}M\right)+j\frac{9}{37}\left(\frac{49}{9}-8M-\frac{1}{3}M^2\right)\right)\ \Omega \qquad (10)$$

Solving (1) and (10):

$$\frac{9}{37}\left(\frac{2}{9}+2M^2-\frac{4}{3}M\right)=0 \Rightarrow M^2-\frac{2}{3}M+\frac{1}{9}=0 \Rightarrow M=\frac{\frac{2}{3}\pm\sqrt{\left(-\frac{2}{3}\right)^2-4\times1\times\frac{1}{9}}}{2}$$

$$\Rightarrow M=\frac{1}{3}\ H \qquad (11)$$

Solving (2) and (10):

$$\frac{9}{37}\left(\frac{49}{9}-8M-\frac{1}{3}M^2\right)>0 \xrightarrow{Using\ (11)} \frac{9}{37}\left(\frac{49}{9}-8\times\frac{1}{3}-\frac{1}{3}\left(\frac{1}{3}\right)^2\right)>0 \Rightarrow \frac{2}{3}>0 \qquad (12)$$

Based on (11) and (12), choice (1) is the answer.

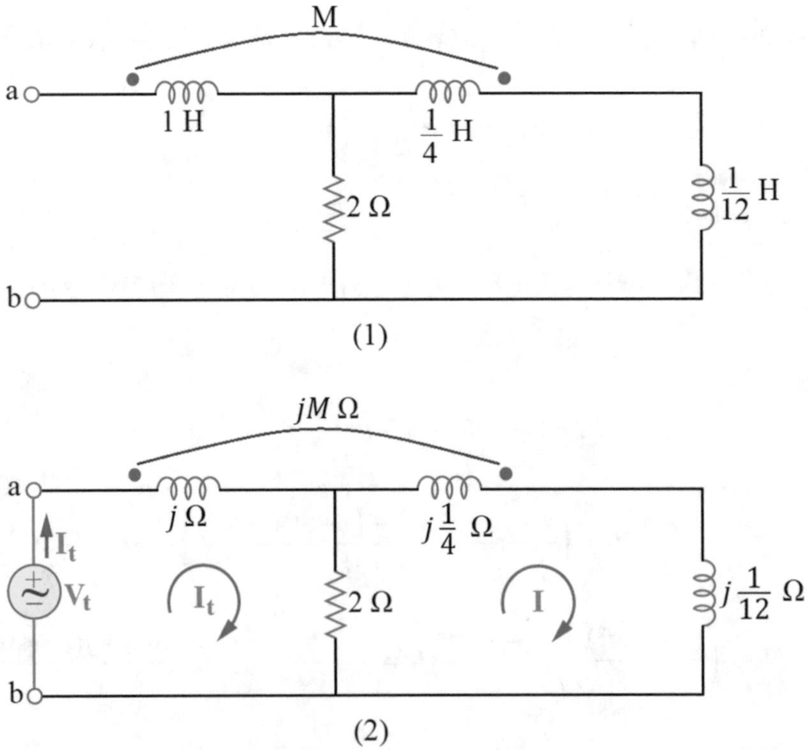

Figure 4.40 The circuit of solution of problem 4.41

4.42. This problem can be solved in time domain, since the circuit is a resistive circuit. Based on maximum average power transfer theorem, to transfer the maximum average power to the resistive load, the resistance of the load (R) must be equal to the Thevenin resistance seen by the load. In other words:

$$R = R_{Th} \tag{1}$$

To calculate the Thevenin resistance, we can connect a test voltage source to the terminal, analyze the circuit, and determine the value of $\frac{V_t}{I_t}$, while the independent power sources are turned off, as is shown in Figure 4.41.2. This problem should be solved by using mesh analysis.

Based on the polarities defined for the primary and secondary voltages and currents of the transformer, we can write:

$$\frac{V_1}{V_2} = \frac{I_2}{I_1} = \frac{4}{1} \Rightarrow \begin{cases} V_1 = 4V_2 & (2) \\ I_2 = 4I_1 & (3) \end{cases}$$

Simultaneously applying KVL and KCL in the right-side mesh (counterclockwise):

$$-V_t - V_2 + 20(I_1 + I_2) = 0 \xrightarrow{Using\ (3)} -V_t - V_2 + 20\left(\frac{I_2}{4} + I_2\right) = 0$$

$$\Rightarrow -V_t - V_2 + 25I_2 = 0 \tag{4}$$

Simultaneously applying KVL and KCL in the middle mesh (clockwise):

$$-V_1 + 30(I_2 - I_t) - V_2 = 0 \xrightarrow{Using\ (2)} -4V_2 + 30(I_2 - I_t) - V_2 = 0$$

$$\Rightarrow -5V_2 + 30I_2 - 30I_t = 0 \tag{5}$$

Simultaneously applying KVL and KCL in the left-side mesh (clockwise):

$$40(I_1 + I_2 - I_t) + V_1 + 20(I_1 + I_2) = 0 \Rightarrow 60I_1 + 60I_2 - 40I_t + V_1 = 0$$

$$\xrightarrow{Using\ (2),(3)} 60\frac{I_2}{4} + 60I_2 - 40I_t + 4V_2 = 0 \Rightarrow 75I_2 - 40I_t + 4V_2 = 0 \tag{6}$$

Solving (5) and (6):

$$-5\left(\frac{-75I_2 + 40I_t}{4}\right) + 30I_2 - 30I_t = 0 \Rightarrow \frac{495}{4}I_2 - 80I_t = 0 \Rightarrow I_2 = \frac{64}{99}I_t \tag{7}$$

Solving (6) and (7):

$$75\left(\frac{64}{99}I_t\right) - 40I_t + 4V_2 = 0 \Rightarrow \frac{280}{33}I_t + 4V_2 = 0 \Rightarrow V_2 = -\frac{280}{132}I_t \tag{8}$$

Solving (4), (7), and (8):

$$-V_t - \left(-\frac{280}{132}I_t\right) + 25\left(\frac{64}{99}I_t\right) = 0 \Rightarrow -V_t + \frac{1810}{99}I_t = 0 \Rightarrow \frac{V_t}{I_t} = \frac{1810}{99} \Rightarrow R_{Th} = \frac{1810}{99}\ \Omega \tag{9}$$

Solving (1) and (9):

$$R = \frac{1810}{99}\ \Omega$$

Choice (3) is the answer.

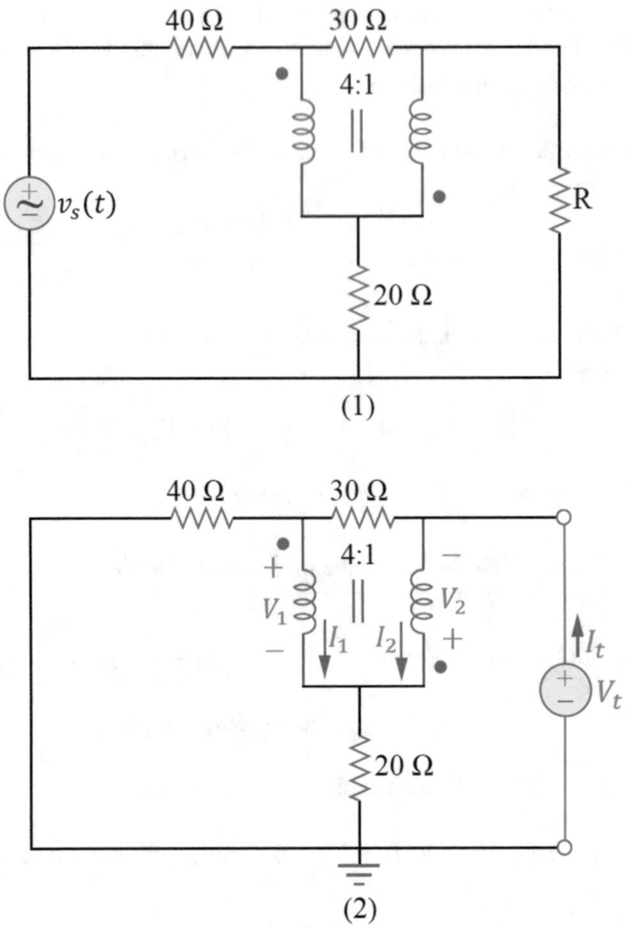

Figure 4.41 The circuit of solution of problem 4.42

4.43. Figure 4.42.2 shows the circuit in frequency domain. The impedances of the components are as follows:

$$\mathbf{Z_C} = \frac{1}{j\omega C}\ \Omega \tag{1}$$

$$\mathbf{Z_L} = j\omega L\ \Omega \tag{2}$$

The resonance frequency of a circuit is the frequency in which the imaginary part of the input impedance (or admittance) of the circuit is zero. Therefore, we need to determine the Thevenin impedance of the circuit and equate its imaginary part with zero to calculate the resonance frequency. In other words, the equation below must be solved:

$$Im\{\mathbf{Z_{in}}\} = 0 \Rightarrow \omega = \omega_0 \tag{3}$$

Herein, we must apply a test source in the terminal to determine the value of $\frac{\mathbf{V_t}}{\mathbf{I_t}}$, to find the Thevenin impedance of the circuit. The problem needs to be solved by using mesh analysis as can be seen in the following.

Based on the polarities defined for the primary and secondary voltages and currents of the transformer, we can write:

$$\frac{\mathbf{V_1}}{\mathbf{V_2}} = \frac{\mathbf{I_2}}{\mathbf{I_1}} = \frac{1}{n} \Rightarrow \begin{cases} \mathbf{V_1} = \dfrac{\mathbf{V_2}}{n} & (4) \\ \mathbf{I_1} = n\mathbf{I_2} & (5) \end{cases}$$

Applying KVL in the middle mesh (clockwise):

$$-\mathbf{V_t} - \mathbf{V_1} - \mathbf{V_2} = 0 \xrightarrow{\text{Using (4)}} -\mathbf{V_t} - \frac{\mathbf{V_2}}{n} - \mathbf{V_2} = 0 \Rightarrow \mathbf{V_2} = -\frac{\mathbf{V_t}}{1+\frac{1}{n}} = -\frac{n\mathbf{V_t}}{n+1} \tag{6}$$

Simultaneously applying KVL and KCL in the lower left-side mesh (clockwise):

$$\mathbf{V_2} + \frac{1}{j\omega C}(\mathbf{I_t} + \mathbf{I_2}) = 0 \xrightarrow{\text{Using (6)}} -\frac{n\mathbf{V_t}}{n+1} + \frac{1}{j\omega C}\mathbf{I_t} + \frac{1}{j\omega C}\mathbf{I_2} = 0$$

$$\Rightarrow \mathbf{I_2} = \frac{j\omega Cn\mathbf{V_t}}{n+1} - \mathbf{I_t} \tag{7}$$

Simultaneously applying KVL and KCL in the lower right-side mesh (clockwise):

$$j\omega L(\mathbf{I_t} - \mathbf{I_1}) + \mathbf{V_1} = 0 \xrightarrow{\text{Using (4), (5)}} j\omega L\mathbf{I_t} - j\omega Ln\mathbf{I_2} + \frac{\mathbf{V_2}}{n} = 0$$

$$\xrightarrow{\text{Using (6)}} j\omega L\mathbf{I_t} - j\omega Ln\mathbf{I_2} + \frac{1}{n}\left(-\frac{n\mathbf{V_t}}{n+1}\right) = 0 \Rightarrow \mathbf{I_2} = \frac{1}{n}\mathbf{I_t} - \frac{\mathbf{V_t}}{j\omega Ln(n+1)} \tag{8}$$

Solving (7) and (8):

$$\frac{j\omega Cn\mathbf{V_t}}{n+1} - \mathbf{I_t} = \frac{1}{n}\mathbf{I_t} - \frac{\mathbf{V_t}}{j\omega Ln(n+1)} \Rightarrow \left(\frac{j\omega Cn}{n+1} + \frac{1}{j\omega Ln(n+1)}\right)\mathbf{V_t} = \left(1 + \frac{1}{n}\right)\mathbf{I_t}$$

$$\Rightarrow \frac{\mathbf{V_t}}{\mathbf{I_t}} = \frac{1+\frac{1}{n}}{\frac{j\omega Cn}{n+1} + \frac{1}{j\omega Ln(n+1)}} = \frac{\frac{n+1}{n}}{\frac{-\omega^2 LCn^2+1}{j\omega Ln(n+1)}} = j\frac{\omega L(n+1)^2}{1-\omega^2 LCn^2} \Rightarrow \mathbf{Z_{in}} = j\frac{\omega L(n+1)^2}{1-\omega^2 LCn^2} \tag{9}$$

Solving (3) and (9):

$$\frac{\omega L(n+1)^2}{1-\omega^2 LCn^2} = 0 \Rightarrow \omega = 0 \tag{10}$$

As can be seen, no non-zero ω can be found as the solution of equation (10). Therefore, no resonance frequency can be detected from terminal a–b.

Choice (4) is the answer.

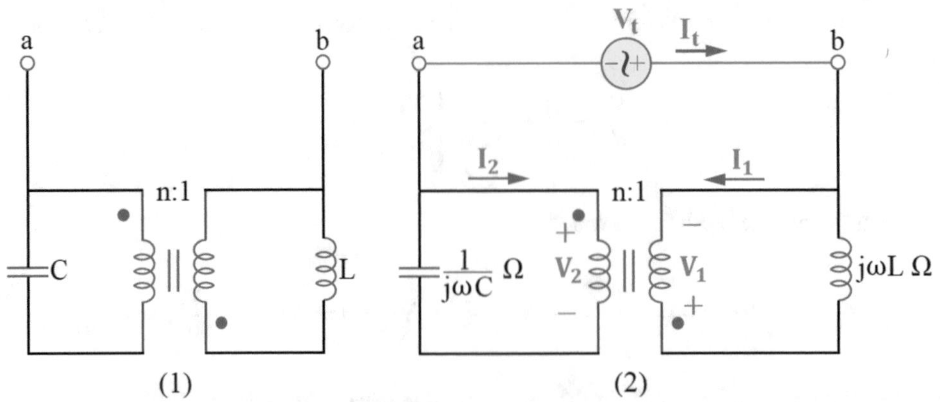

Figure 4.42 The circuit of solution of problem 4.43

Reference

1. Rahmani-Andebili, M. (2020). DC electrical circuit analysis: Practice problems, methods, and solutions, *Springer Nature*.

Index

M. Rahmani-Andebili (ed.), *AC Electrical Circuit Analysis* , https://doi.org/10.1007/978-3-030-60986-3

Printed in the United States
by Baker & Taylor Publisher Services